Aqueous Zinc Ion Batteries

Aqueous Zinc Ion Batteries

Fundamentals, Materials, and Design

Haiyan Wang, Qi Zhang, Yixin Li, and Yougen Tang

WILEY-VCH

Authors

Prof. Haiyan Wang
Central South University
YueLu District
Changsha
CH, 410083

Dr. Qi Zhang
Central South University
YueLu District
Changsha
CH, 410083

Dr. Yixin Li
Central South University
YueLu District
Changsha
CH, 410083

Prof. Yougen Tang
Central South University
YueLu District
Changsha
CH, 410083

Library of Congress Card No.: applied for

British Library Cataloguing-in-Publication Data
A catalogue record for this book is available from the British Library.

Bibliographic information published by the Deutsche Nationalbibliothek
The Deutsche Nationalbibliothek lists this publication in the Deutsche Nationalbibliografie; detailed bibliographic data are available on the Internet at <http://dnb.d-nb.de>.

Print ISBN: 978-3-527-34974-6
ePDF ISBN: 978-3-527-83504-1
ePub ISBN: 978-3-527-83505-8
oBook ISBN: 978-3-527-83506-5

Typesetting: Straive, Chennai, India
Printing and Binding: CPI Group (UK) Ltd, Croydon, CR0 4YY

C9783527349746_150324

Contents

Preface

Rapid economic development has increased the demand for traditional fossil energy sources (e.g. coal, oil, and natural gas), which has induced a series of serious global problems such as acid rain pollution, greenhouse effect, and land desertification. Fossil energy is formed by the natural deposition of ancient organisms over billions of years and cannot be recovered in the short term. To escape from the dilemma of energy depletion, sustainable and environmentally friendly renewable energy sources (e.g. solar, wind, and tidal) have attracted widespread attention in the past two decades. However, there is a critical scientific problem that needs to be solved in practical applications; that is, renewable energy is susceptible to weather and seasonal influences, and its power generation process is discontinuous and uncontrollable. Therefore, an effective energy storage system is needed as a medium device to coordinate the power input and output in the grid, thus improving the tolerance of the grid to renewable energy generation.

In recent years, electrochemical energy storage technology has developed rapidly and is evolving from a miniaturized application for portable electronic devices to an integrated application for large-scale energy storage systems and smart grids. Electrochemical energy storage is less affected by terrain and can store or release electrical energy directly through reversible chemical reactions, which is a very promising energy storage technology with the advantages of high energy density and high energy conversion efficiency. Lithium-ion batteries using organic electrolytes are one of the most widely used electrochemical energy storage systems. However, lithium-ion batteries have met big challenges in large-scale energy storage systems that are more focused on low cost, high stability, and high safety due to the limited lithium resources and the use of unsafe organic solvents. Compared with organic electrolytes, aqueous electrolytes are inexpensive and safe, and their ionic conductivity is greater, enabling the operation of batteries with large loads and high power. Aqueous batteries can completely meet the requirements of large-scale energy storage systems. In recent years, aqueous zinc-ion batteries have developed rapidly due to their high theoretical capacity (820 $mAh\,g^{-1}$), low electrochemical potential (−0.76 V vs. standard hydrogen electrodes), and high natural abundance of zinc resources, which are currently a research hotspot for the application of large-scale energy storage systems.

This book systematically illustrates the basic theory and frontier development of electrochemical theory, key materials, and battery devices for aqueous zinc-ion batteries. Chapter 1 provides a general review of the history of zinc-ion batteries and also provides an overview of the main challenges of cathodes, anodes, separators, electrolytes, and battery devices. Chapter 2 discusses the theoretical fundamentals of aqueous zinc-ion batteries, including the electrochemical reaction mechanism of cathodes and the deposition/dissolution mechanism of zinc anodes. Chapter 3 lists all the promising cathode materials and summarizes their current problems and corresponding optimization strategies. Chapter 4 is a general overview of the development and modification strategy of the zinc anode in aqueous zinc-ion batteries. Chapter 5 introduces the research progress, key problems, and solutions in the design of aqueous electrolyte for aqueous zinc-ion batteries. Chapter 6 illustrates the properties and characteristic parameters of the separator and the strategy to construct high-performance separators for aqueous zinc-ion batteries. Chapter 7 presents the progress in the structure and packaging of full aqueous zinc-ion batteries from the perspective of practical application. Chapter 8 provides advanced characterization tools and theoretical research methods for aqueous zinc-ion batteries. Chapter 9 summarizes the current challenges of aqueous zinc-ion batteries and proposes some future directions for their further development. Thanks to Prof. Yougen Tang, Dr. Qi Zhang, and Dr. Yixin Li, we wrote this book together. Also, many thanks to my group members, Dr. Chao Hu, Zefang Yang, Chunlin Xie, Qi Wang, Yihu Li, Wenbin Li, Huimin Ji, Tingqing Wu, Hao Wang, and Zhiwen Cai, for their participation in the compilation of each chapter. All the above colleagues have been engaged in scientific research in aqueous zinc-ion batteries and have made hard efforts to ensure the quality of this book.

This book aims to provide a comprehensive overview of aqueous zinc-ion batteries in terms of basic theory, frontier science, current status, and development trends of practical applications, which can be used as a reference for science and technology workers engaged in scientific research and technology development in the field of electrochemical energy storage materials and devices. There are inevitably omissions in this book since aqueous zinc-ion batteries are still developing rapidly and new knowledge and research advances are emerging. We hope that experts will offer valuable comments on the additions and revisions.

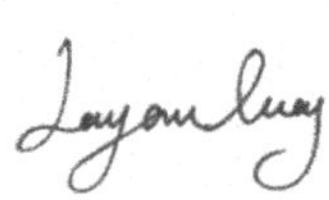

Prof. Haiyan Wang
Hunan Provincial Key Laboratory of Chemical Power Sources
College of Chemistry and Chemical Engineering
Central South University
Changsha, P.R. China

1

Introduction for Aqueous Zinc-Ion Batteries

1.1 History of Aqueous Zinc-Ion Batteries

Aqueous zinc-based batteries can be traced as far back as the voltaic battery, which first used zinc metal as the negative electrode. Subsequently, alkaline zinc-based batteries such as alkaline Zn–MnO_2, Zn–Ni, Zn–C, Zn–Ag, and Zn–Air were developed successively [1]. Among them, alkaline Zn–MnO_2 batteries have dominated the primary battery market since their commercialization. Earlier attempts to develop rechargeable Zn-based batteries were plagued by fast capacity fading and poor coulombic efficiency, mainly due to the uncontrollable growth of Zn dendrites and the formation of insulating, irreversible by-products (e.g. ZnO) in alkaline electrolytes (e.g. concentrated KOH solution). However, these batteries mentioned above cannot be called aqueous Zn-ion batteries (AZIBs) because the reaction mechanism of AZIBs is the plating/stripping of Zn^{2+} at the anode and the intercalation/deintercalation at the cathode in an aqueous solution. The AZIBs differ from the traditional alkaline Zn battery that is based on dissolution/precipitation reactions at the Zn anode ($Zn + 4OH^- \leftrightarrow Zn(OH)_4{}^{-2} + 2e^- \leftrightarrow ZnO + 2OH^- + H_2O + 2e^-$) and distinguished from other batteries with a Zn anode but no intercalation of Zn ions in cathode reactions [2]. The early investigations of AZIBs date back to 1986, when Yamamoto and Shoji first replaced the alkaline electrolyte with a zinc sulfate electrolyte and tested the electrochemical behavior of rechargeable $Zn|ZnSO_4|MnO_2$ batteries [3]. But its research boom was overshadowed by lithium-ion batteries due to lower energy density and poor cycle stability (Figure 1.1a). Studies on rechargeable AZIBs slowed down until 2012, when Kang and coworkers revisited zinc-ion battery chemistries with mildly acidic aqueous electrolytes [5]. As shown in Figure 1.1b, a rechargeable AZIB generally consists of a metallic Zn anode, a Zn^{2+} storage cathode, and a Zn^{2+}-salt electrolyte, operating via the reversible Zn^{2+} intercalation/deintercalation (cathode) and Zn plating/stripping (anode) upon discharging/charging [4]. AZIBs have been intensively investigated as potential energy storage devices on account of their low cost, environmental benignity, and intrinsically safe merits. With the exploitation of high-performance cathode materials, electrolyte systems, and in-depth mechanism investigation, the electrochemical performances of ZIBs have been greatly enhanced. For example, much work has been done on the modification of zinc anodes, and it mainly focuses on issues such

Aqueous Zinc Ion Batteries: Fundamentals, Materials, and Design, First Edition.
Haiyan Wang, Qi Zhang, Yixin Li, and Yougen Tang.

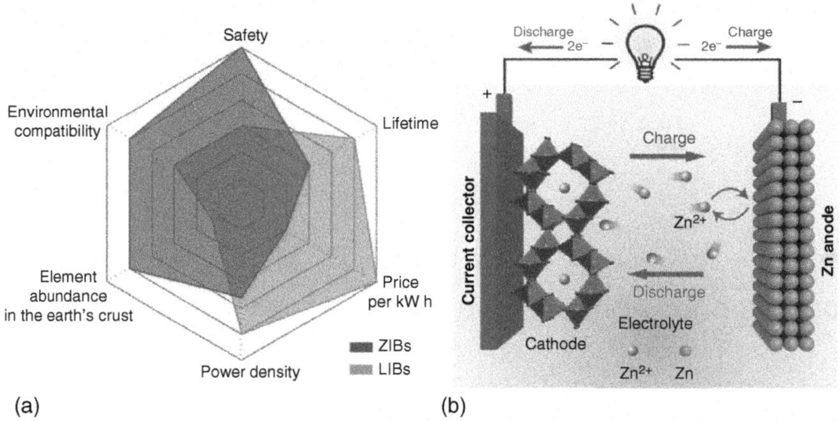

Figure 1.1 (a) Multi-angle comparison of zinc-ion and lithium-ion batteries. (b) Schematic illustration of the working principle of rechargeable zinc-ion batteries. Source: Zhang et al. [4]/Royal Society of Chemistry.

as the inhibition of zinc dendrite growth and the occurrence of side reactions. In terms of electrolytes, in addition to $ZnSO_4$, electrolytes with $ZnCl_2$, $Zn(CF_3SO_3)_2$, and $Zn(CH_3COO)_2$ as the main salts have been developed, while much work has been done in optimizing electrolytes, such as solvents, additives, and concentrations. Cathode materials have also been expanded from the earliest manganese-based materials to vanadium-based materials, Prussian blue and analogs, and conductive polymer materials. The main purpose of these modification strategies is to improve the cycle stability and energy density of AZIBs, which has greatly promoted the development of AZIBs.

1.2 Main Challenges for Aqueous Zinc-Ion Batteries

Recently, AZIBs have attracted much attention due to their advantages of large theoretical capacity, low cost, and environmental friendliness. The research on each component of AZIBs has increased significantly over the past decade [6]. However, the AZIBs have not been widely industrialized because their overall performance is not comparable to that of commercial lithium-ion and lead-acid batteries. The main reason is that the key materials involved in AZIBs, such as electrolytes, separators, anode materials, cathode materials, and current collector materials, cannot meet the cycle life and energy density requirements of practical batteries (Figure 1.2). Herein, the challenges and perspectives for the further development of AZIBs are reviewed, which are instructive for the research toward next-generation batteries for household appliances, electric vehicles, and large-scale energy storage systems.

1.2.1 Cathode

It is important to develop promising cathode materials with excellent electrochemical performance since the overall performance of AZIBs is determined by the

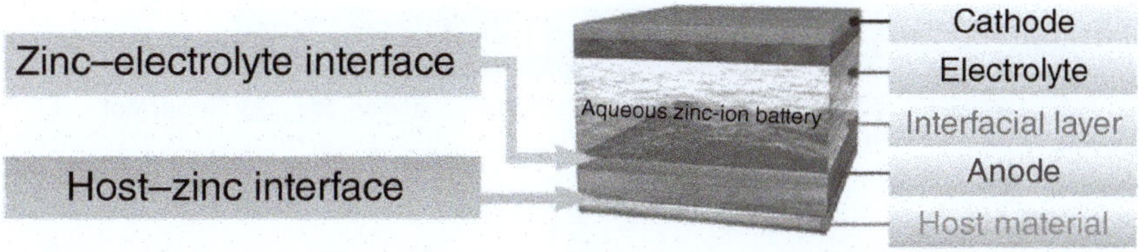

Figure 1.2 The configuration of AZIBs. Source: Zhang et al. [6]/John Wiley & Sons.

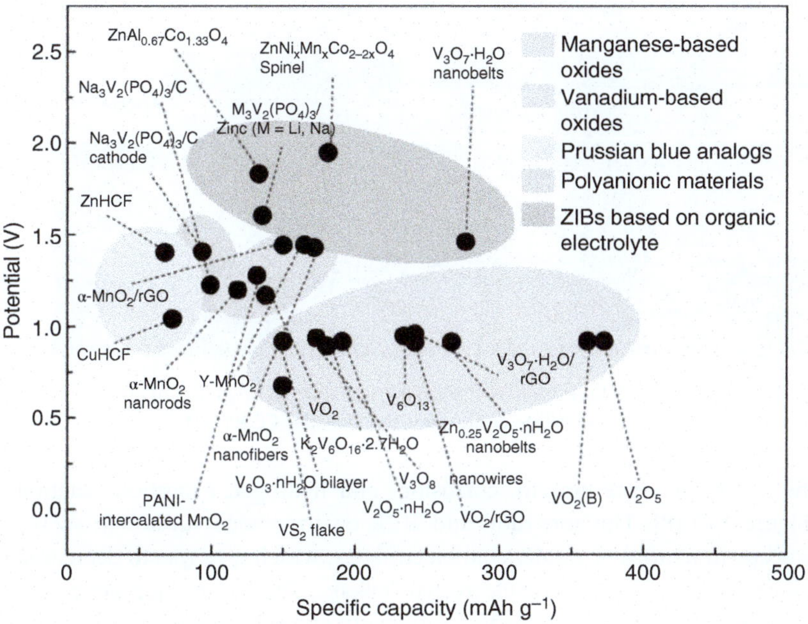

Figure 1.3 Operating voltage vs. specific capacity of various cathode materials currently used for AZIBs. Source: Xu and Wang [7]/Springer Nature/Licensed under CC BY 4.0.

cathode materials. Manganese-based, vanadium-based, and polyanionic materials are the common cathodes applied in AZIB systems (Figure 1.3) [7]. However, these cathode materials exhibit low specific capacity and low-voltage platforms, and their wide application is limited due to the rapid capacity fading caused by cation dissolution, irreversible phase transition, and by-product generation. In addition, the current cathode materials usually cannot meet the requirement of some advanced features for practical AZIBs such as low-temperature performance, low cost, and nonbiotoxicity. Therefore, researchers have attempted to explore the large-scale preparation of commercial cathode materials with outstanding performance to solve the above problems using inexpensive modification strategies.

1.2.2 Anode

Zinc foil is the most common anode material used in AZIBs. However, the electrochemical performances of zinc foils are not sufficient for large-scale applications

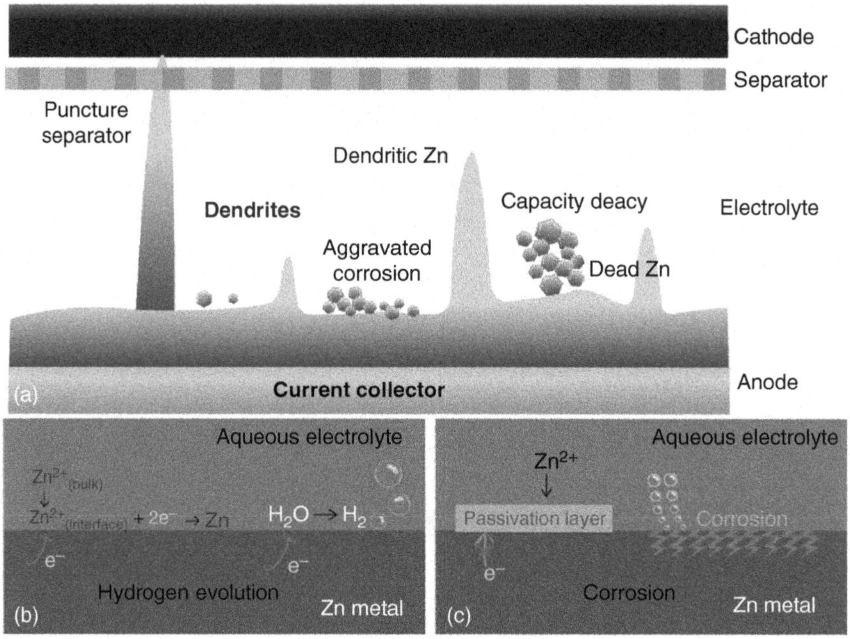

Figure 1.4 The dendrites (a), hydrogen evolution (b), and corrosion (c) of zinc metal anodes. Source: Xie et al. [8]/John Wiley & Sons/Licensed under CC BY 4.0.

due to the severe dendrite growth, corrosion, and hydrogen evolution reaction (HER) (Figure 1.4) [8]. The corrosion and HER on the anode surface in AZIBs generally occur in neutral or mildly acid electrolytes, which give rise to decreased capacity, increased impedance, and electrolyte leakage. The by-products generated from corrosion and HER will hinder the uniformity of ion transmission to induce more dendrites and further exacerbate the corrosion and HER due to the increased specific area of zinc anodes. The zinc foil anode suffers from electrode perforation and joint detachment during the deep charge and discharge processes, which also limits its wide application.

1.2.3 Separator

Glass fibers are widely used as the separator in AZIBs due to their low price, large liquid absorption, and low ion transmission impedance. Glass fibers can significantly improve the ion transport of the cathode material for AZIBs. However, zinc dendrites are easily generated on the zinc anode due to their low strength, irregular pore size distribution, and high zinc affinity (Figure 1.5) [9]. Therefore, the glass fiber is not completely suitable for AZIBs. A large number of studies suggested that Nafion films, nonwoven fabrics, cellulose films, and coating-modified glass fibers can be considered substitutes for glass fibers. However, as there are still some problems with liquid absorption, pore size distribution, strength, and the high cost of these separators, they are not suitable for wide application in AZIBs. In addition, most of

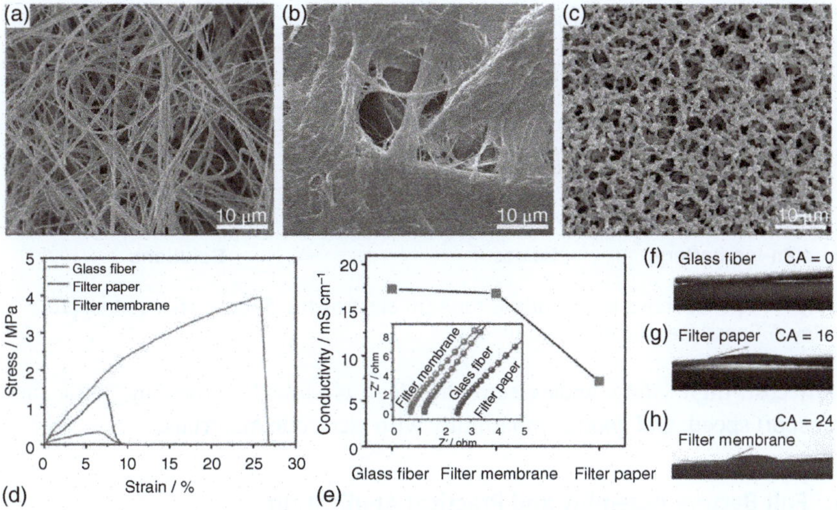

Figure 1.5 Relevant physical properties of the separators. SEM images of (a) glass fiber, (b) filter paper, and (c) filter membrane, (d) stress–strain curves, (e) ionic conductivities (the insert panel is corresponding to EIS curves), (f–h) and water contact angles. Source: Qin et al. [9]/John Wiley & Sons.

the performance evaluations of separators are only carried out in button batteries, and these evaluation results may not be used as test indicators for practical batteries (such as soft pack and box batteries), caused by separators that should possess the strength, wettability, and pore size distribution of the separator, environmental protection, and low cost.

1.2.4 Electrolyte

The design of electrolytes is an important method to improve the performance of aqueous batteries (Figure 1.6) [10]. Therefore, deepening the basic understanding of the solvation structure and interfacial chemistry of electrolytes is of great significance to promote the practical application of ZIBs. Compared with the traditional organic electrolyte battery, the aqueous battery possesses the following advantages, such as avoiding fire disasters caused by battery short circuits and lower cost. Meanwhile, aqueous batteries can be assembled and disassembled in the environment because their components are insensitive to air, which is beneficial to battery manufacturing and recycling. In addition, aqueous electrolytes display higher rate capability and power density than organic electrolytes because the ionic conductivity of aqueous electrolytes (about $0.1\,S\,cm^{-1}$) is much higher than that of organic electrolytes ($1–10\,mS\,cm^{-1}$). However, the presence of water makes the electrochemical window of the common electrolyte only 1.23 V (the organic system has more than 3 V), and the freezing point is high (about $-10\,°C$), which results in lower energy density, obvious battery self-discharge, and poor low-temperature performance for AZIBs. Therefore, it is of great significance to design electrolytes

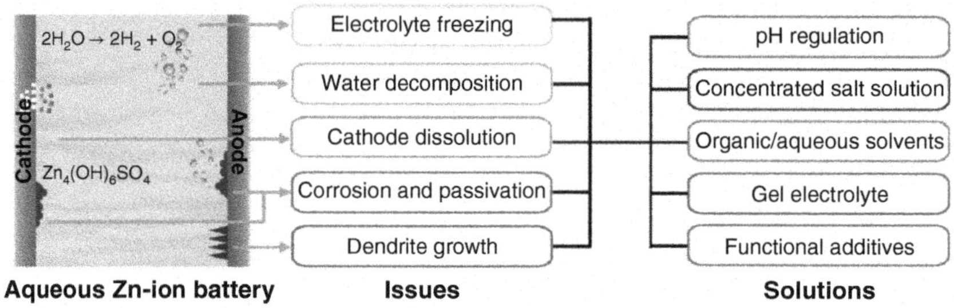

Figure 1.6 The challenges and solutions for AZIB electrolytes. Source: Zhang et al. [10]/ John Wiley & Sons/Licensed under CC BY 4.0.

with low cost, high safety, wide electrochemical window, low freezing point, fast ion transport speed, and good compatibility with electrode materials.

1.2.5 Full Battery Assembly and Practical Application

AZIBs have gradually shown a trend of widespread application due to their low cost, environmental friendliness, intrinsic safety, and relatively high energy and power densities. However, since the energy density, cycle stability, self-discharge behavior, and operating temperature range of AZIBs have not been uniformly optimized, assembling such batteries into commercial batteries still faces serious challenges [11]. The actual energy density of the battery is significantly lower than the theoretical energy density because the coin cells assembled in the laboratory rarely consider the capacity matching of cathode/anode electrodes and the loading of cathode active materials. The problems of low coulombic efficiency, poor cycle stability, and poor low-temperature performance in full cells have not been completely solved. Meanwhile, the widespread application of AZIBs requires a structural design that incorporates the essential characteristics of the battery. For example, aqueous batteries are not sensitive to air, so they can be designed as open batteries, which can replenish electrolytes in time to avoid battery failure caused by gas production and irreversible consumption of electrolytes. The choice of current collectors also has a critical impact on the performance and energy density of AZIBs. The carbon-based and self-supporting electrodes seem to be only suitable for small-scale AZIB devices due to their high price and difficulty in large-scale fabrication and battery assembly. Ni-based and Ti-based metal current collectors are not suitable for wide application due to their high price. Therefore, it is of great significance to develop current collectors (such as special copper foil and foamed stainless steel) that are cheap, show excellent performance, and are suitable for assembling large batteries.

References

1 Hao, J., Li, X., Zeng, X. et al. (2020). Deeply understanding the Zn anode behaviour and corresponding improvement strategies in different aqueous Zn-based batteries. *Energy & Environmental Science* 13: 3917–3949.

2 Song, M., Tan, H., Chao, D., and Fan, H.J. (2018). Recent advances in Zn-ion batteries. *Advanced Functional Materials* 28: 1802564.

3 Yamamoto, T. and Shoji, T.J.I.C.A. (1986). Rechargeable Zn|$ZnSO_4$|MnO_2-type cells. *Inorganica Chimica Acta* 117: 27–28.

4 Zhang, N., Chen, X., Yu, M. et al. (2020). Materials chemistry for rechargeable zinc-ion batteries. *Chemical Society Reviews* 49: 4203–4219.

5 Xu, C., Li, B., Du, H., and Kang, F. (2012). Energetic zinc ion chemistry: the rechargeable zinc ion battery. *Angewandte Chemie International Edition* 51: 933–935.

6 Zhang, Q., Luan, J., Tang, Y. et al. (2020). Interfacial design of dendrite-free zinc anodes for aqueous zinc-ion batteries. *Angewandte Chemie International Edition* 59: 13180–13191.

7 Xu, W. and Wang, Y. (2019). Recent progress on zinc-ion rechargeable batteries. *Nano-Micro Letters* 11: 90.

8 Xie, C., Li, Y., Wang, Q. et al. (2020). Issues and solutions toward zinc anode in aqueous zinc-ion batteries: a mini review. *Carbon Energy* 2: 540–560.

9 Qin, Y., Liu, P., Zhang, Q. et al. (2020). Advanced filter membrane separator for aqueous zinc-ion batteries. *Small* 16: 2003106.

10 Zhang, Q., Yang, Z., Ji, H. et al. (2021). Issues and rational design of aqueous electrolyte for Zn-ion batteries. *SusMat* 1: 432–447.

11 Xu, J. and Wang, C. (2022). Perspective—electrolyte design for aqueous batteries: from ultra-high concentration to low concentration? *Journal of the Electrochemical Society* 169: 030530.

2

Theoretical Fundamentals of Aqueous Zinc-Ion Batteries

2.1 Electrochemical Reaction Mechanism of Cathodes

The energy-storage mechanism of cathode materials remains complicated and controversial in aqueous zinc-ion batteries (AZIBs). To date, the recognized energy-storage mechanisms of zinc batteries are broadly classified into three categories: (i) Zn^{2+}-insertion/extraction mechanism, (ii) co-insertion mechanism, and (iii) chemical conversion reaction mechanism [1]. A brief account of these mechanisms is provided in this section, and readers may refer to other literature for an in-depth mechanistic understanding of charge storage in AZIB.

2.1.1 Zn^{2+}-Insertion/Extraction Mechanism

Among the three mechanisms, the insertion/extraction of Zn^{2+} in the host materials is the most essential mechanism in the AZIBs system, similar to that of the traditional rechargeable lithium-ion batteries (LIBs). In the discharge process, the Zn^{2+} in the electrolyte is inserted into the cathode, and the Zn in the anode loses electrons to produce Zn^{2+}, which maintains the charge balance of the electrolyte. When charging, Zn^{2+} is extracted from the cathode to the electrolyte. It migrates to the anode and collects electrons at the interface, finally diffusing inward in the form of atoms [2]. The cathode materials of the tunnel/layer structure store Zn^{2+} through this mechanism. Reversible Zn^{2+}-insertion/extraction was first proposed to explain the energy-storage mechanisms of ZIBs. Kang and coworkers first reported mild aqueous rechargeable ZIBs [3].

Meanwhile, they proposed that Zn^{2+} can be reversibly inserted into or extracted from α-MnO_2 cathode during discharge/charge processes (Figure 2.1a). And the Zn anode gets dissolving/depositing at the same time. Subsequently, it was proven that the Zn^{2+}-insertion/extraction mechanism would occur in β-MnO_2, γ-MnO_2, and λ-MnO_2 [4]. In addition to reversible Zn^{2+}-insertion/extraction, a reversible phase transition from the tunneled structure (α-MnO_2) to layered polymorphs was observed. This transition is initiated by the dissolution of manganese from α-MnO_2 during the discharge process to form layered Zn-birnessite (Figure 2.1b) [5].

Vanadium-based compounds are mainly layered/tunnel frameworks with a large void space favoring Zn^{2+}-insertion. In a milestone study, a reversible Zn^{2+}-insertion

Aqueous Zinc Ion Batteries: Fundamentals, Materials, and Design, First Edition.
Haiyan Wang, Qi Zhang, Yixin Li, and Yougen Tang.

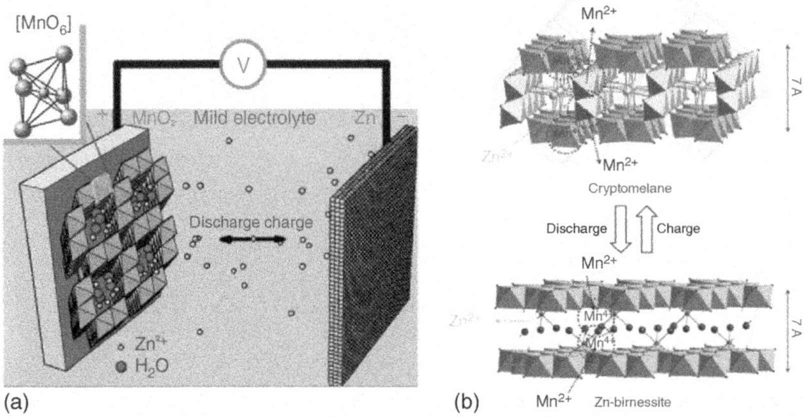

Figure 2.1 (a) Schematics of the chemistry of the zinc ion battery. Source: Xu et al. [3]/John Wiley & Sons. (b) Schematic illustrating the mechanism of zinc intercalation into α-MnO_2. Source: Lee et al. [4]/Springer Nature.

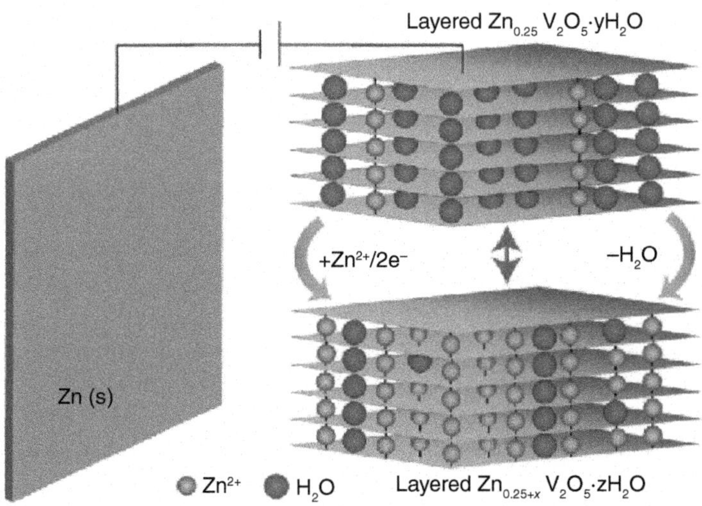

Figure 2.2 Rechargeable $Zn_{0.25}V_2O_5{\cdot}nH_2O$ system. Source: Kundu et al. [6]/Springer Nature.

mechanism of $Zn_{0.25}V_2O_5{\cdot}nH_2O$ was revealed by combining operando X-ray diffraction (XRD) and X-ray photoelectron spectroscopy (XPS) analyses [6]. This corresponds to the following equations (Figure 2.2):

Anode:

$$1.1Zn \leftrightarrow 1.1Zn^{2+} + 2.2e^- \tag{2.1}$$

Cathode:

$$1.1Zn^{2+} + 2.2e^- + Zn_{0.25}V_2O_5 \cdot nH_2O \leftrightarrow Zn_{1.35}V_2O_5 \cdot nH_2O \tag{2.2}$$

Overall:

$$1.1\mathrm{Zn} + \mathrm{Zn_{0.25}V_2O_5} \cdot n\mathrm{H_2O} \leftrightarrow \mathrm{Zn_{1.35}V_2O_5} \cdot n\mathrm{H_2O} \tag{2.3}$$

Alshareef and coworkers reported a simple microwave to synthesize layered metal pyrovanadate nanowires ($Zn_3V_2O_7(OH)_2{\cdot}2H_2O$) with a porous crystal framework [7]. The electrochemical energy storage mechanism can be described by the following equation:

$$\mathrm{Zn_3V_2O_7(OH)_2} \cdot 2\mathrm{H_2O} + 1.9\mathrm{Zn^{2+}}3.8\mathrm{e^-} \leftrightarrow \mathrm{Zn_{4.9}V_2O_7(OH)_2} \cdot 2\mathrm{H_2O} \tag{2.4}$$

Similar Zn^{2+}-(de)intercalation storage processes have been demonstrated in most of the reported vanadium-based cathodes, such as VO_2, $V_5O_{12}{\cdot}6H_2O$, $Na_2V_6O_{16}{\cdot}3H_2O$, $Li_xV_2O_5{\cdot}nH_2O$, and $Ca_{0.25}V_2O_5{\cdot}nH_2O$ [8]. Vanadium-based compounds evolve from V^{5+} to V^{4+} or even V^{3+} through the evolution of the V oxidation state, adapt to the intercalation of Zn^{2+}, and maintain a stable crystal framework. Irreversible phase transitions were also observed in examples of NaV_3O_8-type compounds like $Na_5V_{12}O_{32}$, which experienced structural destruction, phase transition, and thus capacity degradation [9].

2.1.2 Co-Insertion/Extraction Mechanism

The Zn^{2+} insertion/extraction mechanism is the most ideal and acceptable, so the Zn/MnO_2 battery was developed based on the migration of Zn^{2+} ions between the cathode and anode, which was very early reported in 2012. It was found that not only zinc ions but also other ions were involved in the co-insertion/extraction mechanism. Because of the sluggish Zn^{2+} and the strong electrostatic repulsion, other ions or molecules (including H^+, Li^+, and H_2O) in the electrolyte can also be embedded and released when the insertion and extraction of zinc ions are happening in cathode during the process of charge and discharge [10, 11].

2.1.2.1 H^+ and Zn^{2+} Insertion/Extraction Mechanism

For manganese-based cathode materials, Sun et al. [12] first proposed the co-insertion mechanism of H^+ and Zn^{2+} in Zn/ε-MnO_2 batteries, and the insertion of H^+ and Zn^{2+} corresponds to the sloped plateau from 1.8 to 1.35 V in region I and the flat plateau at 1.3 V in region II, respectively (Figure 2.1a). Similar mechanisms of the difference crystallographic (α, δ, and β) are observed in these experiments. The joint charge storage of H^+ and Zn^{2+} delivers high-rate performance and long cycle life in the Zn-δ-MnO_2 batteries [13]. The first step in fast charge storage is non-diffusion-controlled Zn^{2+}-ion-storage mechanism in bulk δ-MnO_2 without significant phase transition. In contrast, the following step reaction is proven to be the diffusion-controlled H^+ conversion reaction in $Zn(TFSI)_2$-based electrolyte (Figure 2.1b). In addition, as evidenced by Gao et al., H^+-insertion can boost the α-MnO_2 electrode [14]. They find that the capacity fading during cycling process is mainly due to the decay of second discharge plateau, which is ascribed to Zn^{2+} insertion (Figure 2.3a). An increased amount of irreversible $ZnMn_2O_4$ is observed on the surface of α-MnO_2, which is less reversible than that of H^+. β-MnO_2

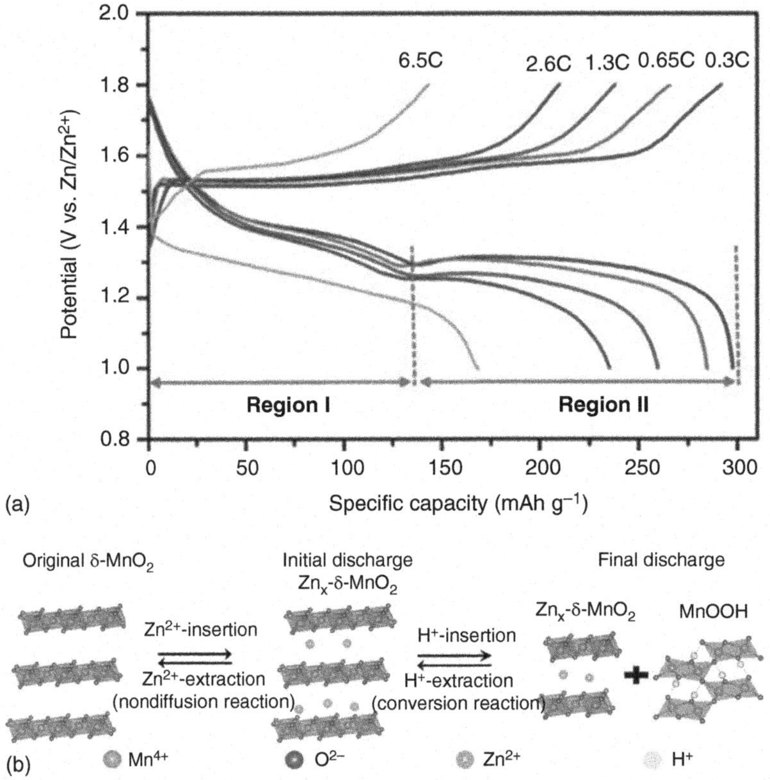

Figure 2.3 (a) Charge/discharge curves at different rates in the first cycle and two discharge regions. (b) Joint non-diffusion-controlled Zn^{2+}-intercalation and H^+-insertion/extraction in δ-MnO_2.

possessing oxygen defects also allows the insertion of H^+, and density functional theory (DFT) computation shows that the β-MnO_2 host structure is much easier for H^+ insertion rather than Zn^{2+} [15] (Figure 2.4).

For Zn/vanadium-based batteries, Wan et al. [16] reported that sodium vanadate hydrate experienced a simultaneous H^+ and Zn^{2+} insertion/extraction process as a cathode, which was mainly responsible for their excellent performance with a capacity of 380 mAh g^{-1} at 4 A g^{-1} (a capacity retention of 82% after 1000 cycles). Furthermore, as shown in Figure 2.1b, Wang and coworkers [17] verified the sequential insertion of H^+ and Zn^{2+} at 1.1–0.71 and 0.71–0.32 V plateaus in Zn/V_2C@CNT battery. Similar co-insertion mechanisms of H^+ and Zn^{2+} are also observed in other vanadium-based battery materials, such as $V_{10}O_{24}\cdot 12H_2O$ [18], $Zn_{0.3}V_2O_5\cdot 1.5H_2O$ [19], and (Ni)VO_2 [20].

2.1.2.2 Zn^{2+}/H_2O Co-Insertion/Extraction Mechanism

Zn^{2+}/H_2O co-insertion/extraction mechanisms usually exist in vanadium-based cathodes because the tunnel structure or layered structure of the cathode is

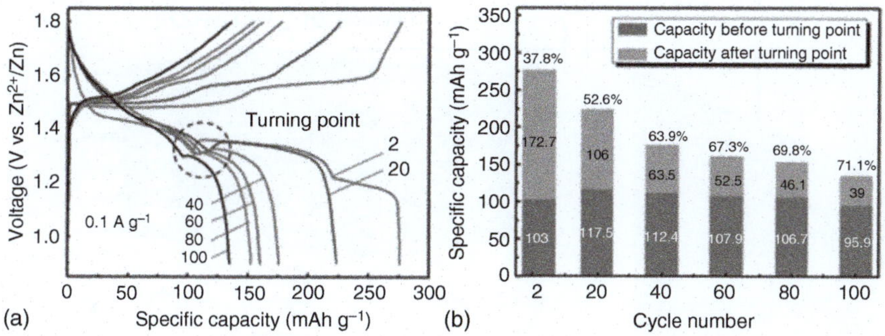

Figure 2.4 (a) Turning point in charge/discharge profiles. (b) Capacity statistics of two discharge plateaus after different cycles.

beneficial to the insertion/extraction of H_2O. Sometimes, the water in such layers of vanadium-based materials provides a pillar effect for stabilizing structures [21, 22].

Kundu et al. [6] reported that a single-crystal layered structure of $Zn_{0.25}V_2O_5{\cdot}nH_2O$ nanobelt possesses two-electron redox containing intercalated divalent cations and water. Interlayer metal ions Zn^{2+} and/or structural water in this layered oxide act as pillars stabilizing the structure (Figure 2.5), providing a capacity retention of more than 80% after 1000 cycles. The water molecules expanding and contracting the layered galleries of $Zn_{0.25}V_2O_5$, allow Zn^{2+}-insertion/extraction in a highly reversible manner, and promise high-rate performance with a capacity of 220 mAh g^{-1} at a 15 C rate.

When used as cathode for AZIBs [23], porous V_2O_3@C material exhibits a capacity of 350 mAh g^{-1} at 100 mAh g^{-1} and capacity retention of 90% after 4000 cycles at 5 A g^{-1}. It is worth noting that the Raman spectra and ex situ XPS demonstrate the co-intercalation of Zn^{2+} and water, and the electrochemical reaction in the Zn//V_2O_3 can therefore be illustrated in Figure 2.6.

2.1.2.3 Li^+- and Zn^{2+}-Insertion/Extraction Mechanism

An aqueous Zn/V_2O_5 rechargeable battery with a Li^+/Zn^{2+} co-insertion mechanism was reported. By XRD, voltammetric, and Raman spectroscopy, it was elucidated that exclusive Li^+-insertion into V_2O_5 occurs up to mid-discharge. Still, the co-insertion of a few zinc ions is likely involved in the second part of the reduction. Besides, Zn^{2+} acts as pillar species, preventing important structural change and hindering the formation of the distorted δ-LiV_2O_5 phase [24].

Except for vanadium-based materials, Prussian blue analogs, for example, iron hexacyanoferrate (FeHCF), can also store and release Zn^{2+} and Li^+ [25]. It is found that Zn^{2+}- and Li^+-insertion/extraction cause crystalline distortion and the reduction of interlayer spacing, respectively (Figure 2.7).

2.1.3 Chemical Conversion of Cathodes

Compared with Zn^{2+} insertion/extraction and co-insertion/extraction mechanisms, the conversion reactions in the battery would tend to provide higher capacity due

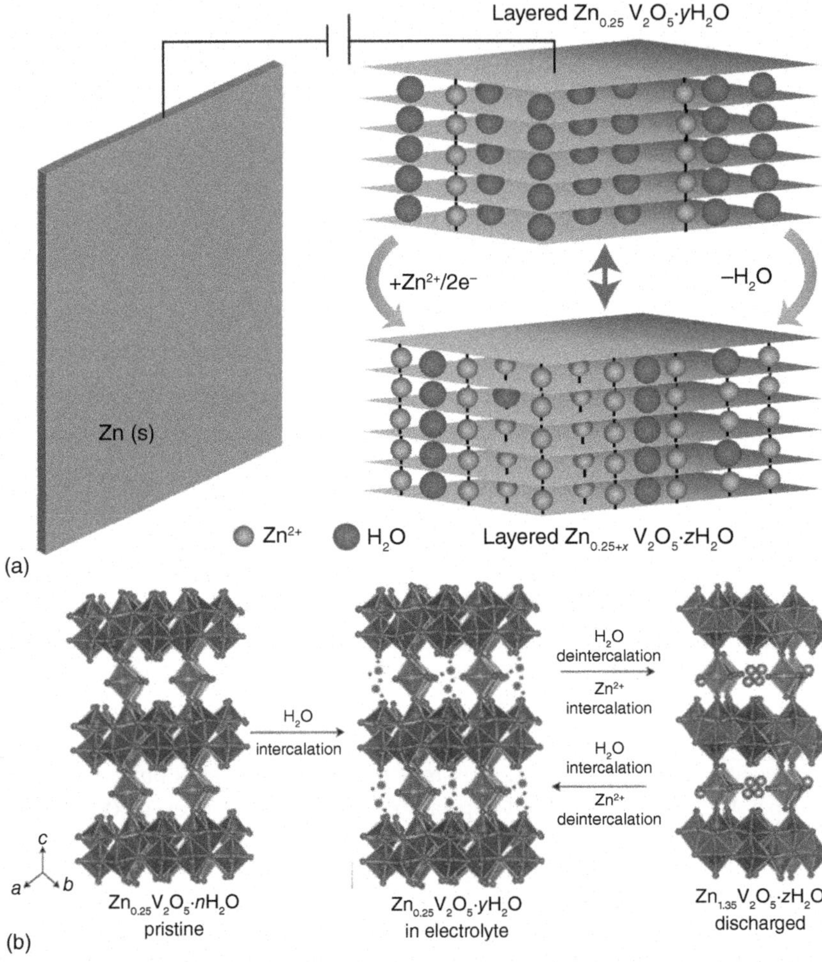

Figure 2.5 (a) Schematic of the Zn metal/$Zn_{0.25}V_2O_5$ battery and the expanded interlayer arrangement of $Zn_{0.25}V_2O_5$. (b) Scheme showing reversible water intercalation into $Zn_{0.25}V_2O_5 \cdot nH_2O$ and the intercalation or deintercalation of the water and Zn^{2+} during discharge/charge.

to direct charge transfer. Hence, it is very promising and effective to develop and design cathode materials with highly reversible conversion mechanisms [11]. Pan et al. [26]. first proved the role of the chemical conversion mechanism in charge storage of MnO_2-based cathodes. During the discharge process, α-MnO_2 reacted with H^+ to form MnOOH. To keep the charge constant in the system, the subsequent OH^- reacts with $ZnSO_4$ and H_2O to form lamellar $ZnSO_4[Zn(OH)_2]_3 \cdot xH_2O$ (ZHS). After charging, the aforementioned products are reduced to the original α-MnO_2, indicating that MnO_2 has reversible electrochemical behavior with MnOOH/ZHS, which can be expressed as follows:
Cathode:

$$H_2O \leftrightarrow H^+ + OH^- \tag{2.5}$$

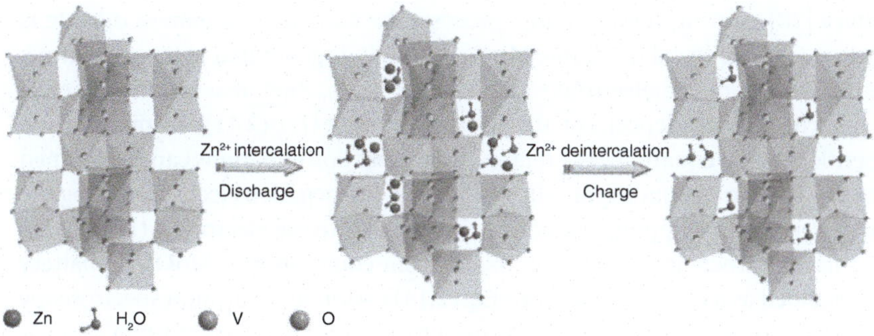

Figure 2.6 Schematic illustrations of Zn^{2+} and H_2O co-insertion mechanisms of V_2O_3 cathode.

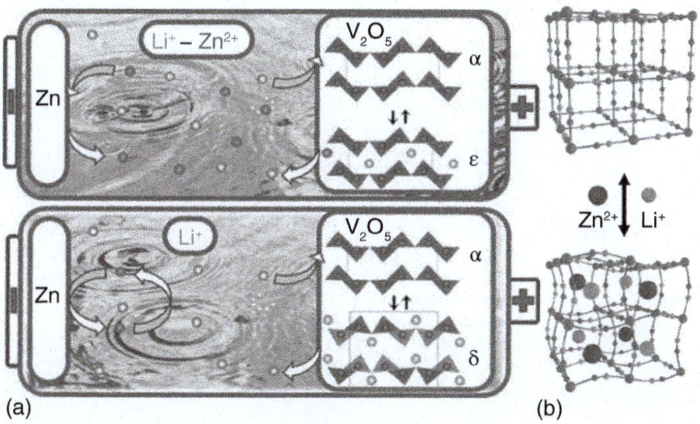

Figure 2.7 (a) Schematic diagram of Zn/V_2O_5 battery with Li^+/Zn^{2+} and Li^+ electrolyte. (b) Schematic illustration of the crystalline structure evolution under the insertion/extraction of Zn^{2+} and Li^+.

$$MnO_2 + H^+ + e^- \leftrightarrow MnOOH \tag{2.6}$$

$$\frac{1}{2}Zn^{2+} + OH^- + \frac{1}{6}ZnSO_4 + \frac{x}{6}H_2O \leftrightarrow \frac{1}{6}ZnSO_4[Zn(OH_2)]_3 \cdot xH_2O \tag{2.7}$$

Anode:

$$\frac{1}{2}Zn \leftrightarrow \frac{1}{2}Zn^{2+} + e^- \tag{2.8}$$

Overall:

$$MnO_2 + \frac{1}{2}Zn + \frac{x}{6}H_2O + \frac{1}{6}ZnSO_4 \leftrightarrow MnOOH + \frac{1}{6}ZnSO_4[Zn(OH_2)]_3 \cdot xH_2O \tag{2.9}$$

Similar to the co-insertion mechanism, the H^+ required for the chemical conversion reaction is generated by water decomposition, and the corresponding generated OH^- leads to the formation of ZHS [27]. There is another explanation for the formation of ZHS and the energy storage of MnO_2 in the charge/discharge process of the battery [28]. During the discharge process, the cathode manganese dioxide is electrochemically reduced to soluble Mn^{2+}, and OH^- is generated at the same time, increasing the pH of the electrolyte (Eq. (2.10)). Thus, Zn^{2+} reacts with the electrolyte to form ZHS and deposits on the electrode surface (Eq. (2.11)). Combining the aforementioned two equations, the total cathodic reaction equations can be obtained (Eq. (2.12)). The corresponding anodic reaction on the negative electrode during the discharge is a typical stripping reaction of zinc metal into the electrolyte (Eq. (2.13)). So, the total electrochemical reaction mechanism equation of α-MnO_2/Zn battery can be described as a reaction equation (Eq. (2.14)). Atomic absorption spectroscopy (AAS) and in situ pH measurement was used to analyze the concentration of manganese and zinc ions in the electrolyte at various stages of the charge–discharge process. Combined with in situ XRD analysis, the key role of pH change in the reaction mechanism of the battery system was further confirmed. Figure 2.8 shows the schematic diagram of the Zn/α-MnO_2 battery discharge process in the $ZnSO_4$ electrolyte.

$$MnO_2 + 2H_2O + 2e^- \rightarrow Mn^{2+} + 4OH^- \tag{2.10}$$

$$4Zn^{2+} + 6OH^- + SO_4{}^{2-} + 5H_2O \leftrightarrow Zn_4(OH)_6(SO_4) \cdot 5H_2O \downarrow \tag{2.11}$$

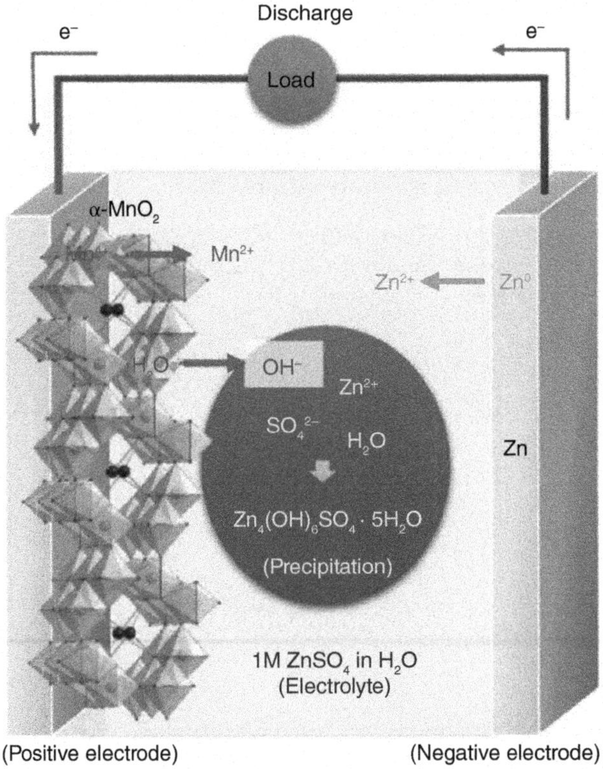

Figure 2.8 Schematic diagram of the conversion reaction energy-storage mechanism for the aqueous Zn/α-MnO_2 battery system. Source: Adapted from Lee et al. [28].

$$3MnO_2 + 8Zn^{2+} + 2SO_4{}^{2-} + 16H_2O + 6e^- \rightarrow 3Mn^{2+} + 2Zn_4(OH)_6(SO_4) \cdot 5H_2O \downarrow \quad (2.12)$$

$$3Zn \rightarrow 3Zn^{2+} + 6e^- \quad (2.13)$$

$$3MnO_2 + 3Zn + 5Zn^{2+} + 2SO_4{}^{2-} + 16H_2O \leftrightarrow 3Mn^{2+} + 2Zn_4(OH)_6(SO_4) \cdot 5H_2O \downarrow \quad (2.14)$$

Considering the important influence of working voltage, Chao et al. [29] gave a new systematic explanation for the above-mentioned disputed reaction mechanism of MnO_2 through the study of a high-voltage Zn//MnO_2 electrolytic battery. The multi-redox reactions of the Zn//MnO_2 battery exhibited strong voltage dependence, as shown in Figure 2.9a, where the discharge process at an operation voltage of 2.2–0.8 V can be divided into three regions: D1 (2.0–1.7 V), D2 (1.7–1.4 V), and D3 (1.4–0.8 V). In the D1 region of high voltage, an electrolytic conversion reaction was dominant, in which the manganese oxide with +4 valence state was reduced and dissolved as manganese ions in +2 into electrolyte; similar (Eq. (2.6)) reactions occurred in the D2 region; and in the D3 region of relatively low voltage, a classic Zn^{2+} intercalation chemistry occurred. It is worth noting that the capacity of the D1 region increases with the cycle number, which is the main contributor to the overall capacity and high voltage. The electrolytic conversion reaction can increase the proportion of the D1 reaction by adjusting the pH value of the electrolyte (by adding appropriate H_2SO_4). As shown in Figure 2.9b, thanks to the

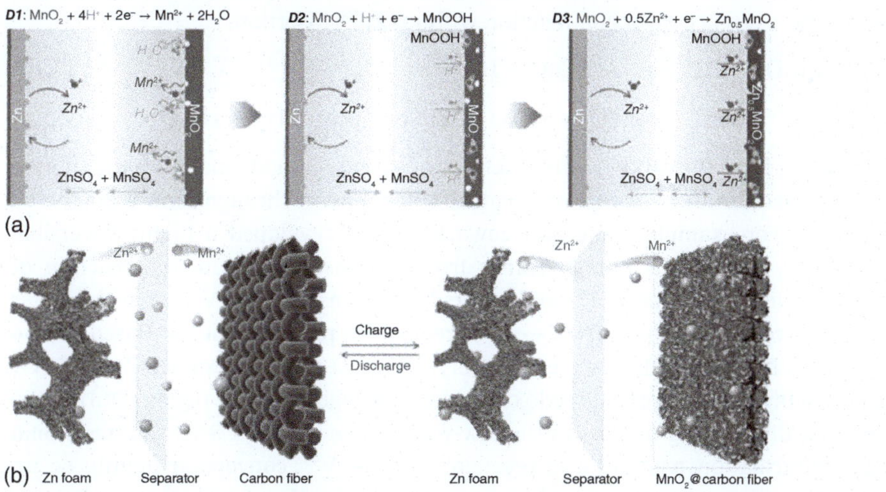

Figure 2.9 (a) The galvanostatic discharge profiles of Zn//MnO_2 battery and the schematic for different reaction mechanisms in the D1, D2, and D3 stages. (b) Schematic representation and charge storage mechanism of the new electrolytic Zn//MnO_2 battery in 1 M $ZnSO_4$ and 1 M $MnSO_4$ electrolyte (During charge, the cations of Zn^{2+} and Mn^{2+} in the electrolyte are deposited onto current collectors of anode and cathode to form, respectively, Zn metal and MnO_2. Zn and MnO_2 dissolution occur at the anode and cathode, respectively, in the discharge.) Source: Adapted from Chao et al. [29]/John Wiley & Sons.

unique dual-electron redox electrolysis reaction of $Mn^{4+} \leftrightarrow Mn^{2+}$ in the D1 region, the prepared Zn//MnO_2 electrolytic battery can achieve a high discharge plateau at about 1.95 V and an imposing capacity.

Reversible cycles

Cathode:

$$MnO_2 + 4H^+ + 2e^- \leftrightarrow Mn^{2+} + 2H_2O$$

$$MnO_2 + H^+ + e^- \leftrightarrow MnOOH$$

$$MnO_2 + 0.5Zn^{2+} + e^- \leftrightarrow Zn_{0.5}MnO_2$$

Anode:

$$Zn \leftrightarrow Zn^{2+} + 2e^-$$

Overall:

$$2Zn + 3MnO_2 + 5H^+ \leftrightarrow 1.5Zn^{2+} + Mn^{2+} + Zn_{0.5}MnO_2 + MnOOH + 2H_2O$$

In addition to manganese-based materials, the AZIBs with Co_3O_4 as the cathode material also have a similar conversion and energy-storage mechanism [30]. Co_3O_4 reacts with H^+ in water and accepts electrons to form CoO (Eq. (2.15)). At the same time, the subsequent OH^- interacts with $ZnSO_4$ and H_2O in aqueous electrolytes to form lamellar ZHS (Eq. (2.16)). Cu^+-based materials can also store Zn^{2+} by direct conversion without Zn^{2+}/H^+ intercalation [31]. As shown in Figure 2.10, due to the higher charge density of Zn^{2+}, the barrier of Zn^{2+} insertion and diffusion based on Zn^{2+} intercalation is much higher than the formation energy of ZnI_2 and Cu metal, so CuI has a direct conversion reaction without Zn^{2+} insertion.

$$Co_3O_4 + 2H^+ + e^- \leftrightarrow 3CoO + H_2O \quad (2.15)$$

$$Zn^{2+} + 2OH^- + ZnSO_4 + xH_2O \leftrightarrow ZnSO_4 \cdot [Zn(OH)_2]_3 \cdot xH_2O \quad (2.16)$$

Different from the cation redox reactions of various cathode materials mentioned above, vanadium-based compounds can also store energy through some anion redox reactions. For example, Fang and coworkers [32] developed a highly reversible Zn//VN_xO_y battery. Its energy comes from the insertion/extraction reaction of typical cations (Zn^{2+} and H^+) and reversible cations ($V^{3+} \leftrightarrow V^{2+}$). On the other hand, the reversible anion ($N^{3-} \leftrightarrow N^{2-}$) redox also provides energy. Based on the $VOPO_4$ cathode and salt-in-water electrolyte, Wan et al. [33] designed Zn//$VOPO_4$ battery with highly reversible redox chemistry in high-pressure region. During the charging/discharging process in the low-voltage region (0.8–1.8 V), the traditional Zn^{2+} intercalation reaction and reversible $V^{5+} \leftrightarrow V^{4+}$ conversion mainly occur. When the voltage window is widened to 2.1 V, the lattice oxygen atoms in $VOPO_4$ undergo an oxygen redox reaction at a high voltage of –1.86 V without the insertion/extraction of zinc ions between 1.8 and 2.1 V (Figure 2.11a). The introduced anionic redox reaction not only increased the average working potential to twice that of the traditional vanadium-based AZIB, but also enhanced the reversible crystal structure evolution of $VOPO_4$ and ensured its superior electrochemical performance.

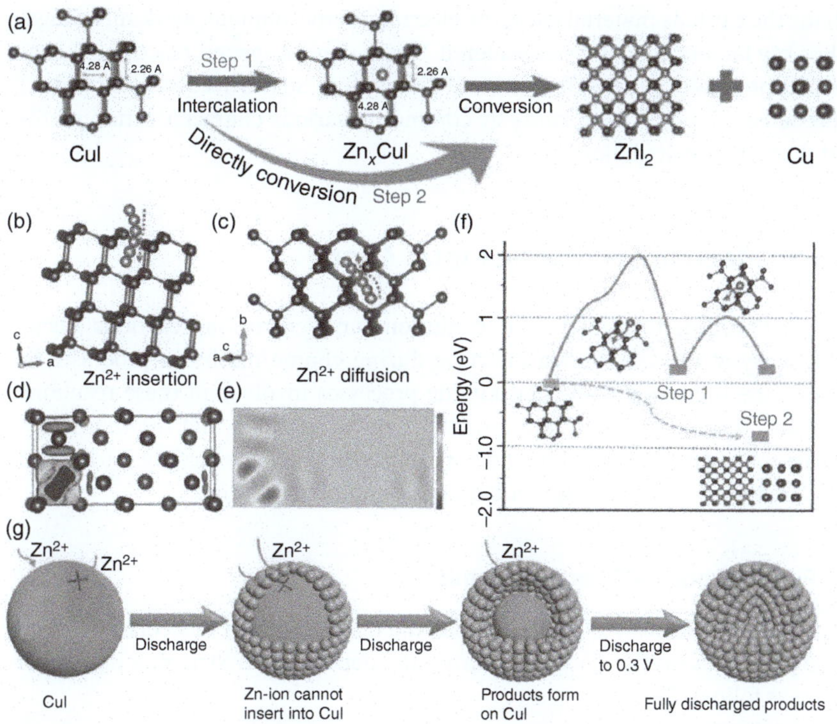

Figure 2.10 (a) The possible Zn-ion storage mechanisms of CuI electrodes. Step 1: Zn^{2+} intercalation followed by the conversion reaction. Step 2: The direct conversion reaction. (b) The simulation model for Zn-ion intercalation into CuI lattice. (c) The simulation model for Zn-ion migration in CuI lattice. (d) The differential charge density of CuI after Zn-ion intercalation. (e) Slice of the electron density difference map to show the unbalanced charge distribution. (f) The energy barriers of Zn-ion intercalation, Zn-ion migration, and the energy changes for the direct conversion reaction. (g) Schematic diagram of Zn-ion storage mechanism and its pathway in CuI electrode, showing the Zn-ion storage can be finished by direct conversion reaction. Source: Hao et al. [31]/John Wiley & Sons.

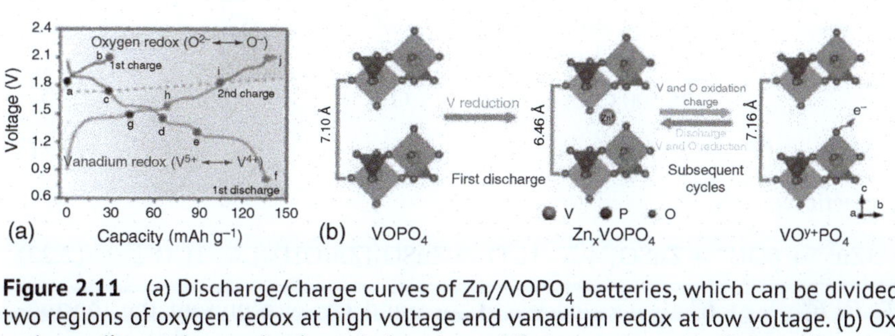

Figure 2.11 (a) Discharge/charge curves of Zn//$VOPO_4$ batteries, which can be divided into two regions of oxygen redox at high voltage and vanadium redox at low voltage. (b) Oxygen and vanadium redox mechanisms during charge/discharge processes. Source: Wan et al. [33]/John Wiley & Sons.

It can be seen that the reaction mechanism of AZIBs is very complex. In addition to studying the cathode material itself, the electrolyte environment, working voltage, and other key factors need to be considered. There may be several reaction mechanisms in the process of battery charging and discharging, so it is necessary to conduct a comprehensive and in-depth study by combining various characterization methods and theoretical calculations.

2.2 The Mechanism of Zinc Metal Anode

Zinc deposition/stripping, hydrogen evolution, and corrosion reactions occur simultaneously at the zinc anode interface during charge/discharge processes in ZIBs. Here, the thermodynamic and kinetic processes involved in these reactions (such as the crystal nucleation and growth of zinc deposition and the thermodynamic basis of hydrogen evolution and corrosion reactions) are mainly introduced to guide the design of zinc anodes with long life and high stability.

2.2.1 Fundamentals of Thermodynamics

Taking the $ZnSO_4$ electrolyte as an example, the main reactions and corresponding electrode potentials on the Zn anode during the deposition or stripping process of AZIBs are as follows [34]:

Zinc deposition reaction:

$$Zn^{2+} + 2e^- \rightarrow Zn \left(E_{Zn^{2+}/Zn} = E^{\theta}_{Zn^{2+}/Zn} - \frac{RT}{2F} \ln \frac{1}{\gamma_{Zn^{2+}} \cdot C_{Zn^{2+}}} \right) \quad (2.17)$$

Electrochemical hydrogen evolution reaction:

$$2H^+ + 2e^- \rightarrow H_2 \uparrow \left(E_{H^+/H_2} = E^{\theta}_{H^+/H_2} - \frac{RT}{2F} \ln \frac{1}{\gamma^2{}_{H^+} \cdot C^2_{H^+}} \right) \quad (2.18)$$

Corrosion hydrogen evolution reaction:

$$Zn + 2H^+ \rightarrow H_2 \uparrow + Zn^{2+} \left(E = E^{\theta}_{Zn^{2+}/Zn} - \frac{RT}{2F} \ln \frac{\gamma_{Zn^{2+}} \cdot C_{Zn^{2+}}}{\gamma^2{}_{H^+} \cdot C^2_{H^+}} \right) \quad (2.19)$$

$$E = 0.76\,V - \frac{8.31\,J\,mol^{-1}\,K^{-1} \times 298\,K}{96\,500\,C\,mol^{-1} \times 2} \ln \frac{1}{(10^{-5})^2} \approx 0.46\,V \quad (2.20)$$

$$\Delta G = -nEF \approx -87.95\,KJ < 0 \quad (2.21)$$

Side reactions:

$$3Zn^{2+} + 6OH^- + ZnSO_4 + X \cdot H_2O \rightarrow ZnSO_4[Zn(OH)_2]_3X \cdot H_2O \downarrow \quad (2.22)$$

R, T, F, $\boldsymbol{E^{\theta}_{Zn^{2+}/Zn}}$, $\boldsymbol{E^{\theta}_{H^+/H_2}}$, $\boldsymbol{\gamma_{Zn^{2+}}}$, $\boldsymbol{\gamma_{H^+}}$, $\boldsymbol{C_{Zn^{2+}}}$, and $\boldsymbol{C\gamma_{H^+}}$, represent the thermodynamic constant (8.31 J mol^{-1} K^{-1}), temperature, Faraday's constant (96 500 C mol^{-1}), zinc standard electrode potential (−0.76 V vs. H^+/H_2), standard hydrogen electrode potential (0 V vs. H^+/H_2), zinc ion activity coefficient, H^+ activity coefficient, zinc-ion concentration, and H^+ concentration, respectively. H^+ will be consumed in the electrochemical hydrogen evolution reaction (2.18) and corrosion hydrogen evolution reaction (2.19), while the residual OH^- will induce side reactions (2.22) at the zinc anode interface.

To analyze the order of zinc deposition (2.17) and hydrogen evolution reaction (2.18), as well as the spontaneity of corrosion reaction (2.19), it is assumed that 1 M zinc sulfate has a pH of 5 and activity coefficients of 1 at a temperature of 298 K. The calculated ΔG of reaction (2.19) is negative, and the E is 0.46 V, which is also equal to the potential of reaction (2.18) minus the potential of reaction (2.17). Therefore, in the absence of overpotential, the HER tends to preferentially occur in the zinc deposition reaction, and the corrosion reaction will proceed spontaneously. Even though $\boldsymbol{E}^{\theta}_{\mathbf{H}^+} > \boldsymbol{E}^{\theta}_{\mathbf{Zn}^{2+}}$, Zn deposition happens ahead of HER due to the high hydrogen evolution overpotential, sluggish hydrogen evolution kinetics, and low activity of H^+ in neutral or mildly acid electrolytes, leading to the reversible deposition/stripping of zinc ions. From the above thermodynamic process, it can be seen that the efficiency of zinc anode deposition/stripping can be improved by increasing the zinc ion activity, reducing the overpotential during the zinc deposition process, adjusting the pH value of the electrolyte, and selecting a substrate with a high hydrogen evolution overpotential [34].

2.2.2 Crystal Nucleation and Growth of Zinc Electrodeposition

From the reaction (2.17), the zinc deposition reaction, determined by the distribution of zinc ions and electrons at the anode interface, is a process of nucleation and growth of zinc ions driven by an electric field in ZIBs. In other words, dendrite-free zinc deposition can be achieved when the electron-ion distribution is more uniform on the reaction interface. The electron-ion exchange process is easier, and the growth direction is more regular on the reaction interface. In this section, current mechanisms and models related to Zn nucleation and growth are introduced to guide the interfacial modification of Zn anodes.

2.2.2.1 Nucleation

The reaction mechanisms of the Zn anode for ZIBs with mild aqueous electrolyte involve the reversible plating/stripping of Zn^{2+} in the anode, which can be summarized as $\mathbf{Zn} \leftrightarrow \mathbf{Zn}^{2+} + \mathbf{2}\boldsymbol{e}^-$ [35]. Similar to the nucleation and growth processes of lithium deposition, the uniformity of zinc deposition is closely related to the nucleation process.[36–38] During the nucleation process, Zn^{2+} needs to transcend a barrier corresponding to the nucleation overpotential, as shown in Figure 2.12a,b. Generally, the more difficult Zn nucleation is at the electrode interface, the larger the nucleation overpotential is. The Zn nucleation process may be driven by the following three factors: electrical field, ion concentration, and surface energy. In terms of the electrical field, due to a result of the "tip effect," the intensity at the tips is much greater than the remaining area, so the Zn near the tips will be subjected to a greater nucleation force (Figure 2.12d). Meanwhile, Zn ion concentration also plays a significant role in the nucleation process because the nucleation barrier

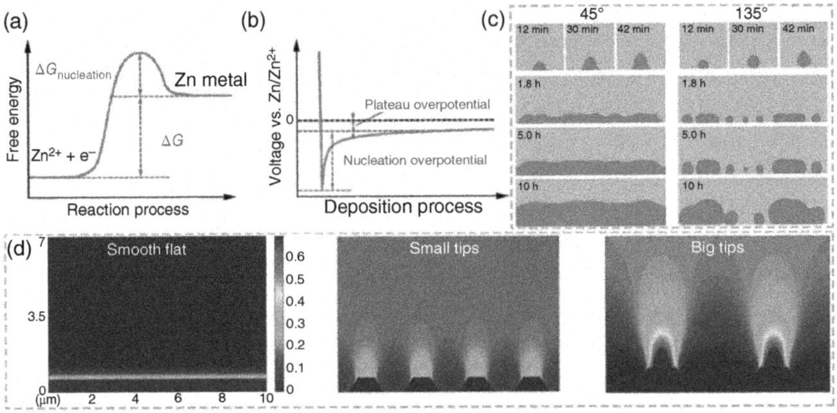

Figure 2.12 (a) Changes in free energy during Zn nucleation. Source: Zhang et al. [35]/John Wiley & Sons. (b) Typical voltage profile during zinc deposition. Source: Zhang et al. [35]/John Wiley & Sons. (c) Simulation of lithium growth on a substrate with 45° and 135° wetting angles. Source: Cheng et al. [38]/American Chemical Society. (d) Electric field simulation of the plane with different roughness. Source: Yang et al. [39]/John Wiley & Sons.

is reduced in regions with higher ion concentrations or faster ion transmission. Another influential factor is surface energy. Surface energy, influenced by zincophilicity (Figure 2.12c), defects, lattice matching degree, and so on, can decrease nucleation barriers by creating abundant nucleation sites, eventually inducing uniform deposition on the surface. It can be speculated that Zn dendrites are mainly caused by partial nucleation due to uneven distribution of electrical field and ions, or a small number of nucleation sites by high nucleation barriers on the interface [40].

The nucleation process of zinc can be described by two typical nucleation models, i.e. continuous nucleation and transient nucleation models. Schairfkeer's classical theoretical model is used to describe these two processes. The I–t curve at constant voltage was measured by chronoamperometry and normalized to represent two processes of nucleation [40]:

$$(I/I_m)^2 = 1.9542(t_m/t)\{1 - \exp[-1.2564(t/t_m)]\}^2 \tag{2.23}$$

$$(I/I_m)^2 = 1.2254(t_m/t)\{1 - \exp[-2.3367(t/t_m)]\}^2 \tag{2.24}$$

Equations (2.23) and (2.24) represent the fitting curves of instantaneous nucleation and continuous nucleation, respectively. t_m and I_m correspond to the maximum current I and the time t at peaks, respectively. The nucleation mode can be determined by how well the experimental curve fits the theoretical curve. As shown in Figure 2.13a, the actual nucleation process is instantaneous at 1.55 V due to the nucleation curve being close to the fitted instantaneous nucleation curve. From the deposition morphology (Figure 2.13b,c), Zn with few nucleation sites and a large nucleus radius undergoes instantaneous nucleation at a current density of 0.2 mA cm^{-2}. However, when continuous nucleation is carried out at a current density of 5 mA cm^{-2}, the number of nuclei is large and the nuclei radius

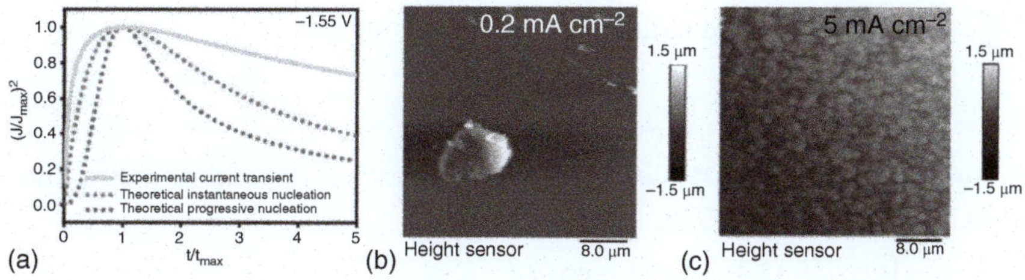

Figure 2.13 (a) Fitting of continuous and discontinuous nucleation curves. AFM images of deposited zinc at current densities of (b) 0.2 and (c) 5 mA cm^{-2}. Source: Hou et al. [41]/John Wiley & Sons.

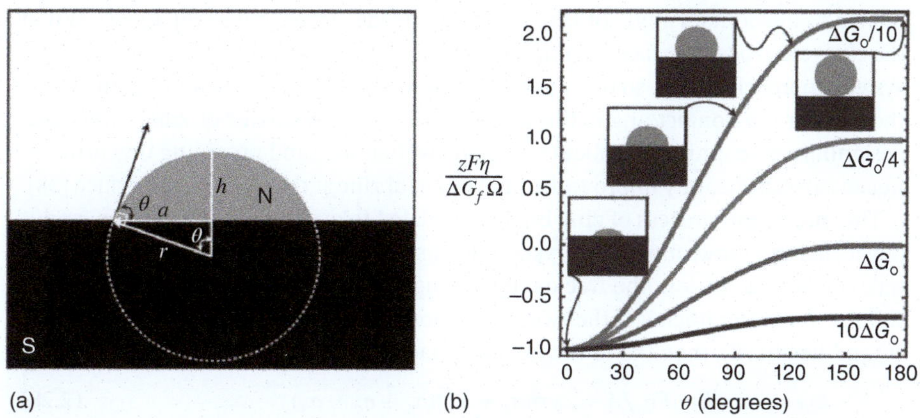

Figure 2.14 (a) The heterogeneous nucleation model. (b) The effects of contact angle and critical Gibbs free energy on nucleation overpotential. Source: Ely and Garcia [42]/IOP Publishing.

is small [41]. It can be seen that the Zn nucleation model during the charging process is closely related to the magnitude of the current, the composition of the electrolyte, and the structure of the electrode interface. In other words, dense Zn deposits can be obtained through a continuous nucleation process under conditions permitted by the polarization voltage.

Heterogeneous Nucleation Heterogeneous nucleation is the crystal nucleation process of zinc at the hetero-conducting interface (Figure 2.14a). At the initial stage of nucleation, the zinc-embryo is unstable and tends to dissolve in the electrolyte or to adsorb at the existing interface. To make the zinc-embryo grow into an independent crystal nucleus, it is necessary to overcome the interfacial energy formed between the zinc-embryo and the electrolyte. The relationship between the growth of the spherical cap nuclei and the Gibbs free energy (ΔG) of the system is described by the following formula [42]:

$$\Delta G = V\Delta G_V + 2\pi r^2(1 - \cos\theta)\sigma_1 - \pi r^2\sigma_1 \cos\theta \tag{2.25}$$

Set $\frac{\Delta G}{\Delta r} = \mathbf{0}$ to get the critical radius r^* of the crystal nucleus:

$$r^* = -\frac{2\sigma_1}{\Delta G_V} \tag{2.26}$$

Substituting (2.26) into (2.25), when the nucleus reaches the critical radius, the required Gibbs free energy ΔG^* is:

$$\Delta G^* = \frac{16\pi\sigma_1{}^3}{3\Delta G_V{}^2} \cdot \frac{(2+\cos\theta)(1-\cos\theta)^2}{4} \tag{2.27}$$

where ΔG_V, σ_1, and r are the volume-free energy released by the growth of the crystal nucleus, the surface tension between the crystal nucleus and the solution, and the embryo radius, respectively. The critical free energy of zinc nucleation is related to the contact angle of zinc on the electrode, according to Eq. (2.27). When the zinc–electrode interface is completely wetted, the critical Gibbs free energy of the system approaches zero, so there is no nucleation barrier. As shown in Figure 2.14b, the smaller the contact angle, the lower the nucleation barrier of zinc. Therefore, designing a zincophilic interface (with good wettability) and providing tiny primary nuclei can significantly improve the nucleation of zinc at the electrode interface [38].

The nucleation process of zinc is closely related to overpotential, and zinc gradually forms and grows along the crystal nucleus epitaxy in the presence of overpotential. During nucleation, the free energy change (ΔG) of the system is a function of the size (r) of the nuclei. If the two-dimensional nucleus is cylindrical with radius (r) and height (h), then the following formula is obtained:

$$\Delta G = -\pi r^2 h\rho nF\eta_c/A + 2\pi rh\sigma_1 + \pi r^2(\sigma_1 + \sigma_2 - \sigma_3) \tag{2.28}$$

Herein, ρ, n, F, η_c, A, σ_1, σ_2, σ_3, are the density of the crystal nucleus, the valence state of the metal ion, Faraday's constant, the overpotential, the relative atomic mass of the metal atom, the surface tension between the nucleus and the solution, the surface tension at the interface between the crystal nucleus and electrode, the surface tension of the electrode and solution, respectively. By calculating the nucleation behavior in the critical state ($\frac{\Delta G}{\Delta r} = 0$), the relationship between the critical nucleation radius r_c and the overpotential can be concluded as follows:

$$r_c = h\sigma_1 \Big/ \left[-\frac{h\rho nF\eta_c}{A} - (\sigma_1 + \sigma_2 - \sigma_3)\right] \tag{2.29}$$

It can be seen that the larger the overpotential during zinc deposition, the smaller the critical nuclei radius (Figure 2.15). The nucleation rate ω, overpotential (η_c), and temperature (T) have the following relationship [41]:

$$\omega = K\exp\left[-\pi\sigma_1{}^2 hLA/(\rho nFRT\eta_c)\right] \tag{2.30}$$

where K, L, and R are the pre-exponential factor, Avogadro's constant, and thermodynamic constant, respectively. It can be seen that the nucleation rate is proportional to temperature and overpotential.

Surface Diffusion and Nucleation Magnesium metal tends to form a dendritic-free morphology in the electrodeposition process compared to zinc metal [38]. The

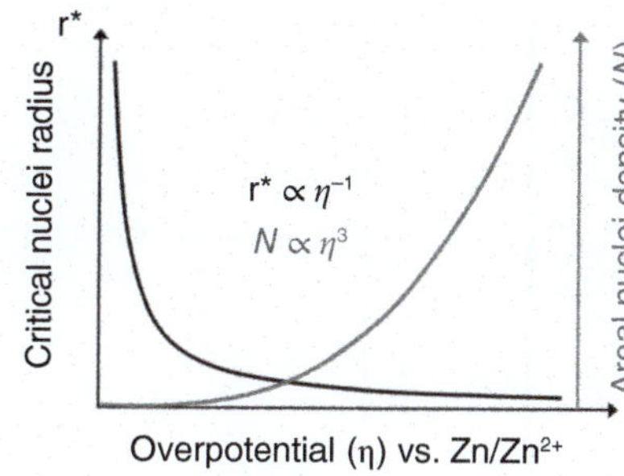

Figure 2.15 Relationship between overpotential and critical nucleus radius. Source: Hou et al. [41]/John Wiley & Sons.

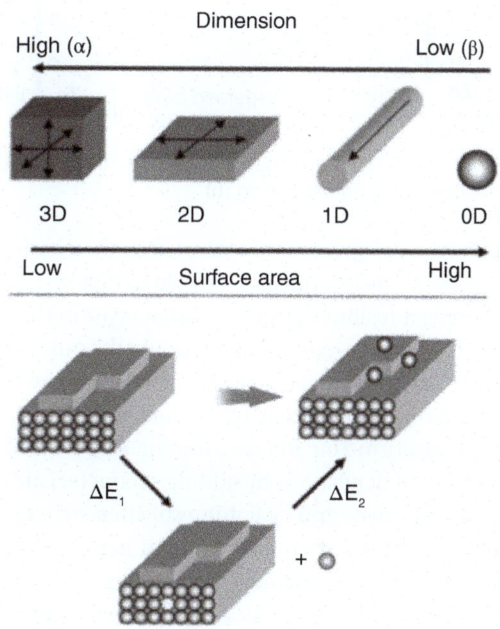

Figure 2.16 Diffusion nucleation behavior of Zn in different dimensions. Source: Cheng et al. [38]/American Chemical Society.

binding force between Mg and Mg atoms is much higher than that between zinc and zinc, so it is more difficult to form a loose shape compared to zinc. The deposition behavior of Mg, Zn, Na, and Li by some researchers indicates that magnesium ions are more likely to be deposited in the vicinity of the deposited location, rather than enriched at a site to form dendrites. Based on the above analyses, magnesium metal tends to form a 3D morphology rather than a 1D dendritic model (Figure 2.16). Therefore, similar to alkali metals, low surface energy and high ion migration barrier are two key factors for the deposition behavior of zinc. Therefore, promoting Zn nucleation can suppress Zn dendrite formation by designing electrode interfaces with low diffusion barriers.

Space Charge-Induced Nucleation Under the attraction of the electric field, in AZIBs, the zinc ions with the solvent sheath migrate from the bulk solution to the anode interface, then remove the solvent molecules to form zinc ions, and finally obtain electrons at the electrode interface to form zinc metal [42]. The deposition/stripping

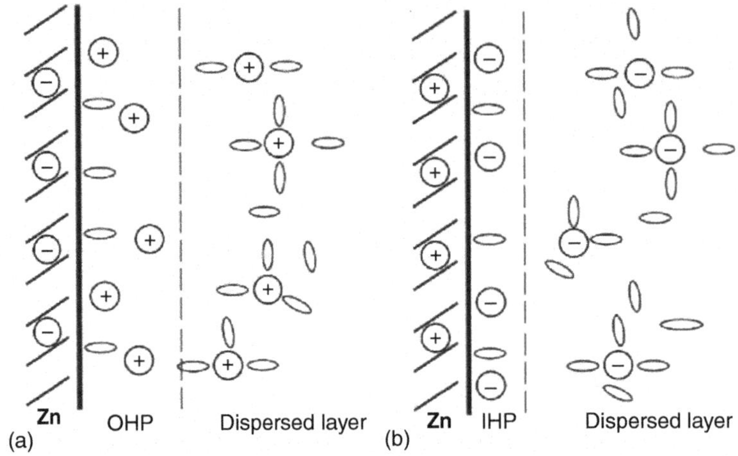

Figure 2.17 Electric double layer structure at the electrode interface during zinc deposition (a) and stripping (b).

behaviors of zinc ions are affected by the distribution of ions and solvent molecules at the electrode interface. As shown in Figure 2.17a, during the zinc deposition process, there is a residual positive charge layer in the solution close to the electrode. These positively charged layers, including the outer compact layer (outer Helmholtz plane, OHP) and the dispersed layer based on the density of the charges, are composed of cations and solvent molecules. Constructing an electric double-layer structure with uniform zinc ion distribution and a suitable transport barrier can achieve uniform zinc deposition by adding characteristic adsorbed cationic and organic molecular additives and designing solvated structures. During the zinc stripping process (Figure 2.17b), there is a residual negatively charged layer in the solution close to the electrode. These negatively charged layers are composed of anions and solvent molecules, which can be divided into two parts: the compact layer (inner Helmholtz plane, IHP) and the dispersed layer. The structure of the electric double layers is affected by the properties of anions, additives, and solvents, which play an important role in the stripping of zinc.

The electric double-layer model can be used to analyze the cause of dendrite formation. "Sand's time," which first studied the hydrogen release in a mixture of copper sulfate and sulfuric acid by J. S. Sand in 1901, is widely accepted to measure the initiation time of dendrite growth in dilute solutions. Currently, this model can also be used to explain the formation of zinc dendrites. When zinc ions are rapidly deposited in a dilute solution, the ion concentration decreases near the electric double layer. Ion depletion in the vicinity of the electrode surface creates a very large space charge and an electric field near the electrode/electrolyte interfaces, leading to a ramified growth of Zn deposits. The time required to form this space charge region, namely "Sand" (τ), can be used to quantify the opportunity for zinc dendrite nucleation. The "Sand" time can be calculated using the following formula:

$$\tau_s = \pi D \left(\frac{C_0 e Z_c}{2J} \right)^2 \left(\frac{\mu_a + \mu_c}{\mu_a} \right)^2 \tag{2.31}$$

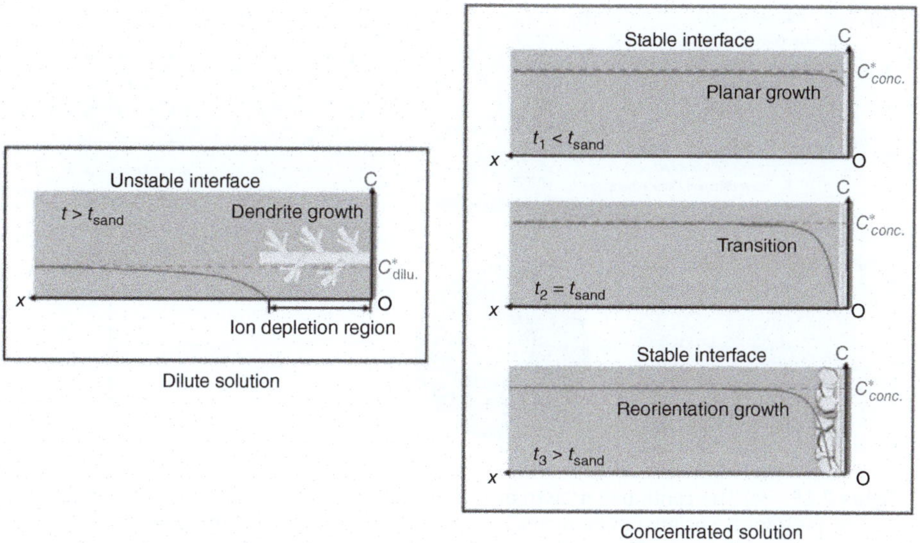

Figure 2.18 Crystal reconstruction and growth behavior during zinc electrodeposition in dilute and concentrated electrolytes. Source: Zheng et al. [43]/American Association for the Advancement of Science.

Among them, D, μ_c, μ_a, e, J, Z_c, and C_0 are the binary diffusion coefficient of ions, anion migration number, cation migration number, elementary charge, current density, cation charge, and initial zinc ion concentration, respectively. The equation shows that high E promotes the growth of dendrites because the higher the J, the shorter the τ. Significantly, the non-uniform ion distribution at the interface can induce the formation of Zn dendrites at low current densities due to the formation of a strong space electric field at the electrode interface. It can be seen that the formation of zinc dendrites is inhibited by using a reasonable current density (J), adjusting the concentration (C_0) of the electrolyte, increasing the migration number of cations (μ_c) in the electrolyte, and regulating the concentration of anions at the electrode interface. As shown in Figure 2.18, the Zn deposition in the concentrated electrolyte is more uniform than that in the dilute electrolyte, and its morphology is controlled by current density (J). In the region of high current density ($t < \tau$), zinc deposition occurs in a planar growth.

2.2.2.2 Crystal Growth of Zinc on Existing Crystal Facets/Nuclei

Surface Diffusion and Lattice Incorporation Zinc is preferentially deposited at sites (such as holes, dislocations, and steps) with higher surface energies, which can release more energy upon binding to zinc atoms. Figure 2.19 presents a schematic illustration of a growing crystal surface that consists of flat regions, the so-called terrace, and steps [44]. Two-dimensional nucleation on an existing terrace creates islands that produce new terraces and steps (the dislocation in crystals offers another source of steps). The steps are usually zigzag due to the presence of kinks. Kink sites are the most stable positions for incoming atoms or molecules to attach because more bonds can be formed there. Kink sites can be created by thermally

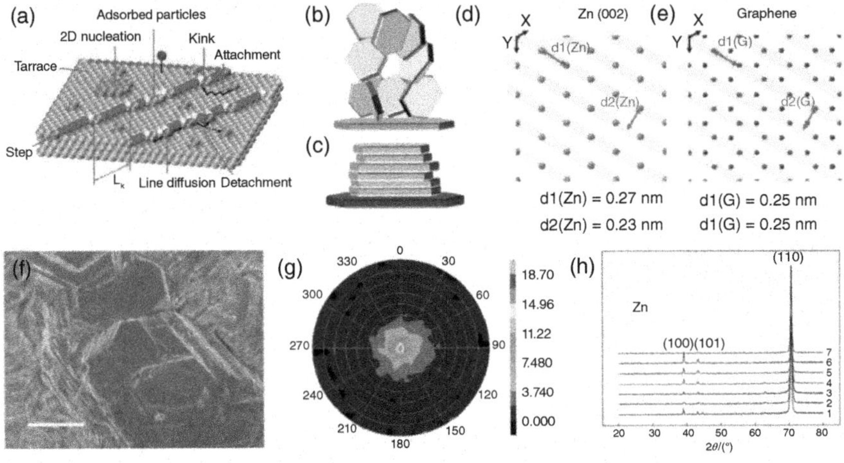

Figure 2.19 (a) The continued growth of zinc on a crystal plane with steps and defects. Source: [44]/Royal Society of Chemistry. Zn deposition behavior (b) without and (c) with epitaxial growth features. Source: Zheng et al. [45]/American Association for the Advancement of Science. (d) Lattice spacing comparison of Zn(002) plane and (e) graphene plane. Source: Adapted from Zheng et al. [45]. Preferential growth of (f, g) Zn(002) plane and (g) Zn(110) plane. Source: Yuan et al. [46]; Du et al. [47]/John Wiley & Sons.

activated detachment or adatom absorption. As the number of molecules attaching to the crystal surface on average is larger than that detaching from the surface in a unit of time, crystal growth occurs. The growth of crystals is represented by steps advancing and the formation of new terraces. During the Zn growth, the surface morphology, which is characterized by qualities such as the density of islands and the width of terraces, will greatly affect the quality of the resulting crystals. Dynamic processes, such as surface diffusion, step diffusion, and mass transfer between terraces, play critical roles in determining surface morphology. These dynamic processes, to a large extent, are controlled by a set of energy barriers that are determined by the strength of bonds between molecules, or, in other words, the interaction between metal atoms. Designing surfaces with defects and steps to direct zinc for rapid growth without crossing high nucleation barriers is an effective approach to homogenizing zinc deposition.

Epitaxial Growth At the initial stage of zinc deposition, the growth of zinc according to the original lattice of the substrate is called epitaxial growth [45]. Epitaxial growth is more obvious when the crystal facet structure of the substrate is slightly different from that of zinc, which can significantly improve the flatness and density of metal deposits and obtain deposits that are less prone to cracking and exfoliation (Figure 2.19b,c). Lattice mismatch δ can be used to predict the epitaxial growth during zinc deposition:

$$\delta = | k(d_s - d_{Zn}) | / d_s \times 100\% \tag{2.32}$$

where d_{Zn}, d_s, and k represent the lattice spacing of the zinc facet, lattice spacing of the substrate facet, and constant coefficients, respectively. In general, the lower the lattice mismatch is, the stronger the tendency for Zn epitaxial growth. If the mismatch between the substrate and Zn is small, the thickness of Zn epitaxial growth can reach more than 4 μm. As shown in Figure 2.19d and e, the degree of mismatch between graphene and Zn(002) planes is only about 7.4%, which proves Zn with strong epitaxial growth tendency on graphene. The epitaxial growth of zinc on graphene was observed by Zheng et al. The epitaxial growth of zinc on MoS_2 and Cu(111) single crystals has also been extensively studied.

Preferred Orientation Growth When epitaxial growth stops, the deposited zinc gradually forms partial twin crystals, and then zinc grows in a disordered crystal orientation. During the later stages, the deposited zinc forms a preferentially oriented layer, which stabilizes the system. For example, since zinc(002) facets grow preferentially in sulfonic acid-based electrolytes, the Zn anode is not prone to dendrites and corrosion (Figure 2.19f,g). However, the addition of amide surfactant to sulfate electrolyte can promote the preferential growth of the Zn(110) facet (Figure 2.19h) [46, 47]. The texture coefficient calculated from XRD data can be used to analyze the preferred orientation of zinc growths:

$$\mathrm{TC}_{\mathrm{hkl}} = \frac{(I_{(\mathrm{hkl})}/I_{0(\mathrm{hkl})})}{\sum(I_{(\mathrm{hkl})}/I_{0(\mathrm{hkl})})} \times 100\% \tag{2.33}$$

Here, $I_{(\mathrm{hkl})}$ is the diffraction intensity of the deposited zinc (hkl) facet, and I_0 is the diffraction intensity of the (hkl) facet in the standard XRD spectrum of zinc powder. In addition to the preferred orientation, average grain size (D) is also an important parameter to describe the zinc deposition behavior, which can be calculated by the Scherrer formula:

$$D = K\lambda/(\beta \cos\theta) \tag{2.34}$$

K, λ, β, and θ are Scherrer constants, X-ray wavelength, half-width of the diffraction peak, and diffraction angle of the diffraction peak, respectively. The dendritic formation and corrosion of Zn foils can be effectively suppressed by tuning Zn deposition with specific preferred orientations and fine grains.

References

1 Wan, F. and Niu, Z. (2019). Design strategies for vanadium-based aqueous zinc-ion batteries. *Angewandte Chemie International Edition* 58: 16358–16367.

2 Shin, J., Choi, D.S., Lee, H.J. et al. (2019). Hydrated intercalation for high-performance aqueous zinc ion batteries. *Advanced Energy Materials* 9: 1900083.

3 Xu, C., Li, B., Du, H., and Kang, F. (2012). Energetic zinc ion chemistry: the rechargeable zinc ion battery. *Angewandte Chemie International Edition* 51: 933–935.

4 Lee, B., Yoon, C.S., Lee, H.R. et al. (2014). Electrochemically-induced reversible transition from the tunneled to layered polymorphs of manganese dioxide. *Scientific Reports* 4: 6066.

5 Zhang, N., Cheng, F., Liu, J. et al. (2017). Rechargeable aqueous zinc-manganese dioxide batteries with high energy and power densities. *Nature Communications* 8: 405.

6 Kundu, D., Adams, B.D., Duffort, V. et al. (2016). A high-capacity and long-life aqueous rechargeable zinc battery using a metal oxide intercalation cathode. *Nature Energy* 1: 16119.

7 Xia, C., Guo, J., Lei, Y. et al. (2018). Rechargeable aqueous zinc-ion battery based on porous framework zinc pyrovanadate intercalation cathode. *Advanced Materials* 30: 1705580.

8 Chen, X., Zhang, H., Liu, J.-H. et al. (2022). Vanadium-based cathodes for aqueous zinc-ion batteries: mechanism, design strategies and challenges. *Energy Storage Materials* 50: 21–46.

9 Ding, Y., Zhang, L., Wang, X. et al. (2023). Vanadium-based cathodes for aqueous zinc ion batteries: structure, mechanism and prospects. *Chinese Chemical Letters* 34: 107399.

10 Yong, B., Ma, D., Wang, Y. et al. (2020). Understanding the design principles of advanced aqueous zinc-ion battery cathodes: from transport kinetics to structural engineering, and future perspectives. *Advanced Energy Materials* 10: 2002354.

11 Chen, D., Lu, M., Cai, D. et al. (2021). Recent advances in energy storage mechanism of aqueous zinc-ion batteries. *Journal of Energy Chemistry* 54: 712–726.

12 Sun, W., Wang, F., Hou, S. et al. (2017). Zn/MnO_2 battery chemistry with H(+) and Zn(2+) coinsertion. *Journal of the American Chemical Society* 139: 9775–9778.

13 Jin, Y., Zou, L., Liu, L. et al. (2019). Joint charge storage for high-rate aqueous zinc–manganese dioxide batteries. *Advanced Materials* 31: 1900567.

14 Gao, X., Wu, H., Li, W. et al. (2020). H^+-insertion boosted α-MnO_2 for an aqueous Zn-ion battery. *Small* 16: 1905842.

15 Han, M., Huang, J., Liang, S. et al. (2020). Oxygen defects in β-MnO_2 enabling high-performance rechargeable aqueous zinc/manganese dioxide battery. *iScience* 23: 100797.

16 Wan, F., Zhang, L., Dai, X. et al. (2018). Aqueous rechargeable zinc/sodium vanadate batteries with enhanced performance from simultaneous insertion of dual carriers. *Nature Communications* 9: 1656.

17 Zhao, Z., Zhao, J., Hu, Z. et al. (2019). Long-life and deeply rechargeable aqueous Zn anodes enabled by a multifunctional brightener-inspired interphase. *Energy & Environmental Science* 12: 1938–1949.

18 Liu, W., Dong, L., Jiang, B. et al. (2019). Layered vanadium oxides with proton and zinc ion insertion for zinc ion batteries. *Electrochimica Acta* 320: 134565.

19 Wang, L., Huang, K.-W., Chen, J., and Zheng, J. (2019). Ultralong cycle stability of aqueous zinc-ion batteries with zinc vanadium oxide cathodes. *Science Advances* 5: eaax4279.

20 Cai, Y., Chua, R., Kou, Z. et al. (2020). Boosting Zn-ion storage performance of bronze-type VO_2 via Ni-mediated electronic structure engineering. *ACS Applied Materials & Interfaces* 12: 36110–36118.

21 Kang, Z., Wu, C., Dong, L. et al. (2019). 3D porous copper skeleton supported zinc anode toward high capacity and long cycle life zinc ion batteries. *ACS Sustainable Chemistry & Engineering* 7: 3364–3371.

22 Lai, J., Zhu, H., Zhu, X. et al. (2019). Interlayer-expanded $V_6O_{13}{\cdot}n$H2O architecture constructed for an advanced rechargeable aqueous zinc-ion battery. *ACS Applied Energy Materials* 2: 1988–1996.

23 Ding, Y., Peng, Y., Chen, S. et al. (2019). Hierarchical porous metallic V_2O_3@C for advanced aqueous zinc-ion batteries. *ACS Applied Materials & Interfaces* 11: 44109–44117.

24 Batyrbekuly, D., Cajoly, S., Laïk, B. et al. (2020). Mechanistic investigation of a hybrid Zn/V_2O_5 rechargeable battery with a binary Li^+/Zn^{2+} aqueous electrolyte. *ChemSusChem* 13: 724–731.

25 Yang, Q., Mo, F., Liu, Z. et al. (2019). Activating C-coordinated iron of iron hexacyanoferrate for Zn hybrid-ion batteries with 10 000-cycle lifespan and superior rate capability. *Advanced Materials* 31: 1901521.

26 Pan, H., Shao, Y., Yan, P. et al. (2016). Reversible aqueous zinc/manganese oxide energy storage from conversion reactions. *Nature Energy* 1: 16039.

27 Wang, J., Wang, J.-G., Liu, H. et al. (2019). Zinc ion stabilized MnO_2 nanospheres for high capacity and long lifespan aqueous zinc-ion batteries. *Journal of Materials Chemistry A* 7: 13727–13735.

28 Lee, B., Seo, H.R., Lee, H.R. et al. (2016). Critical role of pH evolution of electrolyte in the reaction mechanism for rechargeable zinc batteries. *ChemSusChem* 9: 2948–2956.

29 Chao, D., Zhou, W., Ye, C. et al. (2019). An electrolytic Zn–MnO_2 battery for high-voltage and scalable energy storage. *Angewandte Chemie International Edition* 58: 7823–7828.

30 Ma, L., Chen, S., Li, H. et al. (2018). Initiating a mild aqueous electrolyte Co_3O_4/Zn battery with 2.2 V-high voltage and 5000-cycle lifespan by a Co(III) rich-electrode. *Energy & Environmental Science* 11: 2521–2530.

31 Hao, J., Yuan, L., Johannessen, B. et al. (2021). Studying the conversion mechanism to broaden cathode options in aqueous zinc-ion batteries. *Angewandte Chemie International Edition* 60: 25114–25121.

32 Fang, G., Liang, S., Chen, Z. et al. (2019). Simultaneous cationic and anionic redox reactions mechanism enabling high-rate long-life aqueous zinc-ion battery. *Advanced Functional Materials* 29: 1905267.

33 Wan, F., Zhang, Y., Zhang, L. et al. (2019). Reversible oxygen redox chemistry in aqueous zinc-ion batteries. *Angewandte Chemie International Edition* 58: 7062–7067.

34 Xie, C., Li, Y., Wang, Q. et al. (2020). Issues and solutions toward zinc anode in aqueous zinc-ion batteries: a mini review. *Carbon Energy* 2: 540–560.

35 Zhang, Q., Luan, J., Tang, Y. et al. (2020). Interfacial design of dendrite-free zinc anodes for aqueous zinc-ion batteries. *Angewandte Chemie International Edition* 59: 13180–13191.

36 Pei, A., Zheng, G., Shi, F. et al. (2017). Nanoscale nucleation and growth of electrodeposited lithium metal. *Nano Letters* 17: 1132–1139.

37 Gireaud, L., Grugeon, S., Laruelle, S. et al. (2006). Lithium metal stripping/plating mechanisms studies: a metallurgical approach. *Electrochemistry Communications* 8: 1639–1649.

38 Cheng, X.-B., Zhang, R., Zhao, C.-Z., and Zhang, Q. (2017). Toward safe lithium metal anode in rechargeable batteries: a review. *Chemical Reviews* 117: 10403–10473.

39 Yang, Q., Liang, G., Guo, Y. et al. (2019). Do zinc dendrites exist in neutral zinc batteries: a developed electrohealing strategy to in situ rescue in-service batteries. *Advanced Materials* 31: 1903778.

40 Li, Y., Wu, P., Zhong, W. et al. (2021). A progressive nucleation mechanism enables stable zinc stripping–plating behavior. *Energy & Environmental Science* 14: 5563–5571.

41 Hou, Z., Gao, Y., Zhou, R., and Zhang, B. (2022). Unraveling the rate-dependent stability of metal anodes and its implication in designing cycling protocol. *Advanced Functional Materials* 32: 2107584.

42 Ely, D.R. and Garcia, R.E. (2013). Heterogeneous nucleation and growth of lithium electrodeposits on negative electrodes. *Journal of the Electrochemical Society* 160: 662–668.

43 Zheng, J., Yin, J., Zhang, D. et al. (2020). Spontaneous and field-induced crystallographic reorientation of metal electrodeposits at battery anodes. *Science Advances* 6: eabb1122.

44 Zhang, T.H. and Liu, X.Y. (2014). Experimental modelling of single-particle dynamic processes in crystallization by controlled colloidal assembly. *Chemical Society Reviews* 43: 2324–2347.

45 Zheng, K.J., Zhao, Q., Tang, T. et al. (2019). Reversible epitaxial electrodeposition of metals in battery anodes. *Science* 366: 645–648.

46 Yuan, D., Zhao, J., Ren, H. et al. (2021). Anion texturing towards dendrite-free Zn anode for aqueous rechargeable batteries. *Angewandte Chemie International Edition* 60: 7213–7219.

47 Du, N., Shu, W., Zhao, Q. et al. (2013). Preparation of zinc electrodeposits with (110) lattice plane fully preferred orientation and its corrosion resistance. *The Chinese Journal of Nonferrous Metals* 23: 426–433.

3

Cathode Materials for Aqueous Zinc-ion Batteries

3.1 Manganese-Based Cathode Materials

3.1.1 Introduction to Different Mn-Based Materials

Manganese, an abundant transition metal, exists in various valences (Mn^{2+}, Mn^{3+}, Mn^{4+}, Mn^{7+}), which makes the diversification of manganese-based compounds (MnO, MnO_2, Mn_2O_3, Mn_3O_4, $ZnMn_2O_4$, MnS, etc.) [1]. Their electrochemical behavior and performance in energy storage devices are naturally different as a result of their different crystal structures, valence, morphology, and specific surface area [2]. Over the past decades, manganese-based cathode materials of aqueous zinc-ion batteries have become more attractive to scientific researchers. Among them, MnO_2 is dominantly used due to its relatively higher operational voltage (~1.3 V vs. Zn/Zn^{2+}) and theoretical capacity (308 mAh g^{-1}, corresponding to one e^- transfer) [3]. It is constructed from the basic building units (MnO_6 octahedra), which are connected via edges and/or corners. And their linkage manner results in different crystallographic polymorphs, which are tunnel (α-, β-, γ-, and todorokite-) MnO_2, layered (δ-) MnO_2, and spinel (λ-) MnO_2 (Figure 3.1). With the same fundamental building blocks, these structures can easily transform between the various configurations, enabling reversible conversion chemistry during electrochemical cycling [3].

3.1.1.1 Tunnel-Type Structure

As discussed above, α-, β-, γ-, and todorokite-MnO_2 fall into the type with tunnel structure. And they are generally extracted from various minerals. Among them, all the hollandite-type, manjiroite-type, cryptomelane-type, and coronadite-type MnO_2, connected by corner-shared double chains that consist of MnO_6 octahedra (2X2 cross section), can be regarded as α-MnO_2. Besides, Nsutite-type MnO_2 (γ-MnO_2, 1X1 and 1X2 cross sections) is reported as an intergrowth between pyrolusite-type MnO_2 (β-MnO_2, 1X1 cross section) and ramsdellite-type MnO_2 (R-MnO_2, 1X2 cross section). In addition, those built by triple chains composed of MnO_6 octahedra can be classified as todorokite-type MnO_2 and they are usually endowed with the largest tunnel size (3X3 cross-section). Unfortunately, the limited capacity in aqueous zinc ion batteries is the inevitable corollary of these large

Aqueous Zinc Ion Batteries: Fundamentals, Materials, and Design, First Edition.
Haiyan Wang, Qi Zhang, Yixin Li, and Yougen Tang.

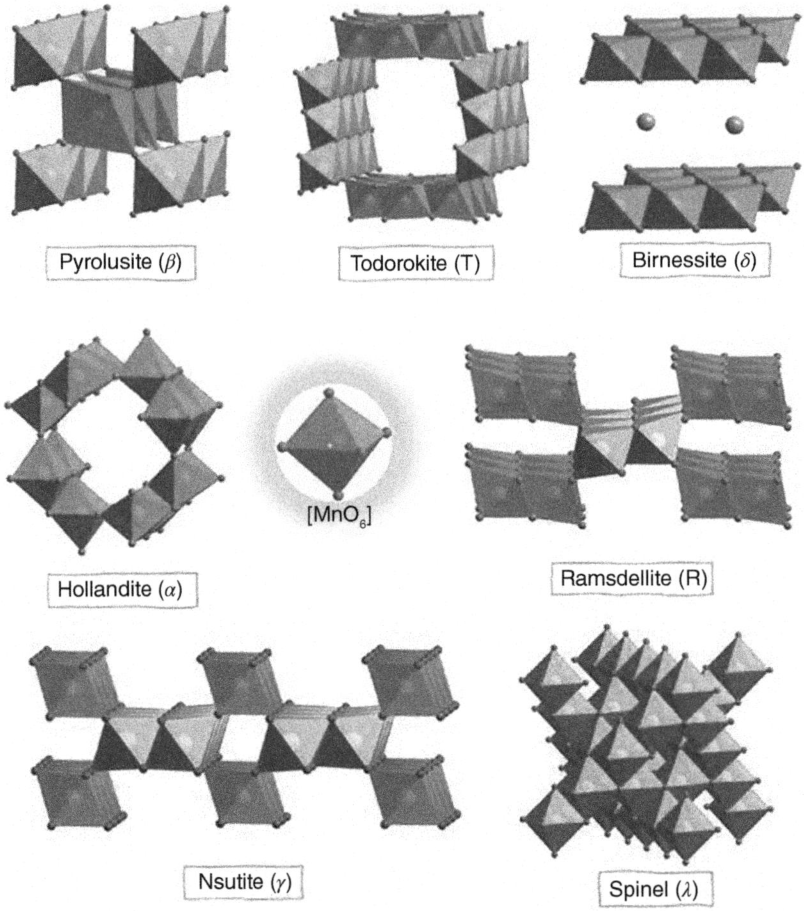

Figure 3.1 Schematic representation of various MnO_2 polymorphs characterized by a tunnel structure (m × n), including α-MnO_2 (2 × 2), β-MnO_2 (1 × 1), γ-MnO_2 (1 × 1) + (1 × 2), R-MnO_2 (1 × 2), δ-MnO_2 (1 × ∞), T-MnO_2 (3 × 3), and λ-MnO_2 (1 × 1). Source: Reproduced with permission from Jia et al. [1]. Copyright 2020, American Chemical Society.

tunnels, because they are likely to accommodate metal ions and water, in which case the average oxidation state of Mn would be reduced [4].

α-MnO_2:

The α-MnO_2 has one-dimensional 2×2 tunnels (4.6×4.6 Å) in its crystal structure. It is the most studied manganese-based material, which enables efficient storage and fast diffusion of foreign cations along the *z*-axis (Figure 3.2a) [5]. According to Kang and coworkers in 2009 [6], Zn^{2+} in neutral zinc sulfate is believed to be embedded in α-MnO_2 to form the $ZnMn_2O_4$ phase during the discharge process, which is evidenced by X-ray diffraction analysis (XRD) (Figure 3.2b). This process could be reversed upon charging (Figure 3.2c). Based on XANES analysis, Kim and coworkers [7] found the tetravalent manganese was completely converted into trivalent when discharging (Figure 3.2d), while the trivalent manganese fully

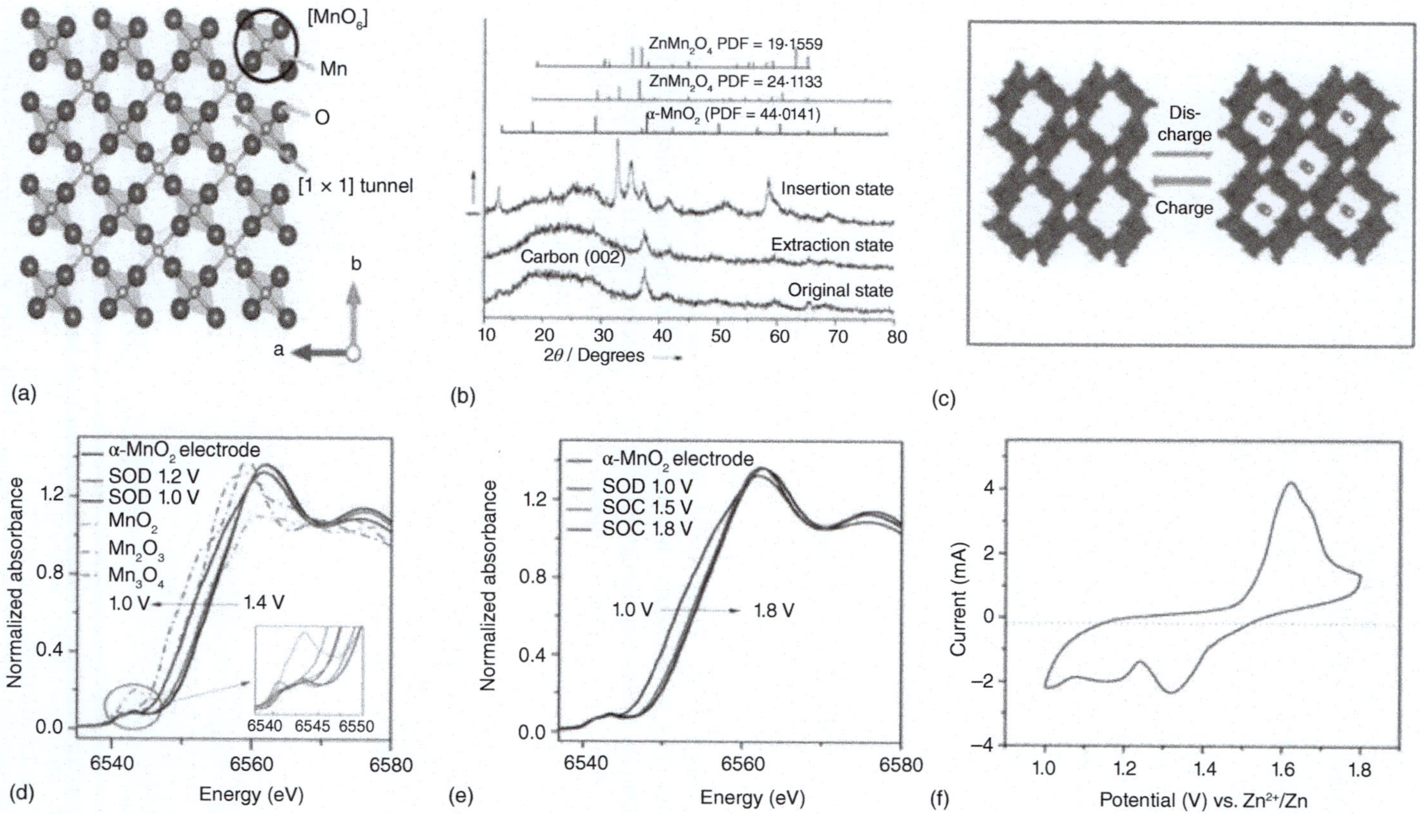

Figure 3.2 (a) Crystal structure of α-MnO_2. (b) The XRD spectrum of α-MnO_2 electrodes in different states (original, extraction, and insertion). (c) The demonstration of Zn^{2+} intercalation in α-MnO_2 cathode. (d, e) Ex-situ XANES test for α-MnO_2 electrode in different discharge (d)/charge (e) states. (f) CV curves of α-MnO_2 electrode at a scan rate of 0.5 mV/s. Source: (a) Adapted from Feng et al. [5]. (b, c) Adapted from Xu et al. [6]. (d–f) Alfaruqi et al. [7]/with permission of Elsevier.

turned back to tetravalent when charging (Figure 3.2e). As demonstrated in the CV curves, there are two distinct peaks located at 1.3 and 1.2 V (Figure 3.2f). Inspired by this, Kim et al. speculated that it may take two steps for Zn^{2+} to insert into the α-MnO_2 cathode. The details could be shown as follows:

$$0.5Zn^{2+} + 2MnO_2 + e^- \leftrightarrow Zn_{0.5}Mn_2O_4 \quad (3.1)$$

$$0.5Zn^{2+} + Zn_{0.5}Mn_2O_4 + e^- \leftrightarrow ZnMn_2O_4 \quad (3.2)$$

Unfortunately, substantial evidence for this mechanism has not been dug out and further exploration is indispensable.

Different from the above, several other discoveries about the working principles of α-MnO_2 have been reported by Oh and coworkers [8] recently. They believe a reversible phase change between α-MnO_2 and layered Zn-birnessite would take place as the Zn^{2+} ions insert/extract (Figure 3.3a). In the middle of a discharge process, ≈1/3 Mn of α-MnO_2 dissolves into the electrolyte, generating Zn-birnessite at the end (Eqs. (3.3) and (3.4)). Upon charging, the original α-MnO_2 structure could be completely recovered (Eq. (3.5)).

$$Mn^{4+}(s) + e^- \leftrightarrow Mn^{3+}(s) \quad (3.3)$$

$$Mn^{3+} \leftrightarrow Mn^{4+}(s) + Mn^{2+}(aq) \quad (3.4)$$

$$Mn^{2+}(aq) \leftrightarrow Mn^{4+}(s) + 2e^- \quad (3.5)$$

As research progressed, they revealed that layered zinc hydroxide sulfate ($Zn_4(OH)_6(SO_4)\cdot 5H_2O$, ZHS) would be formed on the α-MnO_2 surface upon discharging (Figure 3.3b) [9]. This precipitation occurs because unstable trivalent manganese undergoes a disproportionation reaction and is dissolved in the electrolyte during the discharge process, resulting in a gradual increase in the pH value of the electrolyte. This causes zinc hydroxide sulfate to crystallize from the

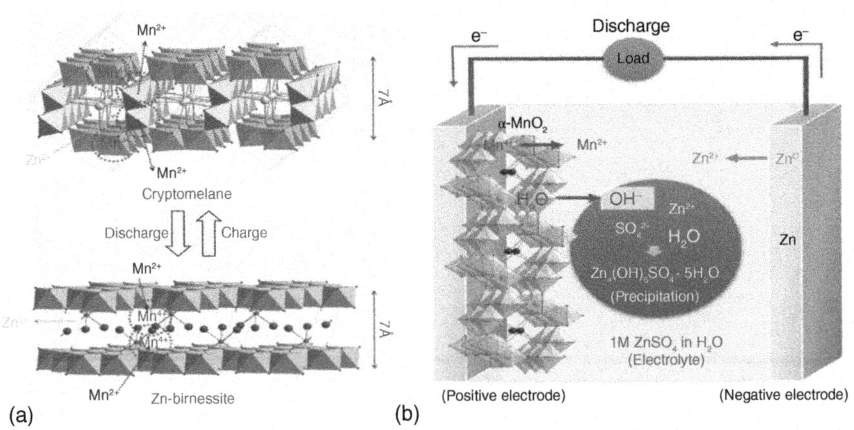

Figure 3.3 (a) The phase change taking place between α-MnO_2 and layered Zn-birnessite upon Zn^{2+} ion insertion/extraction. (b) The formation of layered zinc hydroxide sulfate when discharged in α-MnO_2/Zn cell with aqueous $ZnSO_4$ electrolyte. Source: (a) Lee et al. [8]/Springer Nature. (b) Lee et al. [9]/John Wiley & Sons.

electrolyte on the electrode surface. During the charging process, the pH value of the electrolyte decreases due to the recombination of manganese on the cathode, leading to the dissolution of zinc hydroxide sulfate back into the electrolyte. An analogous phenomenon could also be observed in todorokite. The pathways of these reactions are listed below:

$$MnO_2 + 2H_2O + 2e^- \leftrightarrow Mn^{2+} + 4OH^- \tag{3.6}$$

$$4Zn^{2+} + SO_4^{2-} + 6OH^- + 5H_2O \leftrightarrow Zn_4(OH)_6SO_4 \cdot 5H_2O \downarrow \tag{3.7}$$

Concerning the ZHS mechanism, Liu and coworkers [10] proposed another view based on a different chemical conversion, which is between α-MnO_2 and MnOOH. In this view, α-MnO_2 could react with H^+ from water to form MnOOH and OH^-.

$$H_2O \leftrightarrow H^+ + OH^- \tag{3.8}$$

$$MnO_2 + H^+ + e^- \leftrightarrow MnOOH \tag{3.9}$$

With the OH^-, $ZnSO_4$, and H_2O involved in the following reaction, ZHS would eventually appear on the electrode surface.

$$\frac{1}{2}Zn^{2+} + OH^- + \frac{1}{6}ZnSO_4 + \frac{x}{6}H_2O \leftrightarrow \frac{1}{6}ZnSO_4[Zn(OH)_2]_3 \cdot XH_2O \tag{3.10}$$

β-MnO_2:

With the narrowest (1 × 1) tunnel structure [11], β-MnO_2 makes Zn^{2+} ion insertion a little difficult, which could be significantly improved by tailoring the morphology and structure of β-MnO_2. For instance, β-MnO_2 nanorods with exposed (101) planes exhibit a high discharge capacity (270 mAh g^{-1} at 100 mA g^{-1}) and passable capacity retention (75% over 200 cycles at 200 mA g^{-1}), making the β-MnO_2 acceptable as a cathode of aqueous zinc ion battery [12]. A collection of Zn^{2+} intercalation and conversion reactions was reported to explain the Zn^{2+} storage mechanism in β-MnO_2. Specifically, Zn^{2+} first intercalates into the β-MnO_2 during the initial cycles, and then the phase transition from β-MnO_2 to $ZnMn_2O_4$ occurs after the 30th cycle as well as the zinc sulfate hydroxide hydrate ($ZnSO_4 \cdot 3Zn(OH)_2 \cdot 5H_2O$) precipitates. In the following cycles, the latter $ZnMn_2O_4$ structure is where the reversible Zn^{2+} insertion/extraction process proceeds [12]. However, a different view stated that the phase transition from the tunneled β-MnO_2 to a layered Zn-buserite (B-$Zn_xMnO_2 \cdot nH_2O$) in the initial cycles is irreversible, followed by reversible intercalation of Zn^{2+} into the Zn-buserite structure [13]. The manifesting differences between the beginning and following cycles in the CV curves (Figure 3.4a) and galvanostatic profiles (Figure 3.4b) are strong evidence of the changed Zn^{2+} storage mechanism along with cycling. Similarly, this electrochemical behavior could be observed in α-MnO_2 and γ-MnO_2 as well, indicating that this might be a general reaction mechanism for tunnel-type MnO_2.

γ-MnO_2:

The low-temperature γ-MnO_2 phase, which has stochastically distributed 1 × 1 (2.3 × 2.3 Å) and 1 × 2 (2.3 × 4.6 Å) tunnel blocks, can be regarded as an intergrowth

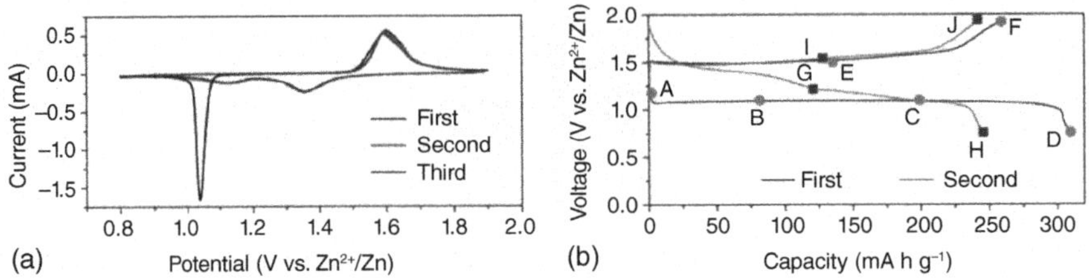

Figure 3.4 β-MnO_2/Zn full batteries: (a) CV curves for the first three cycles. (b) Galvanostatic charge/discharge (GCD) curves for the first two cycles. Source: Zhang et al. [13]/Springer Nature/Public Domain CC BY 4.0.

of R- and β-MnO_2 phases [14]. Alfaruqi et al. first reported the possibility of zinc insertion in the γ-MnO_2 and the electrochemically induced multiphase transformation [15]. As seen in Figure 3.5, the structure of a partial γ-MnO_2 will transform into a spinel-type Mn^{3+} phase ($ZnMn_2O_4$) in the initial stage of discharge. As Zn^{2+} continues to insert into the [1 × 2] tunnels of γ-MnO_2 in the further discharging process, a new tunnel-type Mn^{2+} phase (γ-Zn_xMnO_2) gradually appears. Eventually, a layer-structured L-Zn_yMnO_2 (Mn^{2+} phases) will be formed as a result of some expanded and collapsed tunnels via full zinc-insertion. Note that these three products coexist at full-discharge state, and they will go back to the original γ-MnO_2 when completing a charging process. The discharge reactions are as follows:

$$Zn + 2e^- + 2MnO_2 \rightarrow ZnMn_2O_4 \tag{3.11}$$

$$2Zn^{2+} + 2xe^- + MnO_2 \rightarrow Zn_nMnO_2 \tag{3.12}$$

Additionally, high-valence Mn phases ($ZnMn^{3+}{}_2O_4$) are located on the cathode surface, whereas the inner region is filled with low-valence Mn^{2+} ions. The reasons for this are still ambiguous since the reduction of Mn ions starts

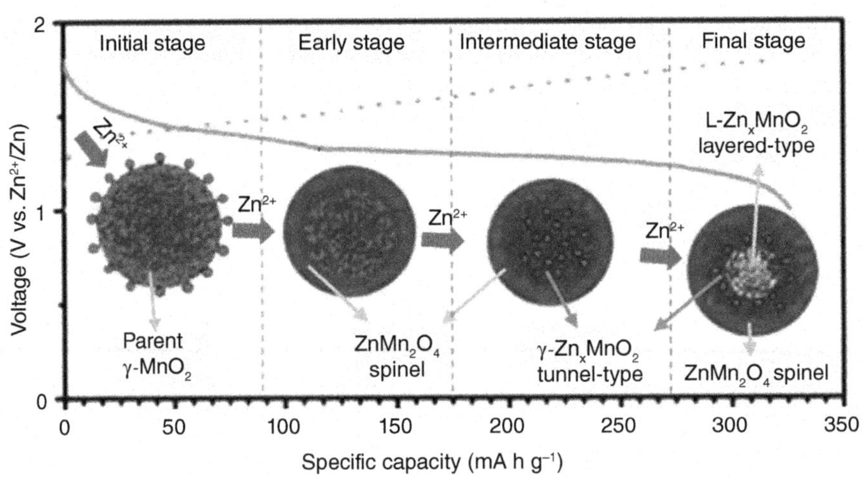

Figure 3.5 The demonstration of multistep phase transformation in the γ-MnO_2 cathode during discharge. Source: Alfaruqi et al. [15]/American Chemical Society.

at the cathode/electrolyte interfaces. With the inspiration of the discovery of V_2O_5–PEDOT electrodes [16], a similar speculation is proposed that the divalent Mn ions on the surface are susceptible to oxidation in an environment containing oxygen.

Todorokite-type MnO_2:

As a result of a large one-dimensional (3 × 3) tunnel structure, todorokite-type MnO_2 is hoped to achieve more Zn^{2+} accommodation and faster Zn^{2+} diffusion [17]. Generally, its structure could be stabilized by the intercalation of metal cations (including Na^+, Mg^{2+}, and Cu^{2+}) and water molecules that can partially shield the charge of Zn^{2+} ions [18]. In theory, with a large tunnel size, this shielding effect can generate a faster Zn^{2+} diffusion within todorokite-type MnO_2 since the electrostatic interactions are alleviated between the Zn^{2+} and the host structure [18b]. This naturally makes sense for the high rate performance and large capacity in contrast to the α-MnO_2 with smaller (2 × 2) tunnels. Unfortunately, its initial discharge capacity in practice is much lower than that of α-MnO_2 (98 vs. 227 mA h g^{-1}) [18b], which should be attributed to those water molecules and metal cations preoccupying the potential Zn^{2+} insertion sites in the tunnels. In short, todorokite-type MnO_2 is promising as cathode material for ZIBs, but its application is critically in shortage of in-depth mechanism exploration.

3.1.1.2 Layered (δ-) MnO_2

Not only MnO_6 octahedra chains but also the sheets assembled by sharing edges of MnO_6 contribute to the construction of layered MnO_2. This layered structure could store various interlayer cations and water molecules, like birnessite and buserite [4, 19].

With a large interlayer distance of 7.0 Å, δ-MnO_2 exhibits great potential as a cathode candidate for ZIBs [19]. In a Zn//δ-MnO_2 battery, DFT calculations demonstrated that the H^+/Zn^{2+} intercalation and conversion appear at about 1.4 and 1.26 V, respectively [20]. However, its capacity degradation is serious upon long-term cycling [21], which is ascribed to the formation of electrochemically inactive $ZnSO_4 \cdot 3Zn(OH)_2 \cdot 4H_2O$ (BZS) on the cathode surface [19a] and the loss of active materials (Eqs. 3.13 and 3.14).

$$MnO_2(s) + H^+ + e^- \rightarrow MnOOH(s)\ (1^{st}\ \text{eletron discharge}) \tag{3.13}$$

$$MnOOH(s) + 3H^+ + e^- \rightarrow Mn^{2+}(aq) \tag{3.14}$$

As shown in Figure 3.6, these side effects mentioned above are allowed by a lower Gibbs free energy barrier at smaller current densities. Therefore, cycling the Zn//δ-MnO_2 battery at a high charge/discharge rate could suppress the kinetic-limiting conversion reactions due to the increased energy barrier [20].

In addition, it is worth noting that no matter what the pristine crystal structures of MnO_2 are, they are inclined to transform into layered birnessite MnO_2 during the Zn^{2+} storage processes [13, 22]. This phase transition is irreversible, and those undesired conversion reactions would inevitably boost the collapse of the MnO_2 crystal structure [8, 20, 22]. Thus, studies to strengthen the layered structure should be carried out in the future.

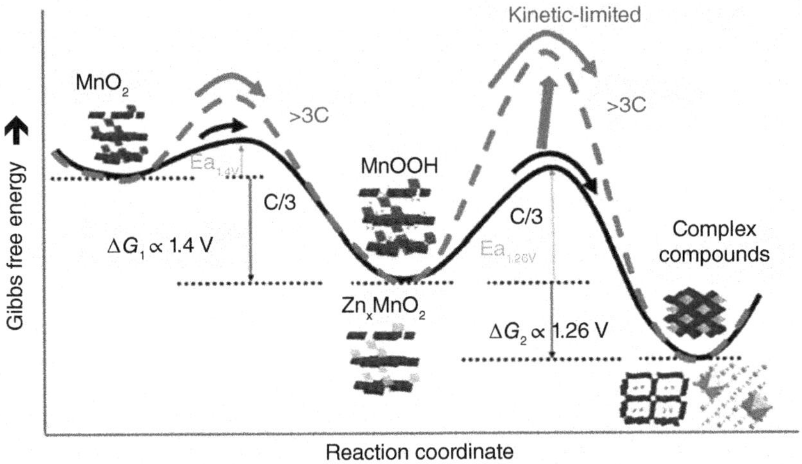

Figure 3.6 Kinetics and thermodynamics of the redox reactions in Zn//MnO_2 batteries at various rates. Source: Li et al. [20]/American Chemical Society.

3.1.1.3 Spinel (λ-) MnO_2

Connected by tetrahedron and octahedron, three-dimensional (3D) tunnels construct spinel MnO_2, featuring favorable ion diffusion. λ-MnO_2 is typical of this structure with Mn and O ions occupying the octahedral 16d and 32e sites. In addition to MnO_2, other Mn-based oxides are built by spinel structure as well, such as Mn_3O_4, $LiMn_2O_4$, and $ZnMn_2O_4$. For Mn_3O_4, the tetrahedral sites are filled with Mn^{2+}; the octahedral sites are filled with Mn^{3+}. In $LiMn_2O_4$ and $ZnMn_2O_4$, Li^+/Zn^{2+} instead of Mn^{2+} is located at the tetrahedral sites likewise [4].

In contrast to those with mesopores (α-, γ-, and δ-MnO_2) [11], λ-MnO_2 is nonporous, which has been proved by its smallest specific surface area ($5.2\,m^2\,g^{-1}$) and total pore volume ($0.0088\,cm^3\,Å^{-1}\,g^{-1}$). This character is held partially responsible for its poor electrochemical activity [11]. Its general synthesis method (i.e. acid leaching from the spinel $LiMn_2O_4$) cripples the electrochemical performance of λ-MnO_2 as well [23]. Because of the acidic environment, material degradation is inevitable, even though the spinel structure can be well preserved after extracting lithium from $LiMn_2O_4$ [24]. To prepare high-performance λ-MnO_2, other synthesis methods are thus desired, such as electrochemical delithiation [24].

Given that the spinel-type $LiMn_2O_4$ has successfully been used as a Li intercalation host, the similar spinel $ZnMn_2O_4$ (ZMO) is expected to be a superior cathode candidate in ZIBs. Researchers reported that its spinel structure could be well maintained during the entire Zn^{2+} insertion/extraction process [25]. However, the large Zn^{2+} diffusion barrier seriously hampers the development of Zn^{2+} ion storage [26]. Note that Zn^{2+} ions are willing to insert into the tetrahedral sites of the ZMO structure instead of the larger unoccupied octahedral sites upon discharging (Figure 3.7a,b). This is the result of a high electrostatic repulsion for Zn^{2+} from the

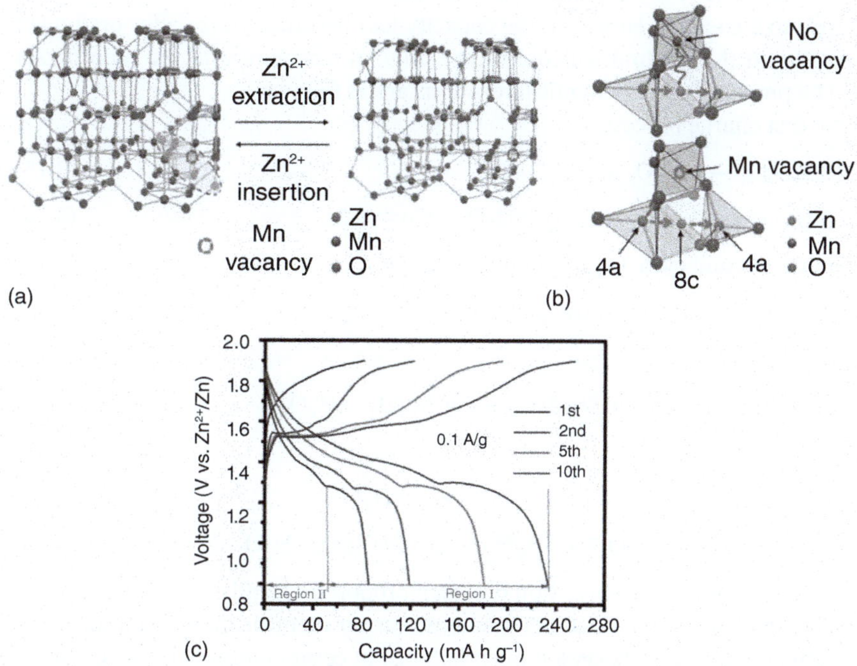

Figure 3.7 (a) The mechanism of Zn^{2+} insertion/extraction in ZMO. (b) Prospective paths for Zn^{2+} diffusion in ZMO with and without Mn vacancies. (c) GCD curves (at 0.1 A g^{-1}) of Mn_3O_4/Zn battery. Source: (a, b) Adapted from Knight et al. [26]. (c) Hao et al. [27]/with permission of Elsevier.

Mn cations in the neighboring octahedral sites when passing through an unoccupied octahedral site.

The tetragonal spinel-Mn_3O_4 ($MnMn_2O_4$) is another spinel-type manganese oxide that has been explored for ZIBs [27]. Its intercalation and conversion reaction process for Zn^{2+} storage is much more complicated in contrast to that in ZMO [27]. At the initial stage of the first charging, a spinel Mn_3O_4 would transform into a layered Mn_5O_8 as a result of the disproportionation reaction of Mn^{3+} (Eq. 3.15). With the participation of the dissolved Mn^{2+} and the intercalated H_2O, the Mn_5O_8 intermediate would turn to a birnessite phase ($MnO_2 \cdot xH_2O$) in the middle of further charging to 1.9 V (Eq. 3.16). Note that, zinc ions do not participate in the first charge process, but they will insert into the birnessite $MnO_2 \cdot xH_2O$ phase upon discharging, generating Zn-birnessite (Eq. 3.18). While that is going on, MnOOH would be formed by the conversion reaction between $MnO_2 \cdot xH_2O$ and H^+, and a simultaneously increased pH of the electrolyte (Eq. 3.20) that triggers the appearance of zinc sulfate pentahydrate (Eq. 3.21). As seen in Figure 3.7c, the charge capacity gradually increases in the first 10 cycles. This is because of the progressive formation of birnessite after the second cycle at which the Zn^{2+} deintercalation from Zn-birnessite (Eq. 3.17) occurs simultaneously with the phase transformation of Mn_3O_4 to birnessite. However, the capacity loss becomes worse after prolonged cycling due to the gradual collapse of the original structure into the amorphous state. The proposed electrochemical reactions are as follows:

In the first charge process,

$$2Mn_3O_4 \rightarrow Mn_5O_8 + Mn^{2+} + 2e^- \tag{3.15}$$

$$Mn_5O_8 + xH_2O \rightarrow 4MnO_2 \cdot xH_2O + Mn^{2+} + 2e^- \tag{3.16}$$

When the second cycle begins, upon charging,

$$Zn_yMnO_2 \cdot xH_2O \rightarrow MnO_2 \cdot xH_2O + yZn^{2+} + 2ye^- \tag{3.17}$$

During charging,

$$MnO_2 \cdot xH_2O + yZn^{2+} + 2ye^- \rightarrow Zn_yMnO_2 \cdot xH_2O \tag{3.18}$$

$$H_2O \leftrightarrow H^+ + OH^- \tag{3.19}$$

$$MnO_2 \cdot xH_2O + H^+ + e^- \leftrightarrow MnOOH + xH_2O \tag{3.20}$$

$$4Zn^{2+} + 6OH^- + SO_4^{2-} + 5H_2O \leftrightarrow Zn_4(OH)_6SO_4 \cdot H_2O \tag{3.21}$$

A novel reversible anion insertion has been proposed recently for the charge storage mechanism in spinel Mn_3O_4. The charge carrier identification is conducted by the electrochemical quartz crystal microbalance in combination with the CV tests. It is found that Cl^- anions conduct a reversible in-/desertion process, as evidenced by a mass change rate of ~35 g/mol e^- accompanied by the redox reaction [28]. In detail, the inserted Zn^{2+} ions are locked in the Mn_3O_4 structure to induce an irreversible phase transformation to $Zn_{0.2}Mn_3O_4$ during the first discharge. Subsequently, the reversible Cl^- anions storage takes place inside the $Zn_{0.2}Mn_3O_4$ in which each trapped Zn^{2+} ion hosts four Cl^- ions, and the final discharged product is $Zn_{0.2}Mn_3O_4Cl_{1.7}$ [28]. This anion insertion mechanism provides new insights into future battery chemistry. As known, a gradual crystallization leads to an irreversible capacity fading. Amorphization of Mn_3O_4 could therefore be regarded as a cathode candidate without the crystallization limiting ion diffusion [29]. But the outlook of the amorphization strategy is uncertain.

3.1.1.4 Other Manganese Oxides

With a rock salt structure, manganese monoxide (MnO) has been successfully selected as electrode material for Li-ion batteries or pseudocapacitors [30]. But, due to the absence of tunnels inside this structure, it was previously considered to be an inactive cathode for ZIBs. Recent works reported that MnO can undergo a phase transformation to a layered MnO_2 to accommodate Zn^{2+} in the first charge process, overturning the previous findings [31]. This transformation could be proved by the X-ray photoelectron spectroscopy (XPS) spectra of Mn 3s (Figure 3.8a) as well. In specific, the binding energy decreases from 5.8 to 4.7 eV after three cycles, suggesting an increased Mn valence state from +2 to +3.9 [31]. Therefore, the newly formed MnO_2 will allow the sequential intercalation of H^+ and Zn^{2+} as an active host in the following cycles, and the $Zn_4(OH)_6SO_4 \cdot nH_2O$ (ZHS) will be simultaneously produced, which is nearly identical to the working mechanism of reported MnO_2 cathode [13, 35]. In other tunnel-less materials, such as Mn_3O_4 [27]

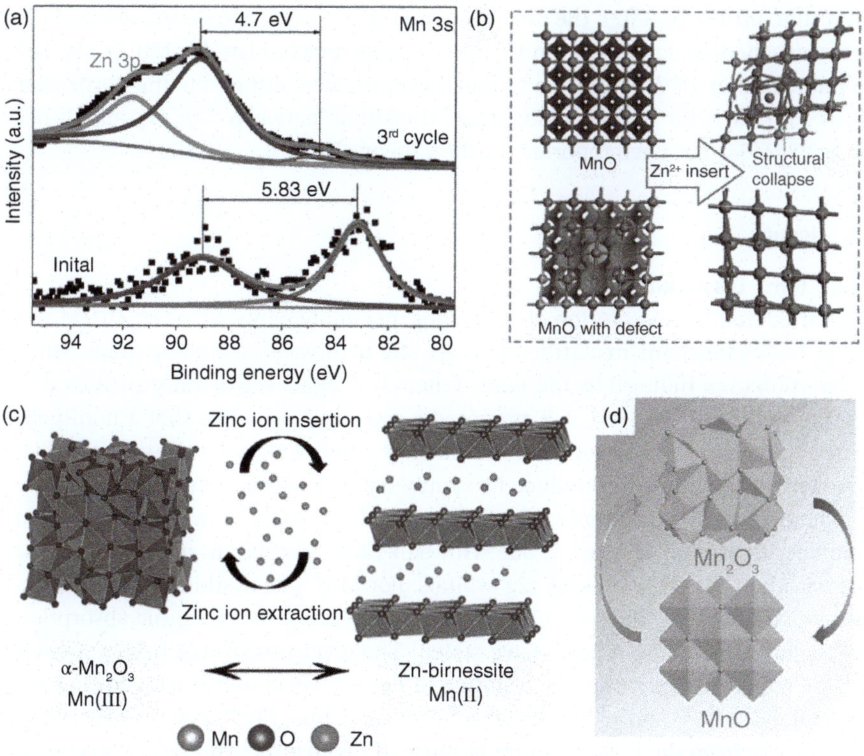

Figure 3.8 (a) XPS curves of Mn 3s in the MnO tested in pre/post-initial cycle. (b) Comparison of MnO charge distribution with/without Mn defects in the middle of Zn^{2+} insertion. (c) The storage mechanism of Zn^{2+} in Mn_2O_3. (d) The phase transition demonstration in the Zn/Mn_2O_3 battery. Source: (a) Reproduced with permission from Wang et al. [31]. Copyright 2019, Elsevier. (b) Reproduced with permission from Zhu et al. [32]. Copyright 2020, Elsevier. (c) Adapted from Cockayne et al. [33]. (d) Reproduced with permission from Feng et al. [34]. Copyright 2020, Springer Nature.

and Mn_2O_3 [36], this phase transformation-induced electrochemical activation process is still applicable.

However, another study took a different view and reported that the introduced Mn defects electrochemically activated the MnO during the first charge [32]. In this case, a facile Zn^{2+} insertion/extraction could be achieved by those Mn defects, providing feasible migration pathways and approachable active sites. This is because electrons usually accumulate around Mn vacancies (Figure 3.8b), attracting Zn^{2+} ions via a strong electrostatic field. Compared with the above phase transformation activated MnO [31], this defect-activated MnO [31] attributes the increased oxidation state of Mn in the initial charging to the extraction of Mn^{2+} from the MnO host, rather than the oxidation of MnO to MnO_2. Nevertheless, these controversial activation processes for MnO are supposed to be in-depth explored with the help of advanced in situ techniques, to unveil novel electrode materials for ZIBs.

As for the bixbyite-structured Mn_2O_3, Mn^{3+} is coordinated in an octahedral manner, while O^{2-} is coordinated with Mn in a tetrahedral manner [33]. Similar to MnO, Mn_2O_3 has a shortage of migration channels for Zn^{2+}. To realize energy storage, the bixbyite Mn_2O_3 would experience a reversible phase transition to a layered zinc-birnessite after the first charge (Figure 3.8c). However, researchers have not reached an agreement on the Zn^{2+} storage mechanism for Mn_2O_3. So far, a co-intercalation of H^+ and Zn^{2+} ions accompanied either by the formation of ZHS [37] or with a reversible phase transition between Mn_2O_3 and MnO (Figure 3.8d) [34] has been reported for the aqueous Zn/Mn_2O_3 battery.

3.1.2 Issues

3.1.2.1 Mn^{2+} Dissolution

It is reported that the dissolution of Mn^{2+} ions inevitably happens in the Zn//MnO_2 systems, but its mechanism clarifications are still inconsistent. Among these, Mn^{3+} disproportionation induced by the Jahn–Teller (J–T) effect is generally attributed to this undesired issue. In 1937, J–T proposed a universal principle that a nonlinear symmetry molecule in a degenerate orbital state (high energy level) was unstable. The molecule will deform to reduce its symmetry and release its degenerate state. This phenomenon is commonly seen in Mn-based materials, in which the valence electron configuration of Mn^{2+} is 3d5 with half-filled 3d orbitals and that of Mn^{4+} is 3d3 with half-filled low-energy t2g orbitals. For Mn^{3+}, its d orbital has four electrons, which are in the high spin t2g3eg1 capacity fading. With atomic absorption (AAS) technology, many reports have stated. The electrons in dz2 have a greater repulsion effect with oxygen ions, resulting in the J–T effect of the octahedron [38], tetragonal distortion, and the d orbital splits again. Thus, Mn^{2+} ions will escape to the electrolyte from the host structure because of the disproportionation process of Mn^{3+} ions to form Mn^{2+} and Mn^{4+} [38]. This will eventually boost capacity attenuation. Note that the dissolution of Mn^{2+} will in turn provoke the disproportionation reaction, resulting in continuous confirmation of this phenomenon by detecting the concentration of Mn^{2+} in the electrolyte [38]. Besides, the dissolved Mn-ions are likely to generate deposition on the anode, which could induce side reactions and thicken the solid electrolyte interphase. But the relative evidence of Mn-ions deposition is still insufficient so far, thus rational experiments are required to uncover this puzzle in the following research. In addition, although the existence of crystal water in Mn-based oxides has been reported to be positive in many types of rechargeable batteries, too much crystal water will promote Mn dissolution from the cathodes of ZIBs. Therefore, suppressing Mn-dissolution is critically significant to enhance the capacity retention of Mn-based materials during long cycling.

3.1.2.2 Structure Instability

Structural instability (including dissolution of $Mn^{3+}/^{2+}$, and phase transition) was generally considered the main reason for the capacity fading of MnO_2-based cathodes in various battery applications, including lithium-ion batteries [39], mild aqueous M-ion batteries (M = Li [40], Na [41], Mg [23], and Zn [21a], etc.), alkaline Zn batteries [42], and so on.

According to Shannon's table, the Mn^{4+}—O bond length (1.93 Å) was predicted to elongate when the Mn was reduced to Mn^{3+} (2.045 Å) or Mn^{2+} (2.23 Å) [43]. In this case, the Mn—O bond was seriously weakened, and the disorder degree of

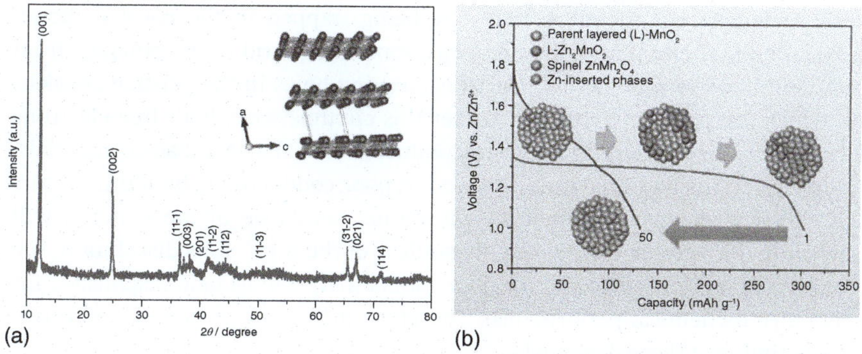

Figure 3.9 (a) XRD curves of layered MnO_2. (b) The mechanism of Zn^{2+} insertion in layered MnO_2. Reproduced with permission from Alfaruqi et al. [21b]. Copyright 2018, Elsevier.

the lattice arrangement increased, i.e. J–T effect mentioned above. On top of that, the intercalated carrier ions exhibited electrostatic repulsion and physical extrusion interactions on reduced $[Mn^{3+}O_6]$ octahedrons or $[Mn^{2+}O_4]$ tetrahedrons. Therefore, the $[Mn^{3+}O_6]$ octahedrons/$[Mn^{2+}O_4]$ tetrahedrons could be easily squeezed out from the crystal lattice of MnO_2 hosts (including both the tunnel-type and layer-type MnO_2) after the insertion of carrier ions. Ultimately, the overall structural instability of MnO_2 induced by the J–T effect could give rise to a large and anisotropic volumetric change, critically undermine the structural integrity, and inevitably block the ion diffusion pathways [44]. The capacity fading of MnO_2 would be exacerbated by this situation without a doubt.

It is reported that, during charge/discharge cycles, Mn-based materials tend to transform into spinel structures after a phase transition [45], whereas the mechanism behind this phenomenon was unclear. Exploring the Zn-storage mechanism of β-MnO_2, Kim and coworkers found that the formation of spinel-phase MnO_2 was because of the shear of oxygen arrangement generated by the interaction between Zn^{2+} ions and Mn^{4+} ions [12]. With further study, they revealed that the sliding of the MnO_2 plane after Zn^{2+} ions insertion contributed to the formation of spinel structure (Figure 3.9a,b) [21b]. In addition, the volume change during Zn^{2+} ions insertion/extraction may also trigger mechanical stress and facilitate phase transformation [46]. The spinel structure was proved to be irreversible, and the storage of Zn^{2+} ions could be impeded by this structure in discharging progress, but how the spinel structure forms during the cycling of MnO_2 is still unclear. Note that, the reversibility of MnO_2 will be significantly reduced at the moment of the formation of the spinel structure. According to the results of ab initio calculations, Manthiram and coworkers reported that Zn^{2+} ions had low mobility in the spinel structure [26]. This spinel structure made Zn^{2+} ions extraction difficult due to the high electrostatic repulsion between the Zn^{2+} ions in the spinel lattice [26].

3.1.2.3 Poor Electrical Conductivity

The electrochemical performance of Mn-based materials is muffled to a large extent because of the poor conductivity of MnO_2 [47]. In general, most electrons in semiconductors cannot move freely by being trapped in the covalent bond. However, due to thermal motion, a few electrons can get rid of the bondage from the covalent bond to become free electrons, leaving holes in the pristine covalent bond, which are positively charged. Under this circumstance, both free electrons and holes become carriers in the semiconductor [48]. For Mn-based materials, fewer free electrons and hole numbers lead to poor conductivity. In addition, part of the electrical energy consumed by the ohmic resistance of the material will also result in a low capacity [37, 49]. Regardless of the Mn^{2+} ions dissolution, the formation of spinel structure and the poor conductivity will be held responsible for the poor electrochemical performance. Therefore, it is essential to design effective strategies aiming at these drawbacks.

3.1.3 Strategies

3.1.3.1 Structure Design

0D–3D Nanostructure:

Over the published papers, nanostructures are normally opted to improve the performance of cathode materials, as is the case with manganese-based materials. Normally, the nanostructure could be divided into four types: (1) 0D: Nanoparticles; (2) 1D: Nanowires, nanotubes, and nanorods; (3) 2D: Nanosheets and nanoplates; (4) 3D: Assembled structures such as nanospheres and nanoflowers.

The nanostructure features a high electrode/electrolyte contact area that could offer sufficient active sites for energy storage. In this case, the transmission path for ions and electrons is shortened, and the enhanced ion diffusion at interfaces improves the electrochemical kinetic process. Benefiting from these, nanostructured manganese-based cathode materials would be endowed with high reversible capacities and exceptional rate performances. With an average size of 6.0 nm, Mn_3O_4 nanodots (Figure 3.10a) were reported as cathodes for ZIBs [50]. Due to the smaller size of nanodots, this material presented a fairly low charge transfer resistance (Figure 3.10b) and a large Zn^{2+} diffusion coefficient of 2.4×10^{-10} cm^2 s^{-1} (Figure 3.10c). Its kinetic process of redox reactions was greatly speeded because the capacitive process made the major contributions to overall capacity (386.7 mAh g^{-1} at 0.1 A g^{-1}) (Figure 3.10d). On top of that, MnO_2 nanosheets could realize sufficient active sites and short ion transmission paths as well [51]. Particularly, the (100) plane of MnO_2 provided 1×1 tunnels with huge exposed areas for H^+ and Zn^{2+} transmission (Figure 3.10e). It exhibited a super rate performance of 115.1 mAh g^{-1} even at 10 C.

Except for the favorable ion/electron transfer kinetics, superior cycle stability could be provided by MnO_2 with nanostructures as well. With a thickness of 1 nm, a single-atomic-layer MnO_2 nanosheet was prepared through an exfoliation process [52]. In contrast to the bulk MnO_2 with the tip-effect caused by sharp corners and irregular morphology, the smooth surface consisting of MnO_2 nanosheets could significantly diminish the concentration gradient in the electrolyte to induce a uniform Zn stripping/plating process (Figure 3.11). As a result, this cathode could keep a capacity of 225 mAh g^{-1} even after 600 cycles at 0.1 Ag^{-1}.

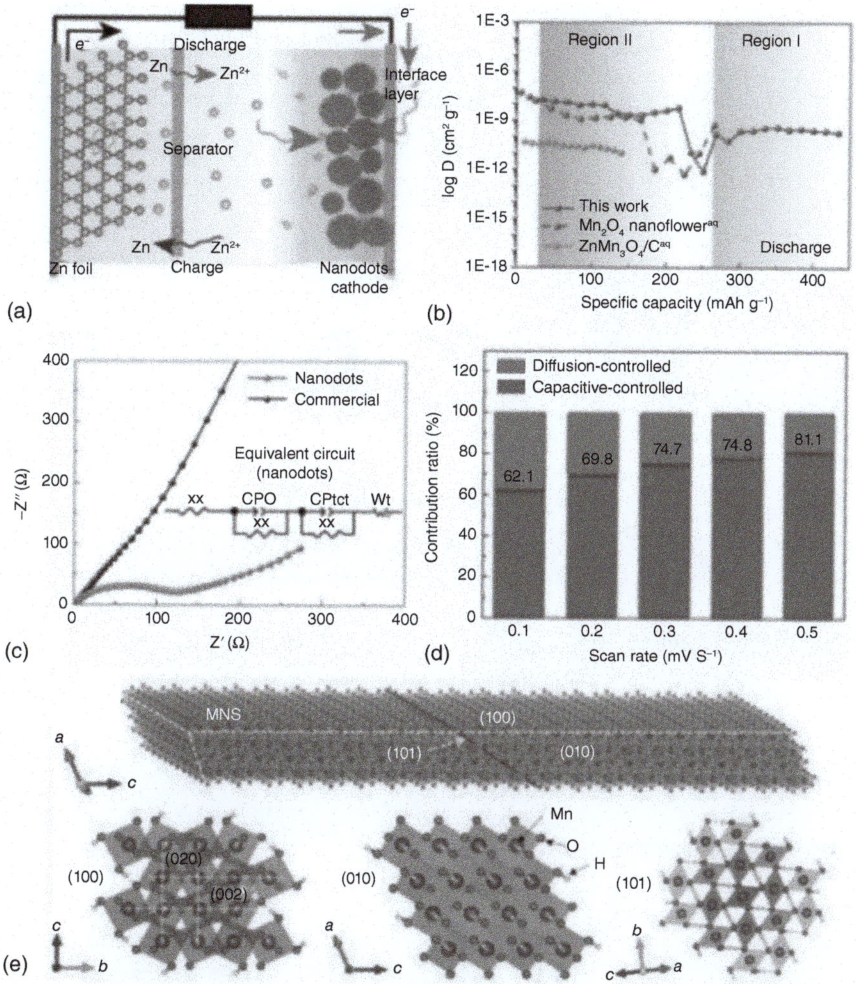

Figure 3.10 (a) Zn^{2+} transport in $Zn-Mn_3O_4$ battery. (b) Zn^{2+} diffusion coefficient. (c) EIS curves of $Zn-Mn_3O_4$ battery. (d) Contribution ratio of capacities at different scan rates. (e) MnO_2 polyhedral models. Source: (a–d) Reproduced with permission from Jiang et al. [50]. Copyright 2019, American Physical Society. (e) Reproduced with permission from Zhao et al. [51]. Copyright 2019, John Wiley & Sons.

Moreover, nanostructures can withstand the volume change during ion insertion/extraction, which could contribute to enhancements in cycle stability as well. Therefore, MnO_2 with 3D nanostructures has been reported in recent years [53]. For instance, through spray drying technology, MnO_2 nanowires with a 3D micro-flower-like structure were successfully synthesized (Figure 3.12a) [53d]. In addition to the abundant active sites and convenient channels for ion/electron diffusion, enormous spatial room to alleviate the volume expansion is the major reason to enable an outstanding performance (Figure 3.12b). Overall, the $Zn//MnO_2$ battery demonstrated a superior rate capacity (shown in Figure 3.12c)

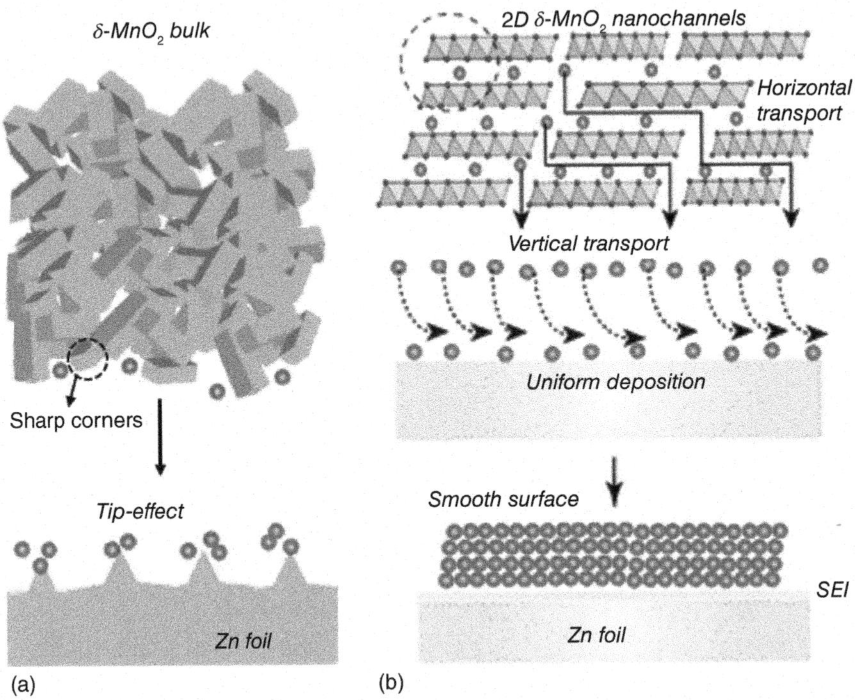

Figure 3.11 Zn^{2+} transport in δ-MnO_2 bulk (a) and MnO_2 nanosheets (b). Source: Adapted from Wang et al. [52].

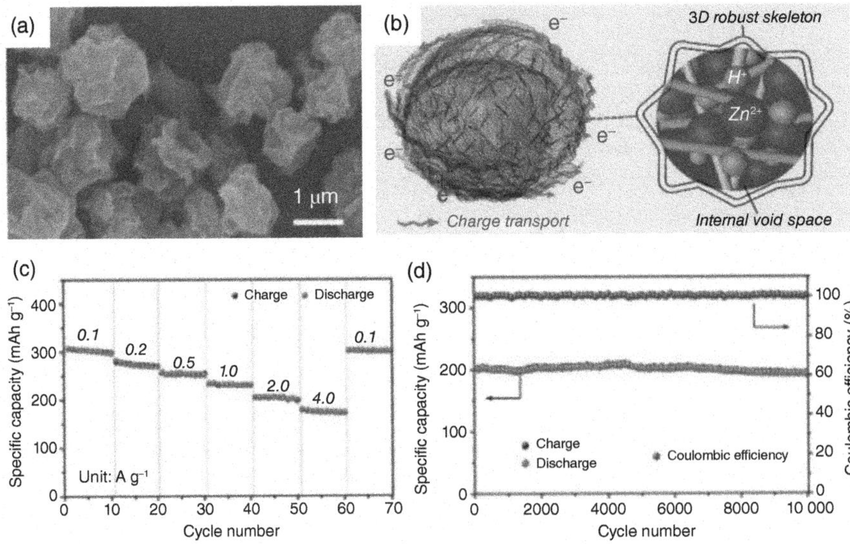

Figure 3.12 3D micro-flower-like MnO_2 Cathode: SEM image (a), schematic diagram (b), rate performance (c), and cycle performance at 2 A g^{-1} (d). Source: Reproduced with permission from Shi et al. [53d]. © 2020, Elsevier.

and exceptional cycle stability of ~97.5% retention even after 10 000 cycles (Figure 3.12d).

Porous and Hollow Structure:

Inspired by the benefits of nanostructures, porous or hollow structures could also achieve sufficient active sites for energy storage and channels for facile ion diffusion. Meanwhile, their innate pores can endure volume change because of the repeated ion insertion/extraction processes. In short, porous or hollow structures are of similar significance to enhance the electrochemical performance of manganese-based cathode materials in ZIBs [49, 54].

As reported [54b], mesoporous Mn_2O_3 (MMO) with a pore size of 3.2 nm (MMO-3.2) showed the morphology of nanoparticles with a diameter of 5 nm (Figure 3.13a) and a high specific surface area of 260 $m^2\ g^{-1}$. Due to a large number of active sites and its capability against volume expansion, the mesoporous MMO-3.2 cathode delivers an excellent rate performance and outstanding cycle stability (Figure 3.13b,c). The hollow structure can be taken as a special type of porous structure. Guo et al. prepared a MnO_2 material with hollow nanospheres, which has a specific surface area of 200 $m^2\ g^{-1}$ [54c]. The hollow MnO_2 shell is around 15 nm thick (Figure 3.13d), extremely shortening the Zn^{2+} diffusion path and avoiding structural collapse caused by volume expansion. In this case, a high specific capacity of up to 405 mAh g^{-1} at 0.5 C (Figure 3.13e) and a high capacity retention could be realized (305 mAh g^{-1} after 100 cycles at 1 C) (Figure 3.13f).

Based on the above, manganese-based cathodes with rational structures undoubtedly offer unique advantages over conventional cathodes due to their typical size effect. That is due to their high specific surface area, fine electron/ion transport channels, and adaptability to volume changes. Therefore, a reasonable structure design is a feasible strategy for improving the performance of manganese-based materials.

3.1.3.2 Compositing with Conductive Materials

As mentioned in 3.1.2.3, poor electric conductivity is a pressing issue for manganese-based materials. Compositing with conductive materials (carbon materials, conductive polymers, etc.) is a corresponding strategy.

Inorganic Carbon Materials:

With high conductivity, a variety of carbon materials (porous carbon, graphene, etc.) have been reported to be applied in manganese-based cathode materials over the past few years [46, 53a, 54c, d, 55]. They are widely used in composite materials, serving as a substrate, network, coating, and so on. Concerning their working mechanism, they can generally improve the electron/ion transport and structural stability of manganese-based cathode materials, which is manifestly promising in ZIBs.

As good substrates, carbon materials are advantageous to the in-situ growth of manganese-based compounds. It is reported that $ZnMn_2O_4$ nanoparticles could be firmly anchored on the porous carbon polyhedrons (ZMO@PCPs) [55c]. Specifically, the porous carbon gave rise to the fast electron transfer kinetics by providing "highways" for electronic transmission (Figure 3.14a). Therefore, the ZMO@PCPs cathode could achieve a remarkable rate capacity of 88.7 mAh g^{-1} at 4 A g^{-1}.

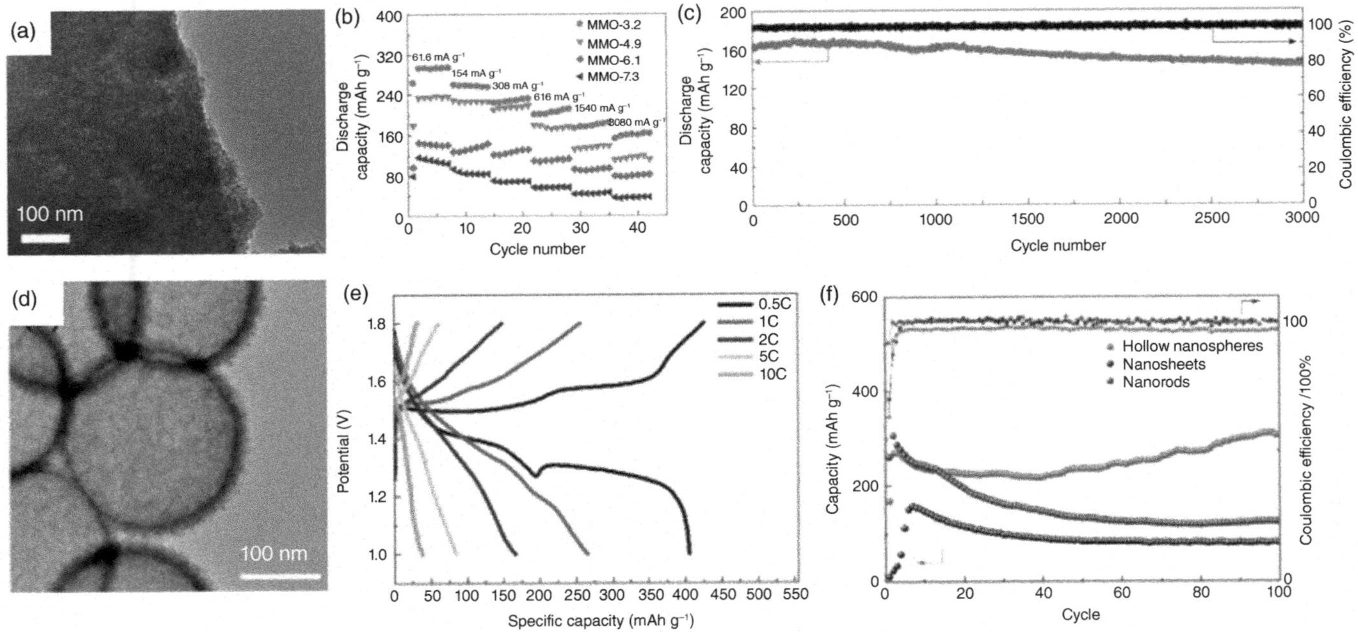

Figure 3.13 MMO-3.2 cathode: TEM image (a), rate performance (b), and cycle performance (c). Hollow MnO_2 cathode: TEM image (d), GCD curves at various current densities (e), and Cycle performance (f). Source: (a–c) Adapted from Wang et al. [54b]. (d) Fu et al. [54d]/John Wiley & Sons.

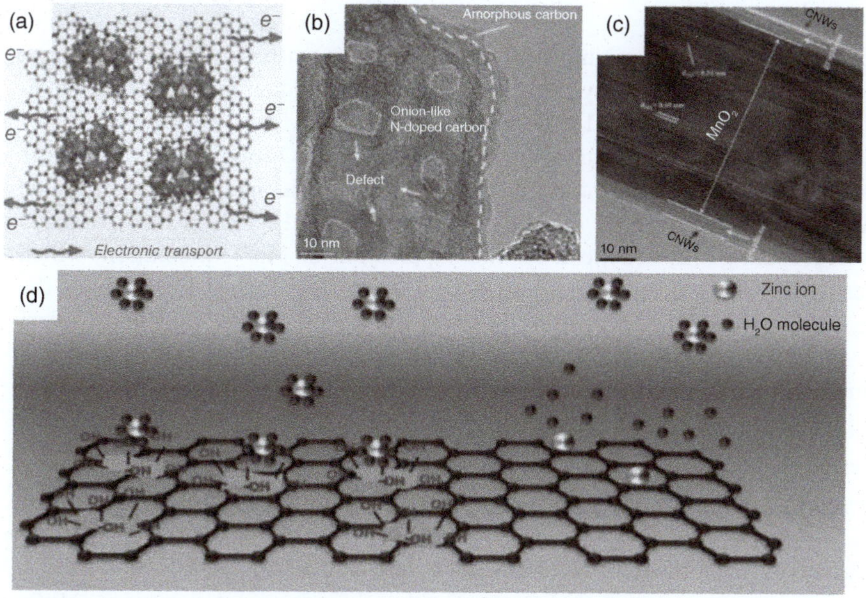

Figure 3.14 (a) The demonstration of electronic transport in ZMO@PCPs. (b) TEM image of MnOx@N-C composite. (c) HRTEM image of MNG. (d) De-solvation behavior of hydrated Zn^{2+} at MNG surface. Source: (a) Reproduced with permission from Yang et al. [55c]. Copyright 2020, Elsevier. (b) Fu et al. [54d]/John Wiley & Sons. (c, d) Reproduced with permission from Fu et al. [55d]. Copyright 2020, Elsevier.

Different from the above, a carbon network with excellent mechanical stability could also secure structural stability in addition to enhancing electric conductivity. An onion-like N-doped carbon composited with MnO_x nanorods (MnO_x@N-C) (Figure 3.14b) was proven to significantly decrease the charge transfer resistance because of the carbon conductive network [54d]. The porous structure was flexible to the volume expansion of ion insertion. As a result, even after 1600 cycles, the battery could deliver stable capacity performance.

Coating manganese-based cathodes with carbon materials not only promotes Zn^{2+} de-solvation but also plays a key role in suppressing Mn^{2+} dissolution. Coated by hydrophobic graphene and amphiphilic cellulose nano-whiskers (MNG) (Figure 3.14c), the prepared MnO_2-based composites achieved a superior rate capacity of 196 mAh g^{-1} at an extremely high current density of 20 C and excellent cycle stability [55d]. As shown in Figure 3.14d, the hydrated Zn^{2+} would be forcedly de-solvated by the proper hydrophobicity of graphene, facilitating Zn^{2+} insertion into host materials. In this situation, the structure could be well preserved because further damage by the insertion of a large number of hydrated Zn^{2+} would disappear. Furthermore, Wu et al. found the graphene coating could effectively inhibit the dissolution of Mn^{2+}, which greatly increased the structural stability during cycling and enabled long-term cycling stability (94% retention after 3000 cycles) [55g].

Conductive Polymers:

As known, conductive polymers are featured by easy synthesis, low cost, and good conductivity. Up to now, polypyrrole (PPy) has been frequently reported to compound with manganese-based materials [37, 56].

As shown in Figure 3.15a, the structural collapse of MnO_2 can be attributed to the dissolution of surface atoms and the high diffusion barrier caused by electrostatic repulsion slowing down the interfacial kinetics. Recently, Huang et al. revealed that PPy coating was beneficial for the suppression of Mn^{2+} dissolution, which is consistent with the above discussion. Due to its high electric conductivity, the sluggish interfacial kinetics were vastly speeded (Figure 3.15b) [57]. In detail, a firm Mn—N bond was constructed between the Mn atom on the surface and the N atom in PPy. This greatly elevated the energy barrier of manganese atoms removal ($E_{\text{a free}} = 0.26\,\text{eV}$) (Figure 3.15c), which is effective in inhibiting Mn^{2+} dissolution.

Others:

Owing to its typical accordion structure with large interlayer spacing, large specific surface area, excellent electrical conductivity, and structural stability, layered MXene is also suitable to composite with manganese-based materials.

Zhu et al. reported the K-V_2C@MnO_2 that is prepared by growing MnO_2 nanosheets on the V_2CTX MXene surface (Figure 3.16a,b) [58]. It is noted that the electrical conductivity of this material was significantly improved. This is because the electrons tend to transfer into the conduction band of MnO_2 due to the lower electron affinity of V_2C (3.84 eV), increasing the density of states of MnO_2 near the Fermi level (Figure 3.16c,d). In addition, the lower Gibbs free energy of Zn^{2+} adsorption on K-V_2C@MnO_2 (0.74 eV) contributed to the fast ion transport kinetics and high reversibility of ion adsorption/desorption (Figure 3.16e). This naturally resulted in a high rate capacity of 87.7 mAh g^{-1} at 15 A g^{-1}.

3.1.3.3 Pre-Intercalation

As discussed above, expanding the interlayer space is desired to reduce the diffusion resistance of Zn^{2+}. Pre-intercalated crystal water, ions, and molecules in the manganese-based compound could serve as a "structural pillar" to realize this and stabilize the host structure of manganese-based cathodes during cycling.

Crystal Water and Molecules:

Since the interlayer space could be enlarged by the pre-intercalated crystal water in the manganese-based compound, the electrostatic interaction between Zn^{2+} and $[MnO_6]$ skeleton would be weakened correspondingly, which is beneficial to the Zn^{2+} diffusion. Nam et al. reported a crystal water pre-inserted MnO_2 layered structure (cw-MnO_2) with a larger interlayer spacing of 7.3 Å than that of pure MnO_2 (7.0 Å) (Figure 3.17a) [59]. They found the cw-MnO_2 was equipped with a lower energy barrier for Zn movement of 0.25 eV, much lower than that of Zn^{2+} in MnO_2 (1.03 eV) (Figure 3.17b), indicating the fast Zn^{2+} diffusion. Moreover, the fleeing Mn^{2+} ions were difficult to form the $[Mn(H_2O)_6]^{2+}$ freely movable in cw-MnO_2 but existed in stable Zn–Mn dumbbell structures (Figure 3.17c). This effectively inhibited the dissolution of Mn^{2+}. Therefore, the cw-MnO_2 cathode was able to present an

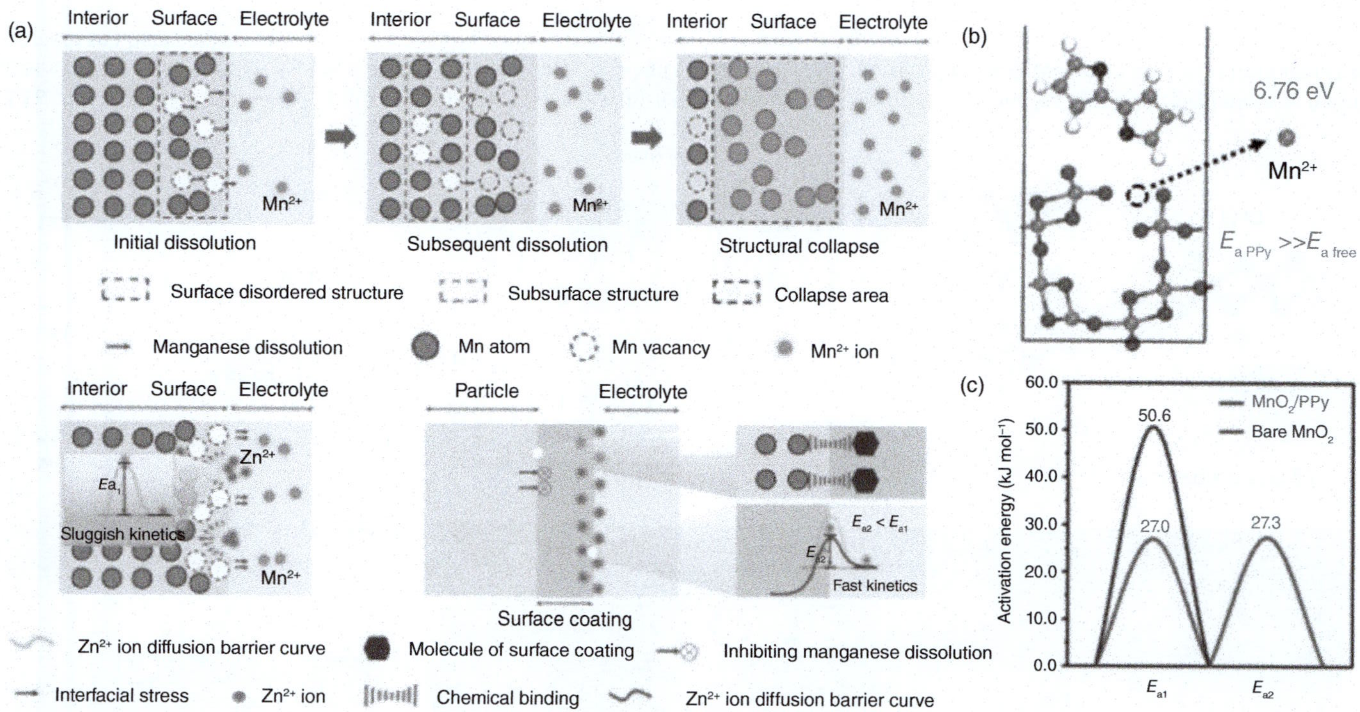

Figure 3.15 (a) The demonstration of the problems and the ideally corresponding strategy for manganese-based cathode materials. (b) Activation energy during charge transfer process for bare MnO_2 and MnO_2/PPy. (c) Removal energy of Mn atom. Reproduced with permission from Huang et al. [57]. Copyright 2020, Elsevier.

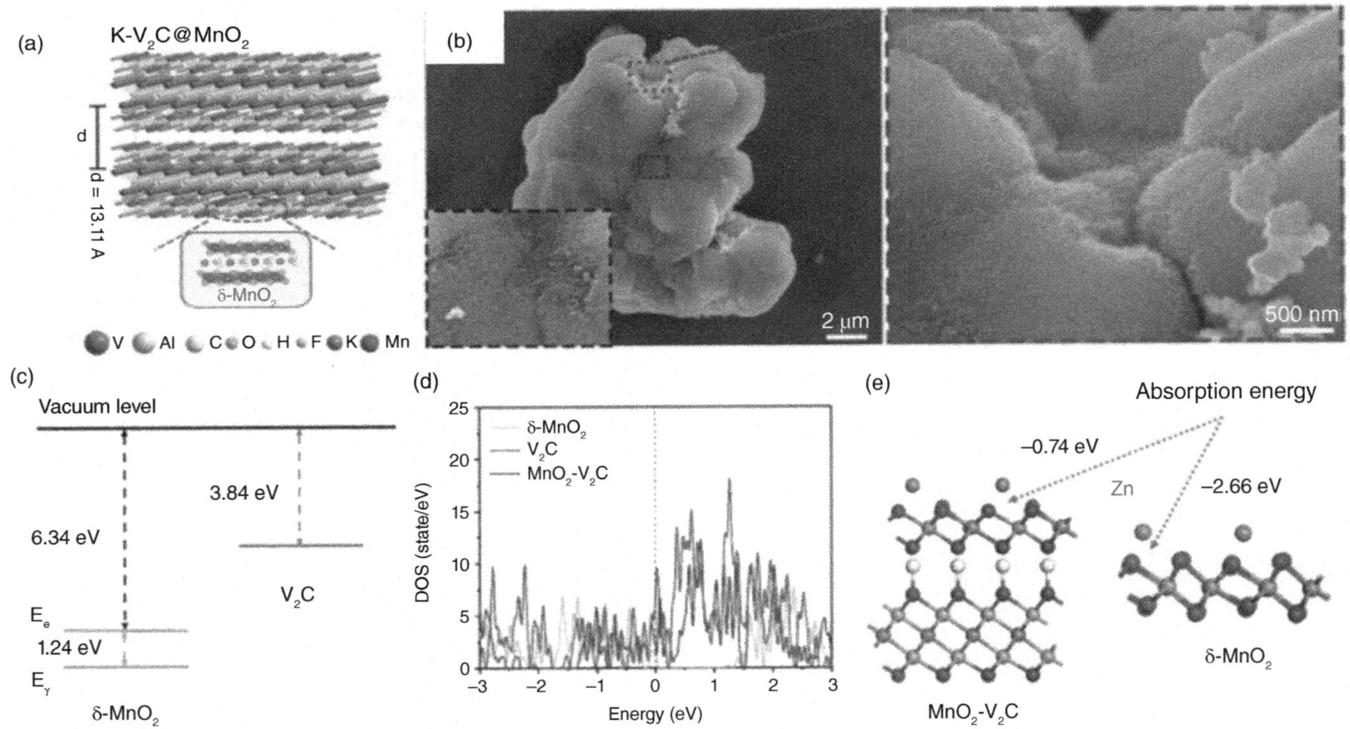

Figure 3.16 K-V_2C@MnO_2: Schematic illustration (a) and SEM image (b). (c) Band edge positions of δ-MnO_2 and V_2C. (d) DOS calculations of δ-MnO_2, V_2C, and MnO_2-V_2C. (e) Absorption energies of Zn^{2+} on the surfaces of δ-MnO_2 and MnO_2-V_2C. Source: (b) Reproduced with permission from Zhu et al. [58].

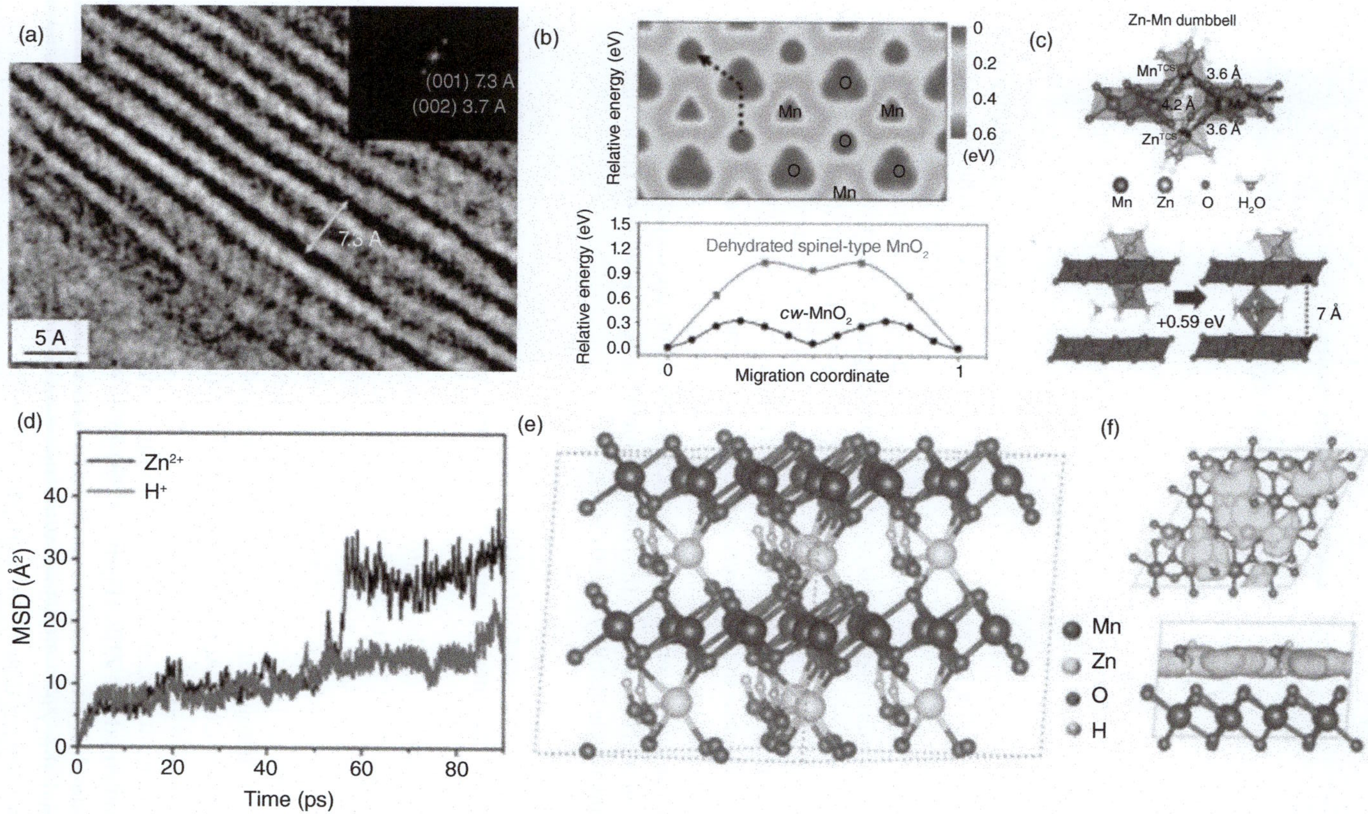

Figure 3.17 cw-MnO_2: HRTEM images (a) and its inside migration pathway and barriers for Zn^{2+} (b). (c) Illustration of Zn–Mn dumbbell structure and Mn-octahedra. (d) Mean square displacements of Zn^{2+} and H^+ in MnO_2. (e) Illustration of Zn^{2+} and H^+ intercalation in MnO_2. Source: (c) Adapted from Nam et al. [59]. (d, e) Jiang et al. [60]/John Wiley & Sons.

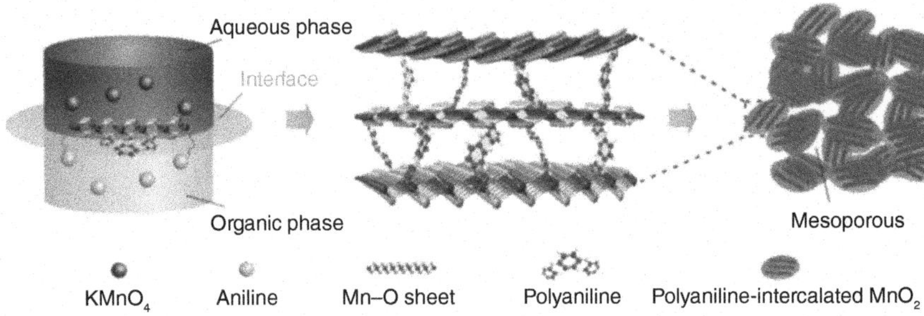

Figure 3.18 Illustration of PANI-intercalated MnO_2. Source: Huang et al. [61]/Springer Nature/CC BY 4.0.

excellent rate capacity of 150.0 mAh g^{-1} at 3 A g^{-1} and superior cycle stability with 75.3% retention after 200 cycles. Through the ab initio molecular dynamics (AIMD) simulation, Jiang et al. found that Zn^{2+} and H^+ could stably co-exist driven by H_2O molecules [60]. At the same time, they could smoothly shuttle among layers of MnO_2 without further damage to the layered structure (Figure 3.17d,e), which made sense for the improved rate performance and long cycle stability up to 10 000 cycles.

Xia and coworkers successfully made polyaniline (PANI) intercalate into MnO_2 via an inorganic/organic interfacial reaction (Figure 3.18) [61]. With the assistance of PANI molecules, the interlayer space of MnO_2 was not only enlarged but also the layered structure was greatly strengthened, avoiding phase change and structural collapse during ion insertion/extraction. Based on the co-insertion/extraction mechanisms of H^+ and Zn^{2+}, the battery achieved a high capacity of 110 mAh g^{-1} at 3 A g^{-1} and a cycle stability of 200 cycles.

Ions:

Similar to crystal water, pre-intercalated ions in the manganese-based compound could expand the interlayer space to accelerate ion transfer and support the host structure.

With Ca^{2+} pre-intercalated into MnO_2, Ca_2MnO_4 (CMO) cathode was reported [62], in which the Ca^{2+} escaped from the CMO and formed a single-component $CaSO_4 \cdot 2H_2O$ (CS) solid electrolyte interface (SEI) film on the surface of CMO in the first charging process (Figure 3.19a). This film further diminished the interface activation energy (Figure 3.19b) to boost the insertion/extraction of Zn^{2+}. Simultaneously, the dissolution of Mn^{2+} could be manifestly suppressed as well, stabilizing the structure of CMO. As a result, the CMO cathode could achieve near-zero capacity fading after 1000 cycles at 1 A g^{-1}.

Apart from cations, anion (PO_4^{3-}) pre-intercalation into MnO_2 was also shown a huge potential for ZIBs (Figure 3.19c) [63]. As reported, the pre-intercalated PO_4^{3-} extended the interlayer space from 0.68 to 0.70 nm (Figure 3.19d), which accelerated the Zn^{2+} diffusion. With the help of vertical multilayer graphene arrays and oxygen defects, the cathode delivered a remarkable high-rate capacity (150.1 mAh g^{-1} at 10 A g^{-1}) and an excellent cycle performance (91.3% retention after 1000 cycles).

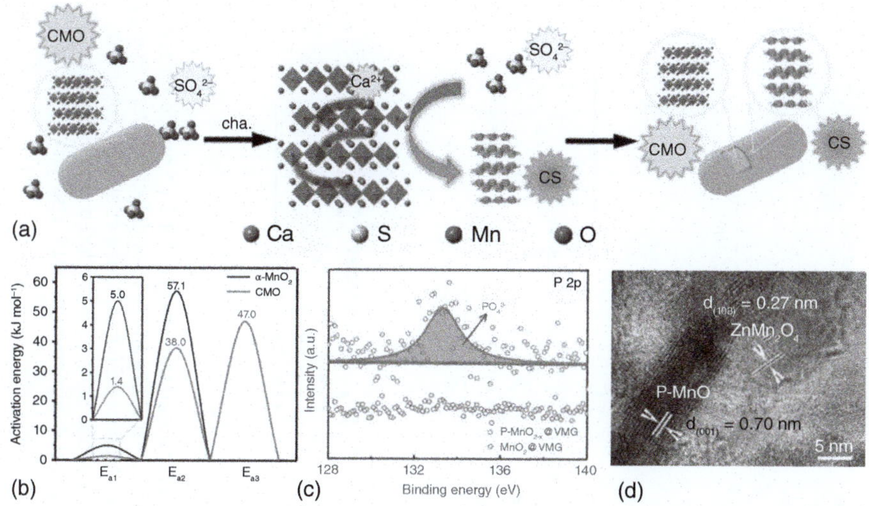

Figure 3.19 (a) The mechanism of CS SEI film formation. (b) The activation energy of MnO_2 and CMO. (c) XPS P 2p spectra of MnO_2@VMG and P-MnO_2-x@VMG. (d) The ex-situ TEM image of P-MnO_{2-x} nanoflakes at a fully discharged state. Source: (b) Reproduced with permission from Guo et al. [62]. Copyright 2019, American Physical Society. (d) Reproduced with permission from Zhang et al. [63]. © 2020/John Wiley & Sons.

3.1.3.4 Defect Engineering

By modifying the crystal structure, defect engineering is expected to reduce the insertion barrier for Zn^{2+} [64]. The defect engineering strategies for manganese-based cathode materials mainly focus on oxygen vacancy, cation vacancy, and cationic doping.

Oxygen Vacancy:

The introduction of oxygen vacancy would commonly improve the thermodynamics of interfacial reactions by growing the number of active sites that promote the transfer of electrons and ions. Xiong et al. found that the oxygen vacancy in oxygen-deficient MnO_2 could enable easier desorption of adsorbed Zn^{2+} because Gibb's free energies of Zn^{2+} adsorption near the vacancies were close to thermoneutral values (0.05 eV) as shown in Figure 3.20a [65]. Subsequently, the ever-occupied active surface areas could accept the following Zn^{2+} adsorption. With the high reversibility of Zn^{2+} adsorption/desorption, the oxygen-deficient MnO_2 cathode showcased a superior rate capacity at 30 A g^{-1} and capacity retention of 84% after 2000 cycles at 5 A g^{-1}.

As explored by Tan et al., the oxygen vacancy affected the electrical conductivity of Mn_3O_4 [66]. With oxygen vacancy involved, the Mn^{3+} charge density was vastly intensified because the octahedral crystal of Mn^{3+} transformed into a pyramidal crystal field (Figure 3.20b). Meanwhile, the bandgap dropped to 0 eV from 0.6 eV. Therefore, the electrical conductivity of Mn_3O_4 was immensely improved. In addition, Fang et al. revealed that introducing oxygen vacancies could unlock the [MnO_6] polyhedral wall and create additional ion channels in the deformed [MnO_6]

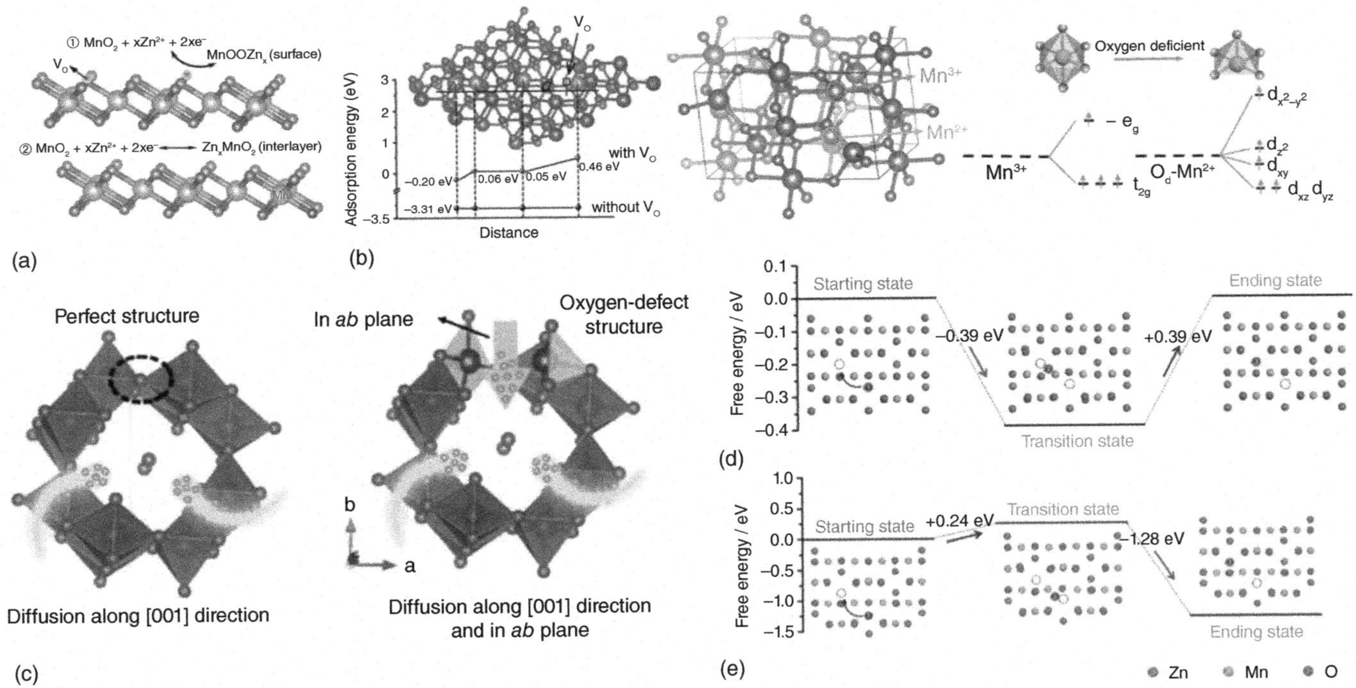

Figure 3.20 (a) Illustration of oxygen-deficient σ-MnO_2 for Zn ion storage. (b) Illustration of the calculated adsorption energies for Zn^{2+} on the surfaces of perfect σ-MnO_2 and σ-MnO_2 with oxygen vacancies. (c) Illustration of H^+ diffusion into KMO with perfect structure and oxygen-defect structure. Energy profiles of Zn diffusion within $ZnMn_2O_4$ (d) and oxygen-deficient-$ZnMn_2O_4$ (e). Source: (a) Reproduced with permission from Xiong et al. [65]. Copyright 2019, John Wiley and Sons. (b) Reproduced with permission from Tan et al. [66]. Copyright 2020, John Wiley and Sons. (c) Reproduced with permission from Fang et al. [67]. Copyright 2019, John Wiley and Sons. (e) Reproduced with permission from Zhang et al. [68]. Copyright 2019, Elsevier.

polyhedral to realize favorable H^+ diffusion (Figure 3.20c) [67]. Also, Zhang et al. simulated the diffusion kinetics of the Zn atom in the $ZnMn_2O_4$ [68]. Oxygen vacancy considerably downsizes the energy barrier of Zn atom migration to near 0.24 eV from 0.39 eV in perfect $ZnMn_2O_4$ (Figure 3.20d,e), indicating its universality in enabling rapid Zn^{2+} diffusion.

These all verify the benefits of oxygen vacancy in promoting ion/electron transfer, contributing to the superior electrochemical performance of manganese-based cathodes.

Cation Vacancy:

According to its application in other ion batteries, cation vacancy could be used to boost the ion insertion into the host because it can debilitate the electrostatic interaction between the ions and the host lattice [69]. In manganese-based cathodes for ZIBs, Mn vacancy is the most common. Through DFT calculations, Zhu et al. reported that the Mn vacancy in $Mn_{0.61}x_{0.39}O$ (x refers to Mn defect) cathode could increase the charge density around the Fermi level to improve the conductivity of MnO [32] (Figure 3.21a,b). Moreover, Mn vacancy could also construct large channels for Zn^{2+} insertion and generate available active sites for Zn^{2+} storage, which would effectively prevent structural collapse during repeated Zn^{2+} insertion/extraction (Figure 3.8b). In combination with the carbon coating, the $Mn_{0.61}x_{0.39}O$ cathode presented a high capacity of 116 mAh g^{-1} at 1 A g^{-1} after 1500 cycles.

In perfect $ZnMn_2O_4$, the strong electrostatic repulsion of Mn cations at octahedral sites obstructs the Zn^{2+} diffusion between adjacent tetrahedrons to a large extent. Nevertheless, as shown in Figure 3.7a,b, introducing Mn vacancy to $ZnMn_2O_4$ could cut down the electrostatic barrier, resulting in a faster Zn^{2+} diffusion process [25]. Thus, the electrochemical performance of cation-defected $ZnMn_2O_4$ has been remarkably enhanced.

Cation Doping:

Cation doping into the manganese-based compounds can change the lattice space of the crystal structure, which can improve the energy storage performance, as discussed before. In a Ni-doped Mn_2O_3 (NM20) (Figure 3.22a) [70], the charge density around the Fermi level was increased thanks to the Ni doping, thereby remarkably enhancing the electronic conductivity of NM20. In addition, the structural stability

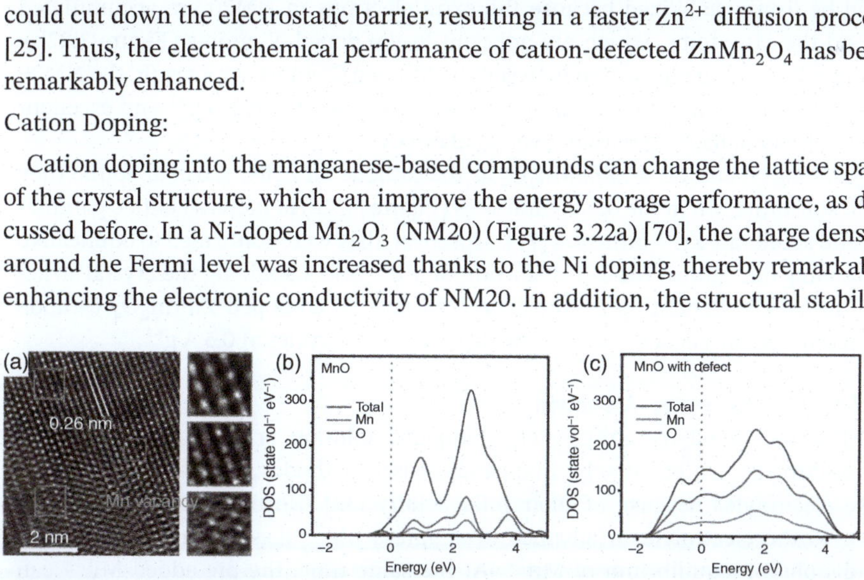

Figure 3.21 (a) HRTEM image of $Mn_{0.61}x_{0.39}O$. Calculated density of states of MnO (b) and $Mn_{0.61}x_{0.39}O$ (c). Source: Reproduced with permission from Zhu et al. [32]. © 2020, Elsevier.

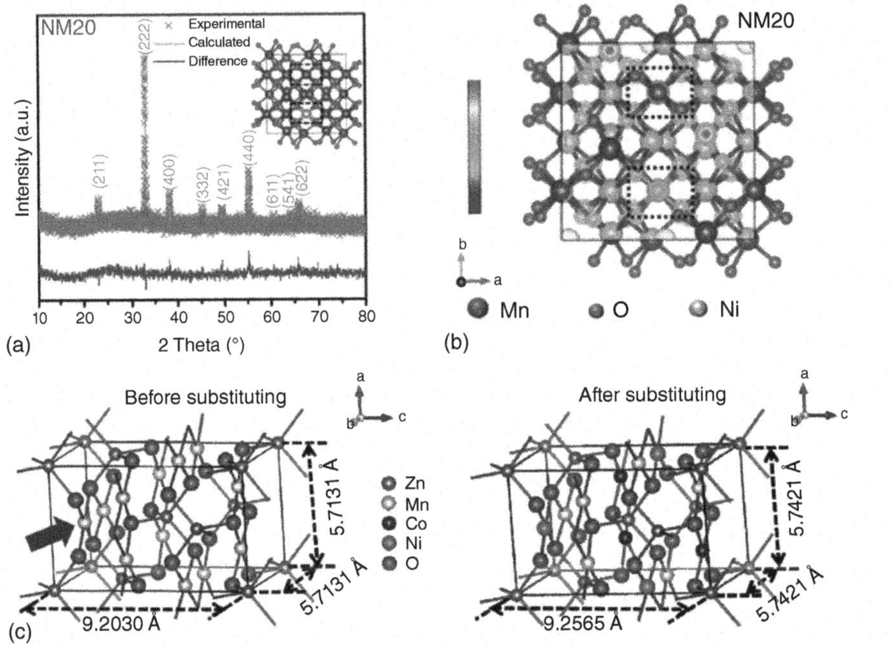

Figure 3.22 (a) Rietveld refinement results of NM20. (b) Charge distribution of Mn_2O_3 and NM20. (c) The schematic illustration of $ZnMn_2O_4$ and co-doped $ZnMn_2O_4$ crystal structure. Source: (a) Zhang et al. [70]/John Wiley & Sons. (b) Adapted from Zhang et al. [70]. (c) Reproduced with permission from Tao et al. [71]. Copyright 2020, Elsevier.

could be firmly stabilized because the extra electrons in NM20 are accumulated around the Mn atoms, which are adjacent to the doped Ni atoms (Figure 3.22b), to generate a strong interaction between the Mn and O atoms. As a result, the NM20 cathode demonstrated a discharge capacity of 132 mAh g^{-1} at 1 A g^{-1} and excellent cycle stability of 85.6% retention over 2500 cycles.

By doping Ni and Co into $ZnMn_2O_4$ as the cathode for ZIBs, Tao et al. found they could substitute Mn in the octahedral sites (Figure 3.22c) to increase lattice parameters and accelerate Zn^{2+} diffusion [71]. Moreover, they were confirmed to be effective in standing out against the J–T distortion of Mn^{3+} and stabilizing the crystal structure ultimately. Benefiting from this, the Ni and Co co-doped $ZnMn_2O_4$ cathode displayed a stable capacity of 117.7 mAh g^{-1} after 600 cycles at 0.5 A g^{-1}.

3.1.3.5 Electrochemical Activation

In ZIBs, the $ZnSO_4$ or $Zn(CF_3SO_3)_2$ aqueous solution is generally selected as the electrolyte. Unfortunately, manganese-based cathodes tend to dissolve into these electrolytes because of their mild acidity [32]. Normally, researchers add Mn^{2+} to the electrolyte in advance to suppress Mn^{2+} dissolution by regulating the dissolution equilibrium of Mn^{2+}. At the same time, the pre-added Mn^{2+} can contribute to the overall capacity of the manganese-based cathodes [72], which could be mainly attributed to the electro-oxidization of Mn^{2+} into active MnO_2 [73]. Inspired by this, manganese-based cathode materials are hoped to be activated via

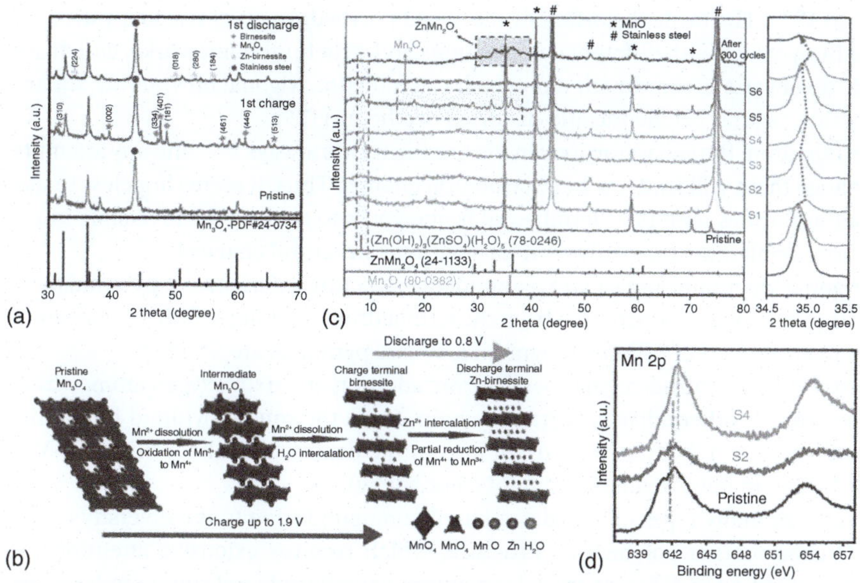

Figure 3.23 (a) XRD curves of Mn_3O_4 at the initial cycle. (b) Reaction progress of Mn_3O_4 in the first cycle. (c) Ex-situ XRD of MnO at the selected state of charge and (d) XPS of MnO tested after 1 cycle and 2 cycles. Source: (a, b) Hao et al. [27]/with permission of Elsevier. (c, d) Reproduced with permission from Li et al. [74]. Copyright 2020, Elsevier.

in situ electrochemical charging. Note that, the activation process is accompanied by phase changes occurring in the initial cycles.

For Mn_3O_4, Hao et al. reported that a spinel-layered phase transition of Mn_3O_4 would be electrochemically induced [27]. In the middle of charging up to 1.9 V, Mn_3O_4 was oxidized to Mn_5O_8 intermediate and subsequently converted into layered Zn-birnessite for Zn^{2+} insertion/extraction, as shown in Figure 3.23a,b. Benefiting from this, the capacity of the Mn_3O_4 cathode gradually escalated to 239.2 from 85.6 mAh g^{-1} at 0.1 A g^{-1}.

Owing to the electrochemical activation, the MnO with divalent manganese, which is electrochemically inactive in principle, becomes feasible for Zn^{2+} storage as well. With the assistance of ex-situ XRD and XPS tests (Figure 3.23c,d), Li et al. found the MnO would transform into Mn_3O_4 for Zn^{2+} and H^+ insertion/extraction in the first 2 cycles [74]. Likewise, commercial MnO particles were oxidized to porous layered-type MnO_2 nanosheets through electrochemical activation [31], indicating that the originally inactive MnO directly used as the cathode material of ZIBs is promising to exhibit excellent electrochemical performance.

3.2 Vanadium-Based Cathode Materials

3.2.1 Introduction to Different Vanadium-Based Materials

Vanadium-based compounds have been extensively applied as zinc storage materials because of their low cost, abundant resources, and multiple oxidation states

(V^{2+} to V^{5+}). The facile distortion of the V—O polyhedron gives rise to a variety of vanadium oxides with different compositions and structural frameworks, which are made up of various coordination polyhedra (tetrahedra, trigonal bipyramids, square pyramids, regular octahedra, and distorted octahedra) [75].

Tetrahedron: the vanadium ions in the tetrahedra are always V^{5+}, and any attempts to change them will lead to a huge structural change. Thus, it seems hopeless to use vanadium oxides with pure tetrahedral frameworks to achieve the reversible intercalation/deintercalation of divalent ions without structural changes.

Trigonal bipyramid and square pyramid: the structures of trigonal bipyramids and square pyramids can be acquired from the tetrahedron by adding a fifth oxygen atom; the vanadium ions of trigonal bipyramids/square pyramids are V^{4+}/V^{5+}.

Distorted octahedron/regular octahedron: addition of a sixth oxygen atom, which comes from another adjacent pyramid chain, turns the square pyramid into a distorted/regular octahedron. Vanadium ions in the distorted octahedron can be V^{4+} or V^{5+}, even in lower oxidation states (+3 or below).

There are many types of vanadium-based compounds due to the diversity of the V—O coordination polyhedron structure, which receives extensive attention for AZIB. According to different crystal structures, vanadium-based compounds can be divided into five types: layer, tunnel, spinel, NASICON, and halite (Figure 3.24) [76].

3.2.1.1 Layered Structure

The layered vanadium-based cathodes have special two-dimensional diffusion channels with large diffusion spaces and adjustable interlayer spacing. This special two-dimensional diffusion channel is conducive to the intercalation/deintercalation of Zn^{2+}, which achieves a high specific capacity for the layered vanadium-based

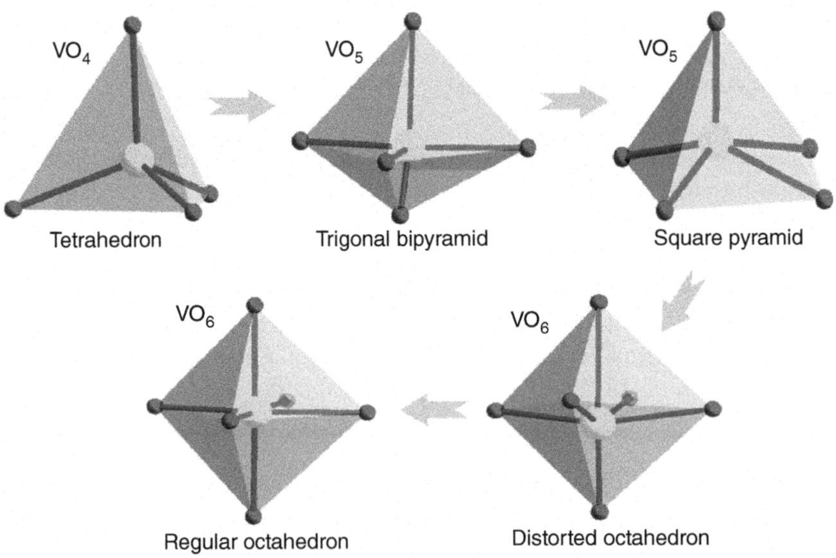

Figure 3.24 Different vanadium coordination polyhedra. Source: Reproduced with permission from Ding et al. [75a]. Copyright 2021, Royal Society of Chemistry.

cathode. Unfortunately, the weak van der Waals force between layers damages the structure during repeated insertion and extraction of Zn^{2+}, resulting in poor battery cycle performance. At present, the layer structure can be regulated by intralayer pillars to obtain stable-layered vanadium-based cathodes [77].

V_2O_5 and $M_xV_2O_5$:

V_2O_5 is a typical layered vanadium-based compound with large interlayer spacing (up to 0.44 nm) and high theoretical zinc storage capacity (589 mAh g^{-1}), in which [VO_5] square pyramid forms a layered structure by co-apex or co-edge [78]. In 1998, V_2O_5 aerogel was reported for the first time to reversibly store Zn ions. However, until 2016, vanadium-based materials became the research hotspot for AZIB cathode. Zhang et al. prepared a V_2O_5 cathode by ball-milling commercial V_2O_5 powder with graphite. On the first discharging, the pristine layered V_2O_5 undergoes structural evolution and transforms into a layered $Zn_xV_2O_5 \cdot nH_2O$; the charging process is reversed. In addition, the morphological evolution from an original bulk V_2O_5 cathode to porous nanosheets was revealed, accounting for the initial capacity increase and enabling a high utilization of active materials [79]. The low electronic conductivity of V_2O_5 and the slow intercalation/de-intercalation of Zn^{2+} seriously hinder its practical application in ZIBs. It is reported that interlayer spacing can be effectively increased by inserting H_2O molecules into V_2O_5 layers. The bilayer V_2O_5 is composed of a VO_5 square pyramid and a VO_6 octahedron, in which intercalated atoms and H_2O molecules exist between the two bilayer layers. Mai and coworkers developed the $V_2O_5 \cdot nH_2O$/graphene (VOG) cathode material, which exhibited extremely high-rate performance (248 mAh g^{-1} at 30 A g^{-1}) and long-cycle stability [80]. The VOG showed an increased interlayer spacing of ~12.6 Å due to the presence of structural water. It was confirmed that structural water played the role of charge shielding for metal ions. The electrostatic interaction between metal ions and the V_2O_5 framework can be reduced by reducing the effective charge, and then effectively promotes the diffusion of Zn ions.

$M_xV_2O_5$:

$M_xV_2O_5 \cdot nH_2O$, a derivative of $V_2O_5 \cdot nH_2O$, is prepared by pre-intercalation of metal cations into the V_2O_5 layer (M = alkali [81], alkaline earth metals [82], and other metals [83]). Metal ion pre-insertion into the structures would improve the electrical conductivity, enhance the ionic diffusion owing to the increase in interlayer spacing, and also enhance structural stability during ion insertion/extraction [84].

Yang developed $Li_xV_2O_5 \cdot nH_2O$ cathode material by the chemical intercalation of Li^+ ions into the interlayer of $V_2O_5 \cdot nH_2O$. The working principle of aqueous ZIBs is presented in Figure 3.25a. The interlayer distance of the (001) facet is enlarged to 13.77 Å when Li ions exist. The galvanostatic intermittent titration technique (GITT) reveals faster Zn^{2+} ion diffusion in $Li_xV_2O_5 \cdot nH_2O$ than that of $V_2O_5 \cdot nH_2O$ (Figure 3.25b). As a result, the LVO-250 demonstrated high-rate capacities and excellent cycling performance [81b]. Afterward, to further improve the performance, K^+-intercalated V_2O_5 (KVO) nanorod with an exposed facet was proposed as a high-performance cathode [81a]. As evidenced from Figure 3.25c,

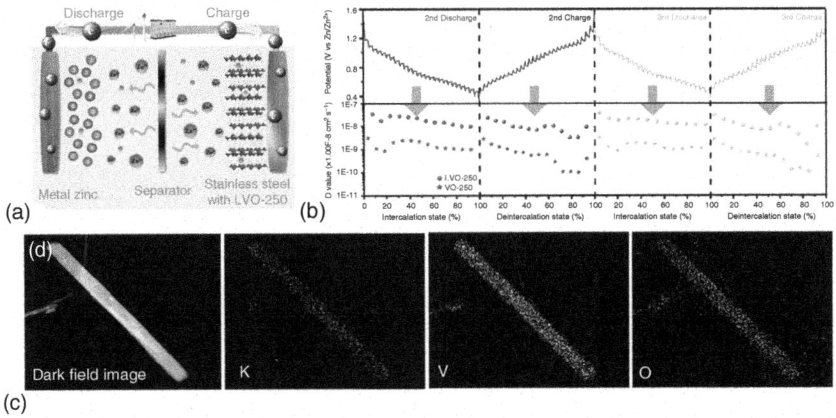

Figure 3.25 (a) The working principle of metallic zinc//LVO-250 aqueous ZIBs, (b) the discharge/charge curves in GITT measurement of LVO-250 and corresponding diffusivity coefficient of Zn^{2+} of LVO-250 and VO-250 samples in the 2nd and 3rd cycles. (c) EDX elemental mapping images of the KVO sample; dark-field image and the corresponding elemental distribution. Source: (a, b) Reproduced with permission from Yang et al. [81b]. Copyright 2018, Royal Society of Chemistry. (d) Reproduced with permission from Islam et al. [81a]. Copyright 2019, Royal Society of Chemistry.

the KVO sample possesses 1D nanorod morphology. The superior performance of the cathode is attributed to its unique exposed layer structure, high conductivity, and low migration barrier for Zn^{2+} migration.

$M_xV_3O_8$:

Another type of vanadate $M_xV_3O_8$ (M = H [85], Li [86], Na [87], NH_4^+ [88], etc. [89]) is also an auspicious electrode material, which has also been extensively studied. LiV_3O_8, NaV_3O_8, and $H_2V_3O_8$ ($V_3O_7 \cdot H_2O$) are composed of VO_6 octahedra and VO_5 square pyramids, and they differ in which corner these two kinds of coordination polyhedra share in Figure 3.26.

Various energy-storage mechanisms have been demonstrated in different cathode materials. For example, the electrochemical mechanism in LiV_3O_8 can be

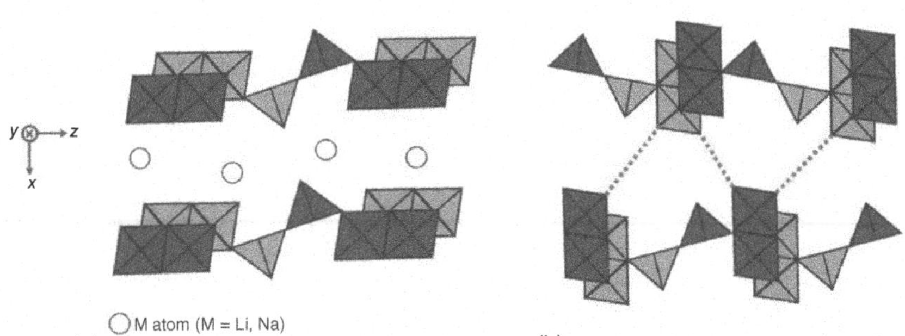

Figure 3.26 Crystal structures of (a) $M_xV_3O_8$ and (b) $H_2V_3O_8$ ($V_3O_7 \cdot H_2O$). Source: Reproduced with permission from Song et al. [75c]. Copyright 2018, John Wiley and Sons.

explained as follows: the initial stage of Zn^{2+}-ion intercalation is attributed to a solid–solution behavior; in the intermediate stage, the coexistence of two phases probably belongs to LiV_3O_8 and $ZnLiV_3O_8$, and the final stage confirms the new Zn-inserted phase ($Zn_yLiV_3O_8$, $y > 1$) [86]. Hu designed a novel ZIB cathode material of $Na_2V_6O_{16} \cdot 1.63H_2O$ (H-NVO) nanowire. During the first discharge process, the Zn^{2+} intercalation was accompanied by the insertion of water molecules into the interlayers. Upon subsequent charge/discharge processes, the Zn^{2+} was reversibly de-intercalated from/intercalated into H-NVO [87b].

The crystal structure of $H_2V_3O_8$ consists of V_3O_8 layers formed by corner- and edge-sharing networks of VO_6 and VO_5, which form two-dimensional slabs in the bc-plane [90]. The water shares an oxygen atom with the VO_5 polyhedron and connects adjacent V_3O_8 layers by hydrogen bonding [91]. Therefore, zinc ions can be easily inserted/extracted with a stable structure due to the elastic buffer spaces between layers based on hydrogen bond vibrations. In addition, the high average valence state (4.67) contributes to a larger specific capacity and more active redox sites. These unique characteristics of $H_2V_3O_8$ make it attractive as an energy-storage material. He synthesized the $H_2V_3O_8$ NWs through a one-step hydrothermal method [85g]. High-resolution XPS spectra of Zn and V elements are analyzed. When discharged to 0.2 V, the insertion of Zn^{2+} into the $H_2V_3O_8$-layered structure is reflected. The intensity of the V^{4+} peak strengthens and the V^{5+} peak weakens in comparison with the initial state, reflecting the reduction of V^{5+} to V^{4+} along with the insertion of zinc ions. Moreover, the addition of graphene was proven to further improve the charge-transfer kinetics of the $H_2V_3O_8$ cathode, especially after long cycling [85c]. A composite material consisting of $H_2V_3O_8$ NWs wrapped by graphene sheets was developed. DFT calculations show that intercalated zinc ions are stable at the center of the vacant site, with slight distortion to neighboring V atoms [85e]. $Zn_xH_2V_3O_8$ is the reversible discharge product, according to various measurements. Recently, Nazar reported the reversible intercalation of up to two Zn^{2+} in layered $V_3O_7 \cdot H_2O$.

Other layer structure:

In addition, $VOPO_4$ [92], $V_6O_{13} \cdot nH_2O$ [93], V_5O_{12} [94], and VS_2 [95] are all layered structures. The researchers have well investigated the energy-storage mechanisms of these different layered materials.

$VOPO_4$, an atypical layered structure, is connected to the PO_4 tetrahedron by a common-angle VO_6 octahedron. Niu and coworkers designed aqueous ZIBs with highly reversible oxygen redox chemistry based on $VOPO_4$ cathodes and water-in-salt electrolytes (Figure 3.27a) [92b]. Owing to the oxygen redox chemistry, the average operating voltage of Zn/$VOPO_4$ batteries was increased to ~1.56 V and provided ~27% additional capacity compared with the case without oxygen redox. Furthermore, the oxygen redox promotes the reversibility of the crystal structure conversion of $VOPO_4$ in the charge/discharge cycles, resulting in excellent rate capability and long-term cycle life. Wang et al. systemically investigated the effect of H_2O in the electrolyte and electrode lattice on the thermodynamics and kinetics of reversible Zn^{2+}-intercalation chemistry based on layered $VOPO_4$ [92c]. They found

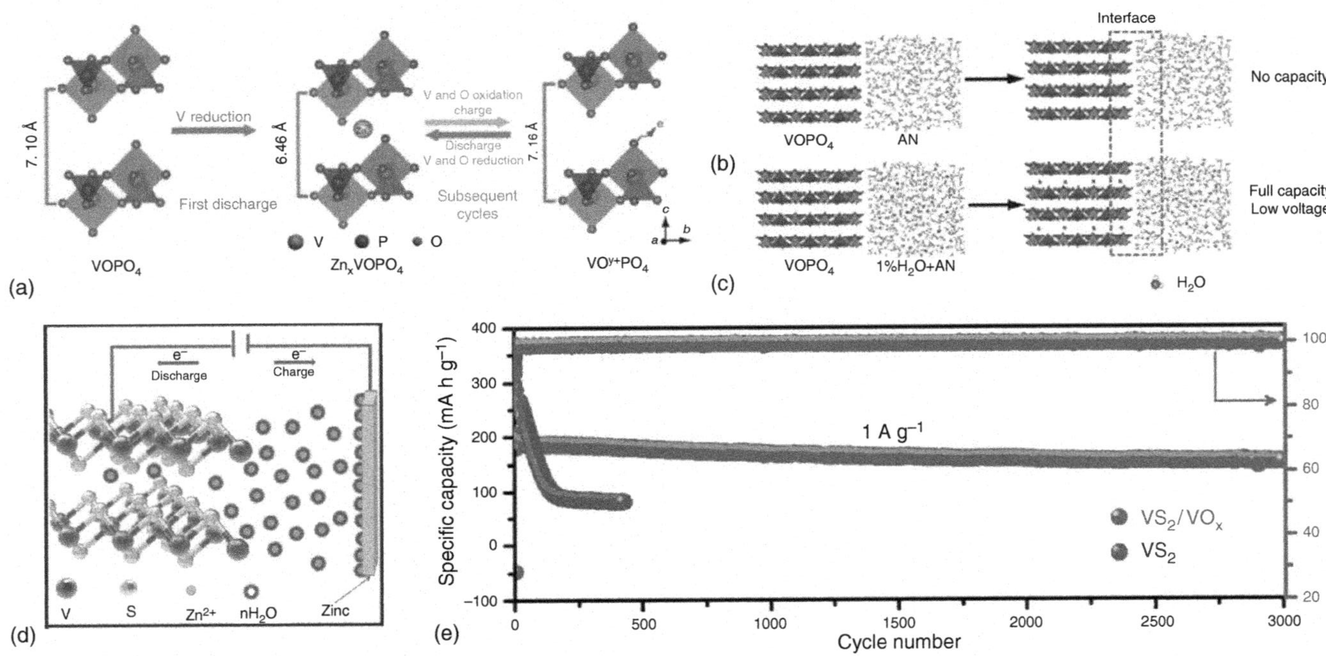

Figure 3.27 (a) Schematic of oxygen and vanadium redox mechanisms during charge/discharge processes. The scheme of $VOPO_4 \cdot 2H_2O$ in the 0.1 M $Zn(OTf)_2$-AN (b) without water and (c) with 1%H_2O. (d) Schematic illustration of the operation mechanism of Zn/VS_2 batteries. (e) The comparison of cycling performance at 1 A g^{-1} of VS_2/VO_2 and VS_2. Source: (a) Wan et al. Ref. [92b]/John Wiley & Sons. (b, c) Wang et al. [92c]/John Wiley & Sons. (d) He et al. [95a]/John Wiley & Sons. (e) Yu et al. [95b]/John Wiley & Sons.

that H_2O at the electrolyte/electrode interface accelerated Zn^{2+} diffusion from the electrolyte to the bulk surface, while H_2O in the bulk electrode altered the working potential (Figure 3.27b,c). When the electrochemical performance of the $VOPO_4 \cdot xH_2O$ cathode in aqueous zinc cells was examined, a serious voltage decay over cycling was observed, suggesting certain instabilities of the material. Hence, Sun et al. reported a high concentration of $ZnCl_2$ salt in the electrolyte to prevent the dissolution of $VOPO_4 \cdot xH_2O$; the cathode showed stable capacity and voltage retentions in 13 M $ZnCl_2$/0.8 M H_3PO_4 aqueous electrolyte, in direct contrast to that in $Zn(OTf)_2$, where the decomposition product VO_x provided the most electrochemical activity during cycling [92a].

Recently, layered transitional metal dichalcogenides (TMDs) have attracted considerable interest as cathode materials for ZIBs (i.e. VSe_2 [96], MoS_2, WS_2 [97], and CoS_2) owing to their outstanding characteristics of graphene-like layered structure, direct bandgap, and fast ion diffusion. VS_2 is a typical family member of TMDs with a hexagonal system, which shows a similar crystal structure to that of graphite lamellar with an interlayer spacing of 5.76 Å. Mai et al. fabricated VS_2 flowers assembled from nanosheets with a thickness of 50–100 nm. The intercalating of Zn^{2+} into the layers of VS_2 can be divided into two steps: the conversion of VS_2 into the conductive $Zn_{0.09}VS_2$, and then into $Zn_{0.23}VS_2$ (Figure 3.27d) [96]. Meanwhile, Du and coworkers developed an in situ electrochemically induced formation of heterostructural VS_2/VO_x material. Thanks to the unique interface properties, the VS_2/VO_x cathode demonstrated enhanced Zn^{2+}-storage performance (Figure 3.27e) [95b].

3.2.1.2 Tunnel-Based Structure

Tunnel-type cathodes have special 1D diffusion channels, high structural stability, and sufficient zinc-ion storage sites. Theoretically, the tunnel structure is an ideal host for storing zinc ions, in which their size and stability play a key role.

It is known that vanadium dioxide (VO_2(B)) is built from distorted VO_6 octahedra, which share both corners and edges, forming tunnel-like frameworks [98]. Yang et al. first demonstrated an intercalation pseudo-capacitance behavior and ultrafast kinetics of Zn^{2+} in VO_2(B) nanofiber. Owing to the abundant tunnel transport pathways and little structural change in Zn^{2+} intercalation, VO_2(B) nanofibers delivered a high reversible capacity, excellent rate capability, and high energy and power densities [98i]. Moreover, it was verified by using first-principle calculations that 0.5 mol of Zn^{2+} can be reversibly (de)intercalated with an average voltage of ~0.61 V (vs. Zn^{2+}/Zn). When Zn^{2+} diffusion occurred between each Zn_{A2} site, the activation barrier was ~586 meV, which indicates that facile Zn^{2+} migration into the crystal structure of VO_2(B) is possible (Figure 3.28a) [98e].

There may be different storage mechanisms for VO_2 cathodes in different types of electrolytes, various voltage ranges, and diverse crystal structures. For instance, Wagemaker demonstrated a new mechanism for the VO_2 cathode toward proton insertion accompanied by Zn-ion storage through the reversible deposition of $Zn_4(OH)_6SO_4 \cdot 5H_2O$ on the cathode surface [100]. It was proposed that the competition between proton and Zn^{2+}-insertion in the VO_2 host is determined

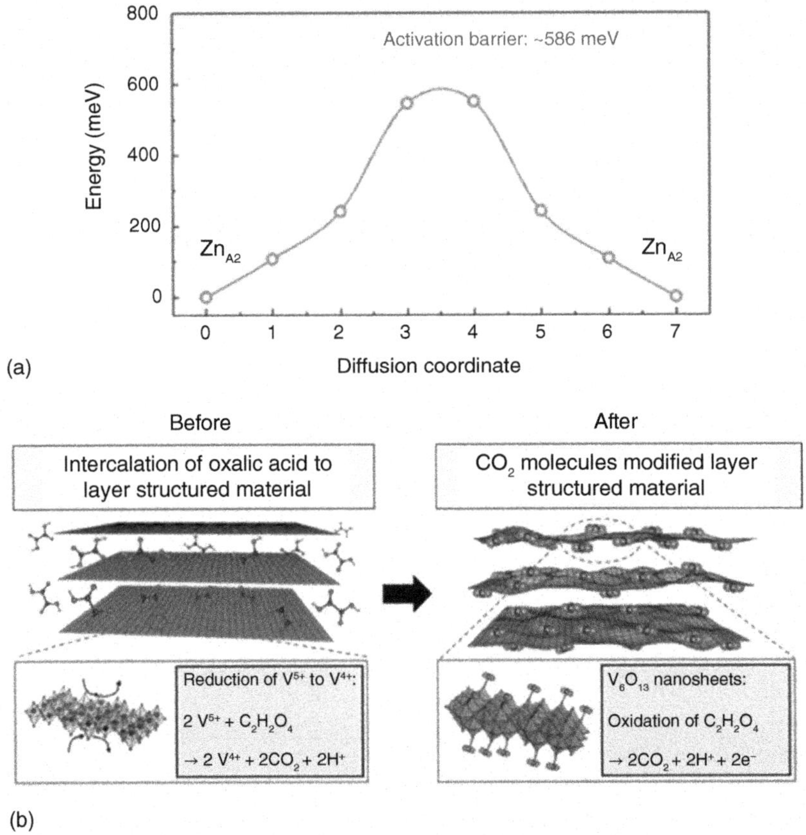

Figure 3.28 (a) Calculated Zn diffusion activation barriers in VO_2(B). (b) Intercalation of oxalic acid to a layer-structured V_2O_5 material (before) and CO_2 molecule-modified layer-structured V_6O_{13} material (after). Source: (a) Adapted from Park et al. [98e]. (b) Adapted from Shi et al. [99g].

by the proton insertion energies and solvation energy of the salt anion, where proton insertion has the advantage of least structural distortion, resulting in long cycle life.

The crystal structure of V_6O_{13} was first studied by Aebi in 1948 and further refined by Wilhelmi et al. in 1971 [101]. The crystal structure comprises edge- and corner-sharing distorted VO_6 octahedra, which form alternating single and double VO_x layers connected via shared corners [93, 99]. V_6O_{13} contains three crystallographically nonequivalent vanadium atoms, V(1), V(2), and V(3), where the vanadium centers are mixed valence with an average vanadium formal charge of $V^{4.33+}$ [99f]. Choi focused on the intercalation behavior of Zn^{2+} within the lattice framework of V_6O_{13}. Interestingly, a series of analyses revealed that water molecules were co-intercalated with Zn^{2+} upon discharge. This intercalation mechanism facilitates Zn-ion diffusion throughout the host lattice and electrode–electrolyte interface via electrostatic shielding and concurrent structural stabilization. Paired with a zinc metal anode, this V_6O_{13} exhibited performance metrics of 360 mAh g^{-1} at 0.2 A g^{-1}, and 145 mAh g^{-1} at a current density of 24.0 A g^{-1}.

Benefiting from the interlayer-expanded strategy through feasible water intercalation and the unique dandelion-like hollow architecture, the $V_6O_{13} \cdot nH_2O$ exhibited a high reversible capacity of 395 mAh g^{-1}. Furthermore, Liang et al. also proposed the conversion reaction mechanism of V_6O_{13}: the intercalated Zn^{2+} ions not only enable the formation of $Zn_{0.25}V_2O_5 \cdot H_2O$ but are also located at the empty sites of both V_6O_{13} and $Zn_{0.25}V_2O_5 \cdot H_2O$ phases. As a result, V_6O_{13} is capable of exhibiting enhanced electrochemical properties, including high capacity and excellent long-term cyclability (206 mA h g^{-1} at 10 A g^{-1} after 3000 cycles). Despite the ease of incorporating simple molecules into the host framework, most are often neglected (other than H_2O) in the development of high performance. Thus, the effect of CO_2 modification was studied by decomposing oxalic acid within a hydrated V_6O_{13} framework (Figure 3.28b) [99g]. Based on the collective simulation and experimental results, the presence of CO_2 drastically lowers the relative energy of Zn^{2+} diffusion through the pathways by forming weak electrostatic interactions between O_{CO2} and Zn^{2+}. The as-synthesized CO_2-V_6O_{13} electrode delivered one of the highest specific capacities reported for vanadium-based oxides of 471 mAh g^{-1}.

Other Tunnel-based structures:

There are many other vanadium-based cathodes with tunnel-type, such as $VO_{1.52}(OH)_{0.77}$ [102], VOOH [103], $Ag_{0.33}V_2O_5$ [104], and $Mo_{2.5+y}VO_{9+z}$ [105]. $VO_{1.52}(OH)_{0.77}$ has oxide ions in the channels along its *c*-axis, with hydrogen atoms attached to the edge-sharing oxygen atoms to form OH groups. The formal valence of vanadium is +3.81 [106]. Myung reported the feasibility of Al-doped $VO_{1.52}(OH)_{0.77}$, which possesses [2*2] tunnel structures [102]. Al preferentially replaces V^{3+}, so that the resulting average oxidation state of V increases with increasing Al content. More importantly, the presence of strong Al—O bonds successfully stabilizes the tunnel structure. The associated electrochemical reaction is driven by $V^{4+}/^{3+}$ redox, which is accompanied by Zn^{2+} intercalation/de-intercalation in the Al-doped $VO_{1.52}(OH)_{0.77}$. In addition, Zhou et al. reported that $Ag_{0.33}V_2O_5$ has a stable tunnel structure and a low mole ratio of Ag/V (Figure 3.29). The $Ag_{0.33}V_2O_5$ demonstrated a highly reversible combination displacement/intercalation mechanism, i.e. Ag^0 nanoparticles would be generated upon the Zn^{2+}-intercalation process and are largely reversible in oxidizing to Ag^+ upon the Zn^{2+} de-intercalation process [104].

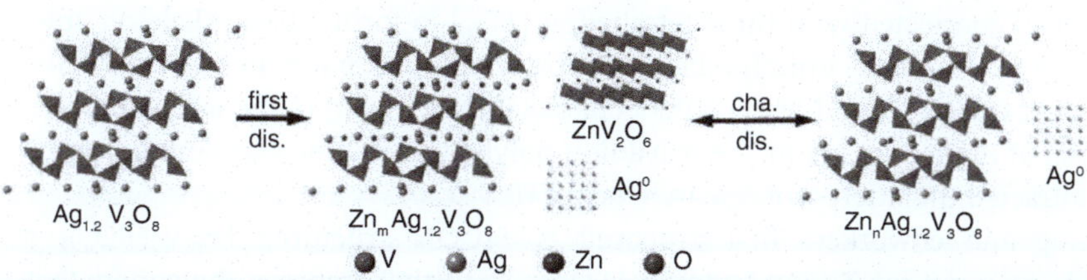

Figure 3.29 The schematic illustration of the electrochemical reaction of $Ag_{1.2}V_3O_8$ during the discharge/charge process. Reproduced with permission from Guo et al. [104]. Copyright 2019, Elsevier.

At present, the performance of tunnel-type vanadium-based cathodes can be improved in two aspects. First, the zinc-ion diffusion rate can be accelerated by regulating the aperture size, and then the rate performance can be improved. Secondly, super-stable cyclic performance can be realized with a stable layer structure.

3.2.1.3 Spinel-Type Structures

Spinel-type structures such as MV_2O_4 (M = Zn [107], Mg [108], etc. [109]) have previously been investigated as cathode materials for various rechargeable batteries. Among the family of spinel oxides, ZnV_2O_4 (ZVO) has drawn significant attention due to its distinct properties.

ZnV_2O_4 has a typical spinel structure with space group symmetry Fd-3m. As shown in Figure 3.30a, Zn atoms occupy the tetrahedral (8a) sites and V atoms occupy the octahedral (16d) sites, forming a standard face-centered cubic (FCC) spinel structure [107c]. Here, a typical spinel ZnV_2O_4 was first investigated as a high-performance cathode material for aqueous ZIBs. The electro-activation process was analyzed via in-situ XRD and ex-situ pair distribution function (PDF) (Figure 3.30b,c), indicating that the electro-activation enhances the surface electrochemical reaction with self-adaptive adjustment of the lattice structure [107d]. These results clarified the micro-crystallization process and the Zn^{2+}-de/insertion mechanism of spinel ZnV_2O_4 at the atomic scale. The electro-activation-induced ZnV_2O_4 delivered an ultra-high specific capacity of 250 mAh g^{-1} at a high rate of 10 C with a remaining reversible capacity of 206 mAh g^{-1} after 1000 cycles.

MgV_2O_4 has the same monoclinic structure as ZnV_2O_4. Tang et al. synthesized urchin-like spinel MgV_2O_4 microspheres by solvent-thermal synthesis. They proposed the Zn^{2+}-storage mechanism of the MgV_2O_4 cathode, containing both a conversion reaction and the formation of a new phase, which gradually transformed from order to disorder during cycling (Figure 3.30d) [108]. Specifically, the battery delivers a high capacity of 272 mAh g^{-1} at 0.2 A g^{-1} and excellent long-term cyclic stability. These remarkable electrochemical performances are attributed to the urchin-like structure with low-valence vanadium, which transforms from order to disorder and produces a new phase during the electrochemical cycling process.

3.2.1.4 NASICON-Type Structure

The general chemical formula of the NASICON vanadium-positive compound is $A_xMV_2(PO_4)_3$(A = Li^+ [110], Na^+ [111], Mg^{2+}, etc. [112]). Specifically, two octahedral [VO_6] connected with three tetrahedral [PO_4] to form a basic skeleton anion $[V_2(PO_4)_3]^{3-}$ "lantern" unit. Its stable 3D host framework has better thermal stability, longer cycle life, and better safety characteristics owing to the strong covalent bonding of oxygen atoms in the polyanion polyhedron. These excellent properties have attracted the interest of researchers in ZIBs.

Huang and coworkers first proposed NASICON-structured $Na_3V_2(PO_4)_3$ as cathode material for Zn-ion batteries [111c]. To further improve the conductivity of $Na_3V_2(PO_4)_3$, graphene-like carbon-wrapped $Na_3V_2(PO_4)_3$ nanoparticles were fabricated. The ex-situ XRD measurements and XPS were carried out to investigate

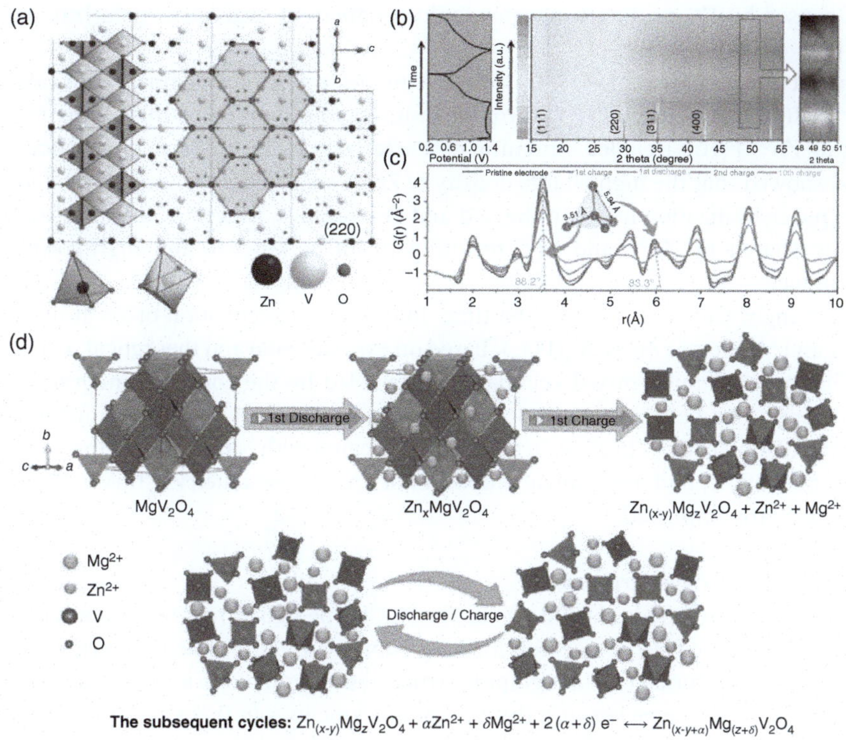

Figure 3.30 (a) Two-dimensional crystal structure of ZnV_2O_4. (b) In situ XRD measurement of the ZnV_2O_4 electrode during the first two cycles, (c) ex-situ PDF analysis of the ZnV_2O_4 electrode. (d) Schematic illustrations of Zn^{2+} storage mechanism. Source: (a) Reproduced with permission from De Juan-Corpuz et al. [107c]. Copyright 2019, American Chemical Society. (b, c) Adapted from Liu et al. [107d]. (d) Reproduced with permission from Tang et al. [108]. Copyright 2020, American Chemical Society.

the mechanism of Zn insertion/extraction into/from the NVP. The transition from $Na_3V_2(PO_4)_3$ to $NaV_2(PO_4)_3$ during charging is accompanied by the oxidation from V^{3+} to V^{4+}. When discharging, Zn ions insert into the $NaV_2(PO_4)_3$ framework, and V^{4+} is reduced back to V^{3+}. The reactions can be written as follows:

The first charge in the cathode:

$$Na_3V_2(PO_4)_3 \rightarrow 2Na^+ + 2e^- + NaV_2(PO_4)_3 \tag{3.22}$$

The rest reactions in the cathode:

$$NaV_2(PO_4)_3 + xZn^{2+} + 2xe^- \rightarrow Zn_xNaV_2(PO_4)_3 \tag{3.23}$$

Later, Mai et al. designed a $Na_3V_2(PO_4)_3$/reduced graphene oxide (rGO) composite, which demonstrated a different mechanism of simultaneous Zn^{2+}/Na^+ intercalation/de-intercalation in a single-component electrolyte [111d]. The battery achieved a high capacity of $114\,mA\,h\,g^{-1}$ at $50\,mA\,g^{-1}$ with a mean discharging platform of 1.23 V. On this basis, replacing some F^- with O^{-2} reduces the induction effect caused by F^-, leading to greater ionic conductivity and lower polarization [112c].

Wang et al. synthesized $LiV_2(PO_4)_3$ by in situ electrochemical extraction of two Li^+ ions from each $Li_3V_2(PO_4)_3$ formula unit, leaving two unoccupied sites for inserting Zn^{2+} [110]. Detailed structural characterization and first-principle calculations showed that the high charge density of Zn^{2+} was effectively accommodated by the p-d hybridization between the V-d and O-p orbitals, as well as the diffusion pathway shared by Zn^{2+} and Li^+, resulting in quick kinetics and high-power performance.

Interestingly, Cui et al., for the first time, found mixed occupation of Zn at both the M1 and M2 sites in $Zn_xNaV_2(PO_4)_3$ based on experimental and theoretical methods (Figure 3.31). Such mixed occupation is enabled by the concerted migration of Na^+/Zn^{2+} ions between M1 and M2, which also activates the M1 site for ion (de)intercalation of Zn^{2+} in NVP. In addition, compared with $Na_3V_2(PO_4)_3$, the electronic conductivity and mechanical properties of Zn^{2+}-intercalated $Zn_xNaV_2(PO_4)_3$ are significantly improved.

3.2.1.5 Rock Salt-Type Structures

Rock salt-type structure is also known as the NaCl-type structure. The lattice is the most densely packed surface center, forming cubic symmetry. Stoichiometric rock salts or sodium chloride crystallographic structures are unable to serve as cathode materials for zinc storage because the diffusion of zinc ions in these closely packed FCC lattices is often limited. Fortunately, a variety of design strategies can be used to expand ion diffusion channels and increase Zn^{2+}-storage active sites [113].

There are many materials with rock salt lattice structures, such as vanadium boride, vanadium phosphide, and vanadium carbide [78]. Vanadium nitride (VN) is a rock salt with a structure similar to NaCl. An important characteristic of VN is its metallic electronic conductivity ($\approx 10^6\ S\,m^{-1}$), which, together with the variety of oxidation states of V, renders it a promising candidate in electrochemical energy storage and conversion (including supercapacitors, batteries, and electrocatalysis) [75b]. Yang and coworkers developed an efficient strategy to unlock the electrochemical activity of rock salt vanadium oxynitride via the substitution of low-valent oxygen for high-valent nitrogen, then forming disordered rock salt with abundant vacancies/defects due to the charge-compensating function [113g]. Subsequently, Zhou and coworkers reported the insertion/extraction reaction and pseudo-capacitance-like surface redox reaction mechanisms. The energy storage is caused by a simultaneous reversible cationic ($V^{3+} \leftrightarrow V^{2+}$) and anionic ($N^{3-} \leftrightarrow N^{2-}$) redox reaction, which is the main reason for the high reversibility and structural stability of VN_xO_y [113i].

The rock salt-structured vanadium-based compounds further enrich the vanadium-based cathode system for AZIBs. In addition, the rock salt-structured vanadium-based cathode exhibits great potential for Zn storage by flexibly using defect regulation strategies and cleverly exploiting the unique pseudocapacitive behavior of the rock salt structure.

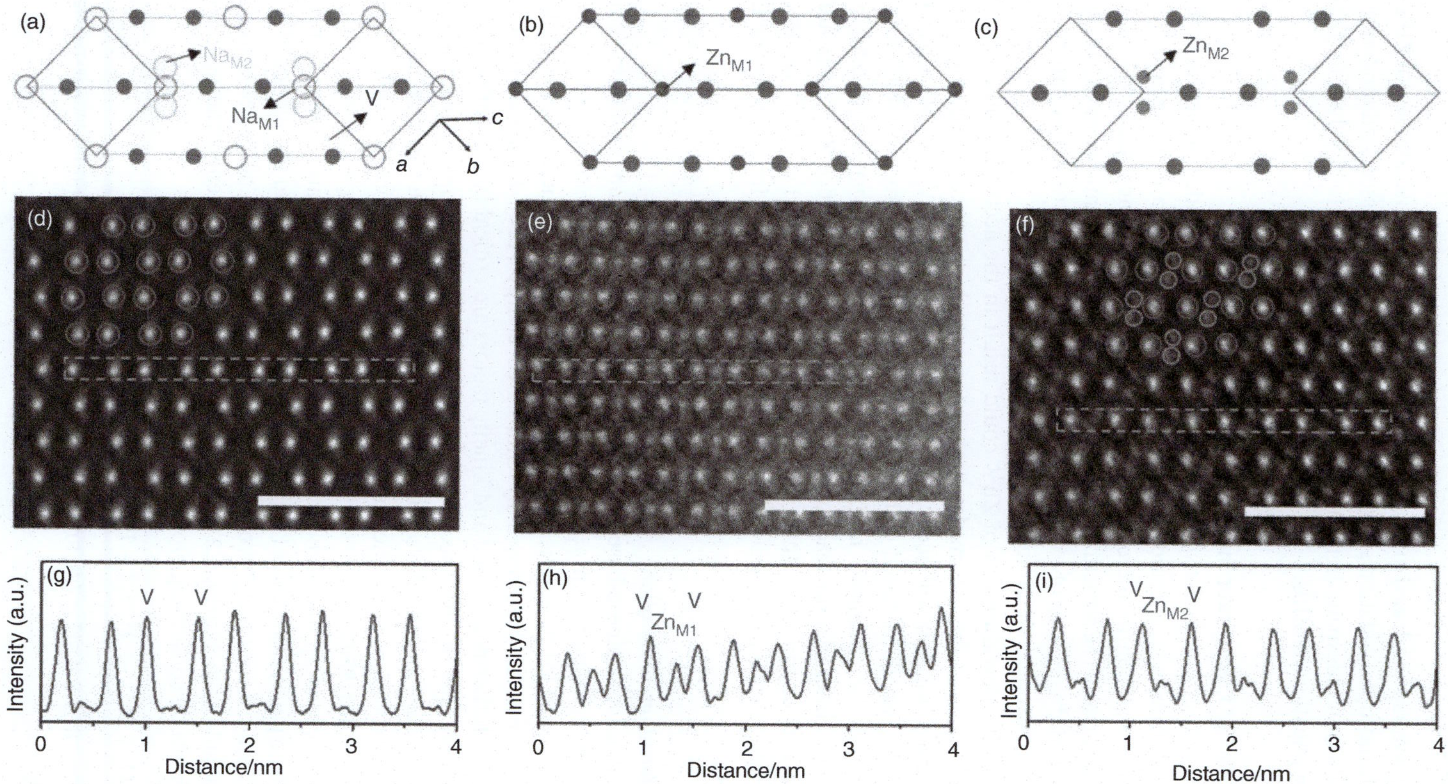

Figure 3.31 (a–c) Schematic diagrams of the structure for $Na_3V_2(PO_4)_3$ (a) and $ZnNaV_2(PO_4)_3$ (b, c) along the [221] projection. (d–f) STEM-HAADF images of $Na_3V_2(PO_4)_3$ (d) and $ZnNaV_2(PO_4)_3$ (e, f) along the [221] projection. (g–i) Line profiles along the STEM-HAADF images of $Na_3V_2(PO_4)_3$ (g) and $ZnNaV_2(PO_4)_3$ (h, i). Scale bars are 2 nm. Source: Reproduced with permission from Hu et al. [111a]. © 2020, John Wiley & Sons.

3.2.2 Issues

Even though vanadium-based materials are an important family of cathodes for ZIBs due to their multiple oxidation states, large amounts of supply, associated moderate rate capability, and lifespan, there are still many issues that need to be solved for the practical applications of ZIBs.

3.2.2.1 Effect of Electrostatic Interactions

Zn^{2+} ions with high charge density will render stronger electrostatic interaction with the host structure, which could result in slow Zn^{2+} diffusion and go against the reversible intercalation/deintercalation in host structures to some extent [93]. As a result, it could cause the structural collapse of cathodes. Recent studies have shown that the ionic diffusion rate of oxide cathodes is slower than that of sulfide cathodes because of the stronger electronegativity of O^{2-} [114].

Moreover, the slow diffusion of Zn^{2+} could result in the accumulation of Zn^{2+} in the lattice. Once the accumulated zinc ions reach a critical mass, an irreversible phase transition will occur. The typical new-phase $Zn_3V_2O_7(OH)_2 \cdot 2H_2O$ appears in the vanadium-based cathode during the discharge process, which may be ascribed to the enhanced interaction between Zn^{2+} and vanadium-oxygen layers and the intercalation reaction with water molecules. Since Zn^{2+} ions are difficult to extract from the ZnO_6 octahedra and $Zn_3V_2O_7(OH)_2 \cdot 2H_2O$ is electrochemically inactive, the electrochemical behavior would be impaired to some extent [90c].

3.2.2.2 Vanadium Dissolution

Similar to those manganese-based cathodes, mass dissolution is one main problem for vanadium compounds in aqueous electrolytes due to the strong polarity of water molecules [115]. Theoretically, vanadium dissolution is a thermodynamically favorable process, as determined by Gibbs free energy [116]. However, in real situations, both thermodynamic and kinetic factors are critical factors influencing the dissolution rate. In the aspect of thermodynamics, the amount of water at the electrode interface and the dissolubility of the electrolyte are the two determining factors for dissolution. From the viewpoint of dissolution kinetics, it is related to time/temperature [95a].

In general, vanadium-based materials are slightly soluble in water, giving them a characteristic pale yellow color (Figure 3.32a) [117]. According to the pH-log[V^{5+}] diagram (Figure 3.32b), VO_2^+ would be the stable existence form of V-species in an aqueous solution when pH < 3.2 and [V^{5+}] < 10^{-1} M (pH dependent), leading to an increase in pH by consuming protons [118].

$$V_2O_5 + 2H^+ \rightarrow 2VO_2^+ + H_2O \tag{3.24}$$

However, at $3.8 \leq pH < 7.8$ and [V^{5+}] < 10^{-4} M, the most stable V-species is $VO_2(OH)_2^-$:

$$V_2O_5 + 3H_2O \rightarrow 2VO_2(OH)_2^- + 2H^+ \tag{3.25}$$

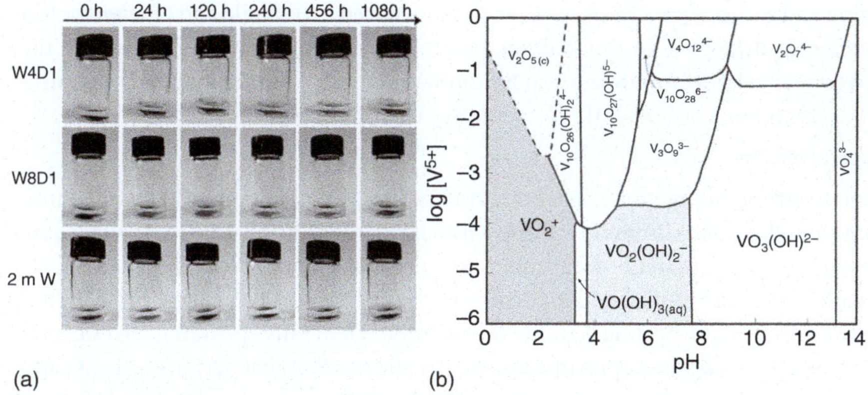

Figure 3.32 (a) Optical images of V_2O_5 dissolution in different electrolytes. (b) The pH-log [V^{5+}] diagram in the V_2O_5-H_2O system. Source: (a) Dong et al. [117]/Royal Society of Chemistry. (b) Reproduced with permission from Lu et al. [118]. Copyright 2020, John Wiley and Sons.

The formed $VO_2(OH)_2^-$ can further react with Zn^{2+} in the electrolyte, yielding $Zn_3V_2O_7(OH)_2 \cdot 2H_2O$ (ZVO) via the following reaction:

$$2VO_2(OH)_2^- + 3Zn^{2+} + 3H_2O \rightarrow Zn_3V_2O_7(OH)_2 \cdot 2H_2O + 4H^+ \qquad (3.26)$$

On the one hand, the dissolution of vanadium-based materials would directly lead to capacity deterioration owing to the loss of active materials. On the other hand, the dissolved transition metal ions would be deposited on the anode surface to increase the battery resistance [119]. Vanadium dissolution is time-dependent, and a longer duration would cause more severe dissolution [120]. Consequently, the previously reported vanadium-based cathodes often show ultra-stable cycling capability with thousands or even tens of thousands of cycles at a high current density but very poor cycling lifespan at a low current density [121]. To this end, it is a huge challenge to prolong the cycling lifespan of vanadium-based cathodes at low current density [122].

3.2.3 Modified Strategy

The Zn^{2+}-ion storage performance of vanadium-based cathodes can be augmented by (i) rational defect engineering, (ii) interlayer intercalation, (iii) morphology optimization, (iv) composite material, and (v) electrochemical activation.

3.2.3.1 Defect Engineering

Defect engineering has gradually become a recognized strategy as a means to afford new functional properties of materials by altering the surface chemistry, electronic structure, or coordinated mode of the pristine materials in a wide range of fields, such as catalysis and energy storage [123]. In particular, defect engineering has been widely employed in the modification of vanadium-based materials for ZIB applications, with very motivating improvements in the performance of batteries [124].

Doping and vacancy creation have been demonstrated to be facile and efficient defect engineering strategies. In this section, defect engineering strategies for tuning the electrochemical performance of vanadium-based electrode materials will be introduced, combined with several representative examples.

Oxygen Defects:

Due to strong interactions between divalent Zn^{2+} and host materials, vanadium-based cathodes offer a limited cycle life with high cathode utilization in AZIBs. These trapped Zn^{2+} ions gradually accumulate during cycling, creating physical barriers on the cathode surface and eventually reducing the number of active sites for Zn^{2+} storage, thus resulting in cycling stability degradation. Generally speaking, the effects of the oxygen vacancy strategy on optimizing the electrochemical performance of cathode materials can be classified into the following: (i) increased energy-storage sites; (ii) enhanced ion diffusion and electron transfer kinetics; and (iii) greater structural stability [124a].

For example, compared with oxygen-deficient V_6O_{13}, the relatively defect-free structure (V_6O_{13}) has a much higher Gibbs free energy of desorption, pointing to a rather difficult Zn^{2+} release. The oxygen-deficient V_6O_{13} structure has a thermoneutral Gibbs free energy of Zn^{2+} desorption, suggesting effortless Zn^{2+} release by the cathode. In addition, it had been reported that the incorporation of oxygen vacancies into VO_2 had a similar effect. For example, the oxygen-deficient VO_2 possessed a more moderate adsorption energy (−0.039 eV) than that of perfect VO_2 (−0.198 eV) (Figure 3.33a) [98h]. Furthermore, the electrical structure can be regulated through defect engineering. As shown in Figure 3.33c, the uniform charge distribution in the perfect structure was broken after introducing the oxygen defects, and the electrons around the vacancy-defect sites were prone to accumulate. In general, when regular Zn^{2+}-insertion occurs into the ab plane of V_6O_{13} (Figure 3.33b,d), the oxygen-deficient sites in Od-VO may open up extra insertion channels to allow Zn^{2+} diffusion along the *c* axis, hence greatly boosting the reaction kinetics and electrochemical reactivity [125].

Besides, Wang and coworkers proposed that introducing oxygen defects into potassium vanadate (KVO) could promote solid-state diffusion in the host structure [126]. The significantly lower diffusion barrier (0.63 eV) in the oxygen-defective KVO further confirms that the presence of oxygen defects can enable much faster reaction kinetics along the ab-plane, which effectively enhances the diffusion capacity of zinc ions, accordingly attaining superior rate performance and stable cycling capability. As a result, the as-prepared oxygen-defective KVO exhibited superior rate performance (166 $mAh\,g^{-1}$ even at 20 $A\,g^{-1}$) and outstanding cycling stability with a 95% capacity retention over 1000 cycles.

Although oxygen vacancies greatly affect the electrochemical properties of electrodes, there is still a lack of a suitable preparation strategy with controllable oxygen vacancy (Vö) incorporation, and the impact of oxygen vacancy concentration on electrochemical performances remains unclear. Given this, Sun and coworkers developed a facile spontaneous polymerization strategy to achieve tunable Vö concentration in V_2O_5 [123c]. The introduction of poly(3,4-ethylene

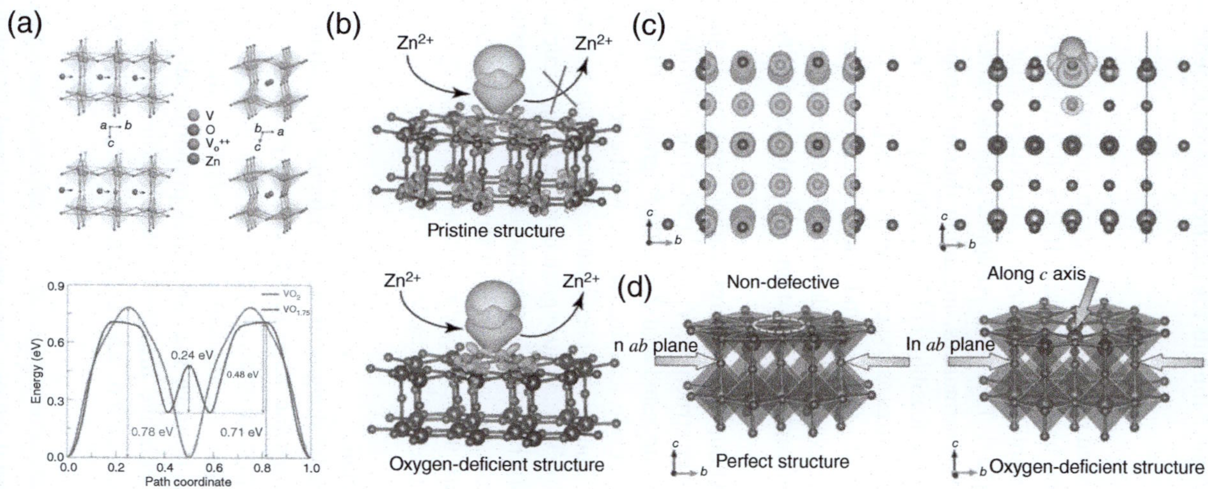

Figure 3.33 (a) Diffusion energy barriers of VO_2 and $VO_{1.75}$. (b) Schematic illustration of Zn^{2+} adsorption/desorption for perfect V_6O_{13} and oxygen-deficient V_6O_{13}, (c) charge distribution, and (d) corresponding structure of perfect vanadium oxide and oxygen-deficient vanadium oxide. Source: (a) Reproduced with permission from Li et al. [98h]. Copyright 2020, American Chemical Society. (b–d) Reproduced with permission from Liao et al. [125]. Copyright 2020, John Wiley and Sons.

dioxythiophene) (PEDOT) not only leads to the formation of Vö in V_2O_5 but also results in a larger interlayer spacing. According to the results, the V_2O_5-PEDOT with 20.3% Vö concentration showed superior electrochemical stability. Based on the theoretical calculations and experimental results, excessive Vö in V_2O_5 will lead to an increase in the band gap, which inhibits the electrochemical kinetics and charge conductivity.

Doping:

Previous studies have revealed that cation doping can energetically promote ionic bonding, thus stabilizing the layered structure. It is well known that Al doping greatly contributes to the stability of the crystal structures of $LiCoO_2$, $LiNiO_2$, and $LiNi_{1-x}Co_xO_2$ in LIBs [127]. Therefore, Al-doped cathode materials generally exhibit improved electrode performances, including thermal properties. For the first time, Jo et al. suggested the feasibility of $VO_{1.52}(OH)_{0.77}$ and Al-doped $V_{1-x}Al_x O_{1.52}(OH)_{0.77}$ with tunnel structures as the cathode for ZIBs [102]. The substitution of V^{3+} by Al^{3+} formed strong Al—O bonds and helped stabilize the tunnel structure during cyclic testing. As a result, the Al-doped $VO_{1.52}(OH)_{0.77}$ demonstrated a substantial improvement in capacity, retention, and rate capability in zinc-ion batteries.

To elucidate the effect of Al-doping in different cathode systems, Li and coworkers synthesized an Al-doped layer structure, $V_{10}O_{24} \cdot 12H_2O$ as cathode material for AZIB [123a]. XRD analyses suggested that $Zn(H_2O)_6{}^{2+}$ groups diffused along the (001) planes of the $V_{10}O_{24}$ electrode, and the material will be activated after the first discharge/charge progress. The strong electrostatic interaction between divalent zinc ions and crystal structure would cause great damage to the structure and lead to fast degradation of capacity. Fortunately, structural stability was greatly enhanced by Al-doping. Benefiting from the stable layered structure, the material exhibited an excellent cycling life with 98% capacity retention after 3000 cycles at $5\,A\,g^{-1}$, which is much higher than the 65% in pure $V_{10}O_{24} \cdot 12H_2O$.

3.2.3.2 Interlayer Intercalation

For vanadium-based materials, sluggish transport kinetics are usually caused by the narrow interlayer spacing and large diffusion energy barrier, which greatly reduce the electrochemical performance. Interlayer intercalation generally involves the introduction of metal ions or molecules/polymers into the layer to widen the interlayer spacing and improve the structural properties of target materials. In this section, three effective interlayer intercalation strategies, including ion intercalation, H_2O intercalation, and conductive polymer intercalation, are introduced.

Ion Intercalation:

Until now, a series of metal cations have been developed to stabilize the layered vanadium-based hosts, relying on their large interlayer spacing and cation pre-intercalation to achieve robust structures. Many polyvalent and monovalent ions have been successfully applied, such as Zn^{2+}, Ca^{2+}, Mg^{2+}, Mn^{2+}, Ni^{2+}, Li^+,

Na^+, and K^+ [81b, 88a, 128]. Specifically, the introduction of metal ions enables the vanadium-oxide layers to transform into VO_5 square pyramidals and VO_6 octahedra, thus leading to the *c*-axis extension, which synergistically constructs a more stable bilayered V_3O_8 structure along the *b*-axis. During the charging/discharging processes, it demonstrated higher stability than the structure only containing VO_5 square pyramidals. $Zn_{0.25}V_2O_5 \cdot nH_2O$ nanobelt cathodes with Zn^{2+} intercalation and structural water retention were first reported by Nazar and coworkers [128d]. In aqueous ZIBs, exhibiting high capacities (>300 mAh g^{-1}) and long-term cycling stability (>1000 cycles) due to their enhanced reaction kinetics and firmly layered structure caused by the indigenous Zn^{2+}.

To capitalize on these features, Xia proposed a double-layered calcium vanadium oxide bronze ($Ca_{0.25}V_2O_5 \cdot nH_2O$, CVO) as an aqueous ZIB intercalation cathode (Figure 3.34a–c) [128b]. It is more suitable compared to $Zn_{0.25}V_2O_5 \cdot nH_2O$ (ZVO) for the following considerations: (i) large interlayer spacing due to the larger size of CaO_7 polyhedra than the ZnO_6 octahedra pillars, assuring the fast insertion/extraction of Zn^{2+}; (ii) the smaller molecular weight and density of CVO probably give higher gravimetric and volumetric capacity; and (iii) higher electrical conductivity of CVO than ZVO. The CVO cathode delivered a high capacity of 340 mAh g^{-1} at 0.2 C, good rate capability, and long cycling life (96% retention after 3000 cycles at 80 C). Moreover, the intercalation of high-amount metal ions may reduce the specific capacity of V_2O_5-based materials. Yang et al. successfully achieved the chemical intercalation of small amounts of transition metal ions (Fe^{2+}, Co^{2+}, Ni^{2+}, Mn^{2+}, Zn^{2+}, and Cu^{2+}) into the interlayer of V_2O_5, which can effectively improve the ions' intercalation kinetics, electronic conductivity, and structural stability of V_2O_5-based composites, thus improving the electrochemical performance (Figure 3.34d) [128a].

Moreover, researchers discovered that low-cost alkaline metal cations play the same "pillar" role as transition metal cations in stabilizing vanadate structures [129]. K^+ with a larger ionic radius (0.138 nm) illustrates a better relief of electrostatic resistance for Zn^{2+} transfer than Li^+ (0.076 nm) and Na^+ (0.102 nm) [130]. Tian reported a structural engineering method by incorporating K^+ into hydrated vanadium pentoxide (KVOH) [128c]. In contrast to previously reported work, K^+ introduction leads to a new phase of KVOH with faster ion diffusion kinetics and better long-term cycling stability. This work establishes an understanding of the role of K^+ incorporation in KVOH that goes beyond its conventional categorization as an agent for interlayer spacing adjustment, reflecting its importance in maintaining structure flexibility for effective Zn^{2+}-insertion/extraction even at high rates. In conclusion, pre-intercalated metal ions in vanadium-based materials have been proven to be an effective method to regulate their electronic structure, electronic conductivity, and Zn^{2+} diffusivity, realizing the great improvement in electrochemical performance applied to AZIBs. However, the crystallographic complexity of the double-layer V_2O_5 host lattice and the challenges associated with evaluating the precise role of various pre-intercalated ions have limited systematic improvements in the design of these cathode materials.

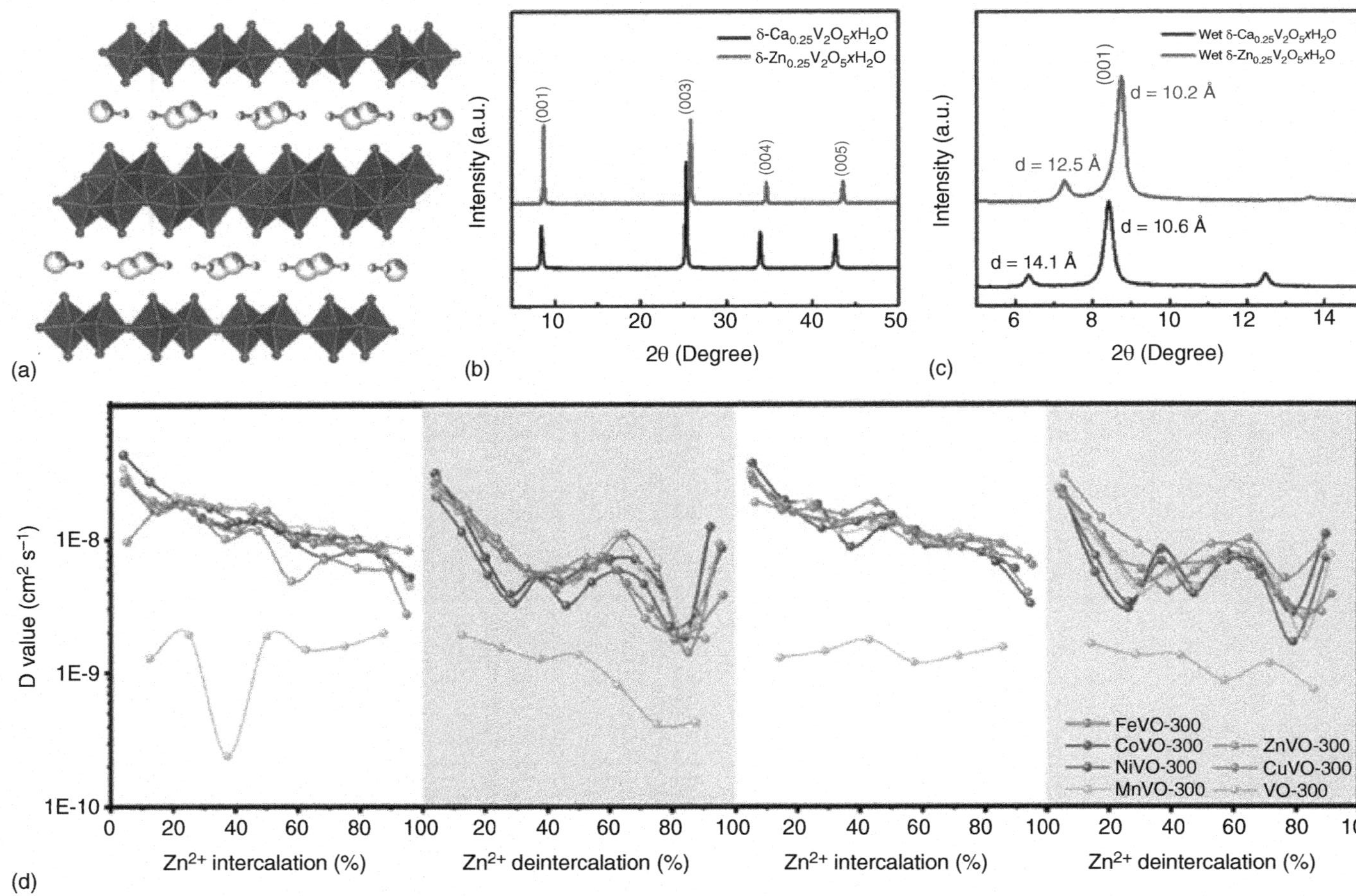

Figure 3.34 (a) Crystal structure viewed along the *b*-axis of CVO, (b,c) Typical XRD pattern of as-synthesized CVO. (d) Corresponding diffusivity coefficient (D) of Zn^{2+} of TVO-300 and VO-300. Source: (a–c) Reproduced with permission from Xia et al.[128b]. Copyright 2018, John Wiley and Sons. (d) Reproduced with permission from Yang et al. [128a]. Copyright 2019, Elsevier.

H_2O Intercalation:

In previous studies, structural water intercalation modification has been regarded as an effective strategy for tuning the electrochemical performance of vanadium oxide materials. During cycling, the solvating H_2O works as a charge shield for the metal ions (Al^{3+}, Mg^{2+}, Li^+, etc.), reducing their effective charges and hence their interactions with the host frameworks [131]. This strategy has been investigated to enhance the capacity and rate capability of Li^+, Na^+, and Mg^{2+} batteries [132]. Gu et al. found that the hydrated form of vanadium pentoxide ($V_2O_5 \cdot nH_2O$) with better chemical stability than crystalline V_2O_5 can reversibly uptake more than two Li^+ ions due to the expanded interlayer space. With a layered structure with double V_2O_5 sheets, $V_2O_5 \cdot nH_2O$ with water molecules contained in the oxide network leads to good ion mobility owing to the electrostatic shielding of the metal ions. Recently, structural water in $Zn_{0.25}V_2O_5 \cdot nH_2O$ was proven to exchange with Zn^{2+} reversibly, resulting in good kinetics and rate performance. For example, the critical role of structural H_2O on Zn^{2+} intercalation into bilayer $V_2O_5 \cdot nH_2O$ was demonstrated. The results suggested that the H_2O-solvated Zn^{2+} possesses largely reduced effective charge, thus reducing electrostatic interactions with the V_2O_5 framework and effectively promoting its diffusion. Benefiting from the "lubricating" effect, the aqueous Zn battery system displays a specific energy of 144 Wh kg^{-1}, a capacity of 295 mA h g^{-1} and a power density of 6.4 kW kg^{-1} (based on the cathode and 200% Zn anode) [80].

Moreover, crystal water molecules and defects were introduced into the VO_2 tunnel structure simultaneously [133]. The lattice defects break the strong electrostatic interaction between ions and host structure, offering a rapid ion diffusion pathway. The crystal water plays the role of stabilizing the channel structure and working on the charge-shielding effect by reducing the interaction with the surrounding framework. The results illustrated that the HVO is beneficial to ultrafast cation diffusion, and the kinetics were not restricted by diffusion, thus achieving the intercalation pseudocapacitance (Figure 3.35).

Conducting Polymer Intercalation:

More effective strategies for interlayer spacing regulation have recently been explored by using conductive polymers as intercalated pillars [88b, 134]. For example, Liu developed an in-situ PANI intercalation strategy to promote the Zn^{2+} (de)intercalation kinetics in V_2O_5 (Figure 3.36a–d) [134a]. The intercalation of PANI not only expands the Zn^{2+} diffusion channels but also maintains their structural stability as interlayer pillars. In particular, its unique π-conjugated structure acts as an electron reservoir, shielding the electrostatic interactions between Zn^{2+} and the V_2O_5 host. PEDOT, having been verified to display greatly enhanced stability compared to PPy and PANI, has attracted extreme interest for application in supercapacitors in recent years. Yang and Xia's groups synthesized PEDOT-intercalated layered V_2O_5 nanoflakes and PEDOT-$NH_4V_3O_8$ nanobelts, respectively [88b, 134b]. Coupled with the conductive nature of intercalated polymers, promotions of electron/ion transport as well as structural integrity, have

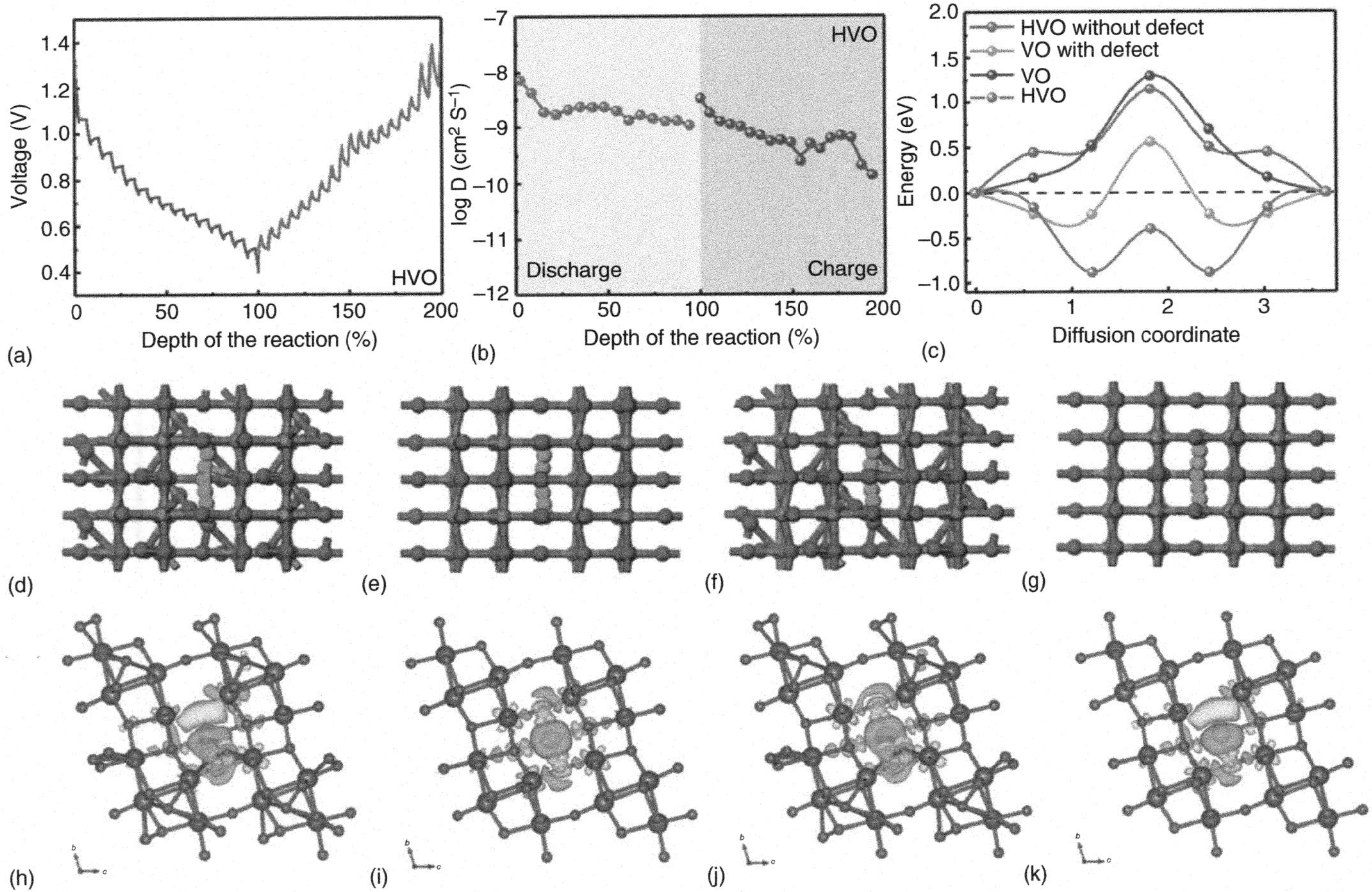

Figure 3.35 (a, b) The GITT tests and corresponding ion diffusion coefficient of HVO cathode. (c) The corresponding diffusion energy barrier profiles. (d–g) The migration paths of Zn ion diffusion and (h–k) differential charge density calculation. Source: Reproduced with permission from Liu et al. [133]. Copyright 2020, John Wiley and Sons.

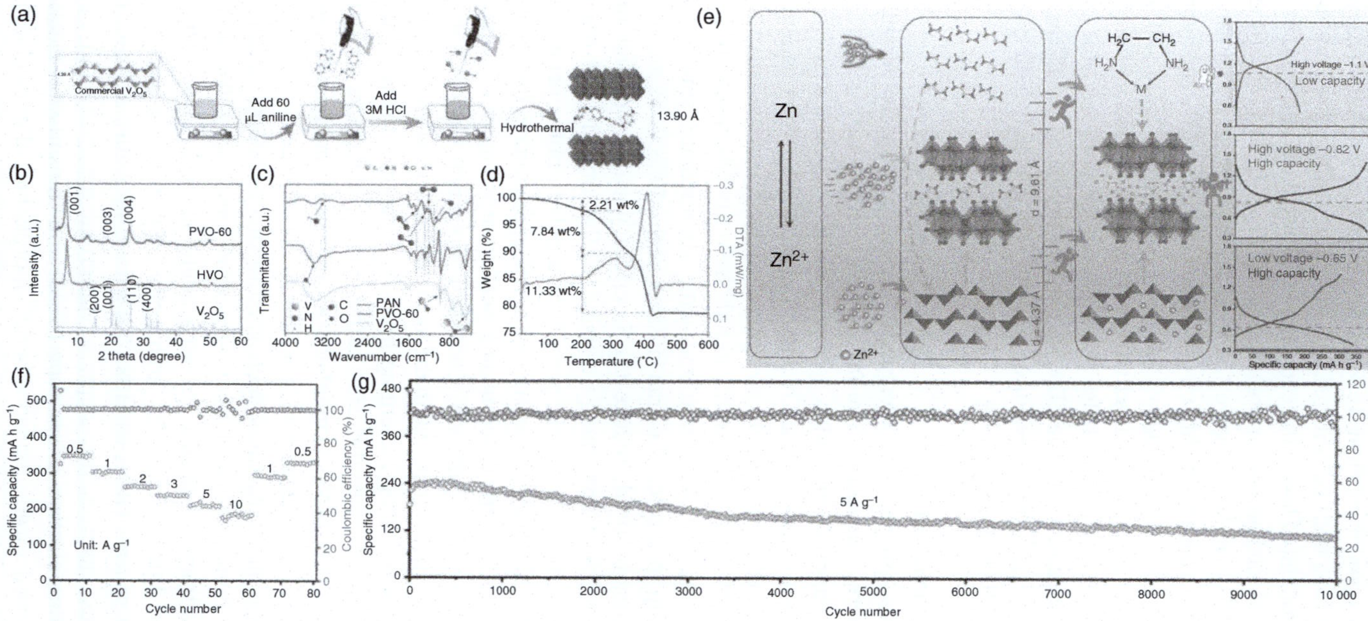

Figure 3.36 (a) Illustration of the synthesis process of PANI-intercalated V_2O_5 nanosheets. (b) XRD patterns and (c) FTIR spectra of PANI-intercalated V_2O_5, as-prepared $V_2O_5 \cdot nH_2O$ (d) The TG and DTA curves of PVO-60. (e) Schematic illustration of the organic (EDA)-inorganic hybrid cathode with high capacity and high operating voltage. (f) The rate capability of EDA-VO. (g) Long-term cycling performance of EDA-VO electrodes at 5 A g^{-1}. Source: (a–d) Reproduced with permission from Liu et al. [134a]. Copyright 2020, John Wiley and Sons. (e–g) Reproduced with permission from Ma et al. [134c]. Copyright 2022, John Wiley and Sons.

been achieved in these cathode materials for the AZIBs, thereby exhibiting notably improved high-rate capability and cycling stability.

However, the organic components just worked as pillars to expand the interlayer spacing instead of exploiting their advantages. Therefore, the strategy does not fundamentally solve the problem of low voltage in vanadium-based cathodes. Ma et al. reported an organic (ethylenediamine)–inorganic (vanadium oxide) hybrid cathode, $C_2H_8N_2V_7O_{16}$ (denoted as EDA-VO) (Figure 3.36e) [134c]. Benefiting from the high capacity of the vanadium oxide and the high working voltage of the EDA, the hybrid cathode provides a high operating voltage of 0.82 V. In addition, the EDA can not only increase the layer spacing of the vanadium oxide, with improved mobility of Zn ions in the V-O layered structure, but also participate in the storage of zinc ions as a bidentate-chelating ligand. Thus, the EDA-VO cathode presented a high capacity (382.6 mA h g^{-1} at 0.5 A g^{-1}) and superior cycling stability (capacity fading only 0.047‰ per cycle after 10 000 cycles at 5 A g^{-1}) (Figure 3.36f,g).

3.2.3.3 Morphology Optimization

Morphology control has also been considered an effective strategy for designing advanced cathode materials in AZIBs. The rational structural design is expected to endow the cathode materials with high specific surface area, rich active sites, short ion/electron transport path, and fast reaction kinetics [135]. The frequently-used strategies include reducing particle size and designing porous and multi-dimensional structures [98c, d, 136]. Nanomaterials have the advantages of high specific surface area, excellent electronic and ionic conductivity, small mass, volume, etc., which help to shorten the transmission path of electrons and ions, accelerate the transmission speed, reduce the interface polarization, and thus improve the rate performance and energy density [137]. Normally, the nanostructure can be classified into 0D, 1D, 2D, and 3D materials from the perspective of spatial growth [99h, 135b, 138]. In recent years, researchers have investigated the use of cathode materials with different nanoscale morphologies in aqueous rechargeable zinc batteries. For instance, Cao et al. prepared a novel ancient Chinese coin-shaped VO_2 structure based on the localized Ostwald ripening-guided dissolution-regrowth process [98c]. Benefiting from the structural merits, a large exposed surface and abundant mass diffusion pathways were created, accelerating zinc-ion transfer at the surface of VO_2 nanoplates. As a result, it could present high specific capacity and excellent cycle stability for ZIBs as the cathode.

In addition, complex hollow structures, such as multi-shelled hollow structures, yolk-shell structures, rambutan-like hollow structures, and bowl-like hollow structures, have also attracted extensive research attention [98d, 136a, 137b, c]. Zhou reported a novel intricate complex architecture, i.e. bubble-encapsulated double-shelled hollow spheres (BDHSs) [138]. As shown in Figure 3.37a–d, the synthesis of the BDHSs involves two steps: solvothermal treatment and calcination process. Scanning electron microscopy (SEM) and transmission electron microscopy (TEM) showed a well-retained spherical morphology after calcination. Moreover, rougher surfaces of the shells (Figure 3.37e,f) and enlarged voids at the surface protrusions (Figure 3.37g,h) were observed. This layered structure not only

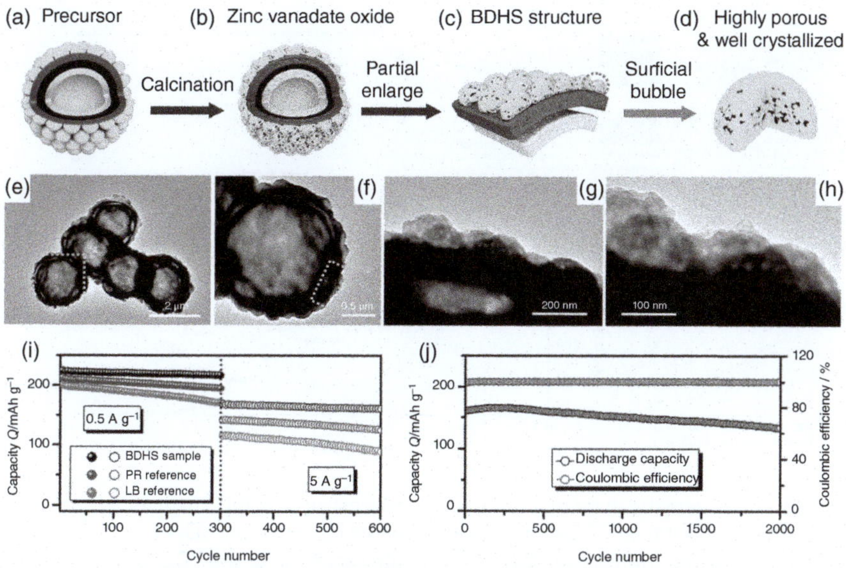

Figure 3.37 (a–d) Schematic illustrations of the precursor from solvothermal treatment. (e) Low-resolution TEM image of the products. (f) The enlarged observation of one sphere and the partially enlarged image of the outside shell (g), (h) enlarged image of the surface bubble layer and HRTEM image of the nanoparticles. (i) Cycling properties of all the samples at both 0.5 and 5 $A\,g^{-1}$. (j) Long-term high-rate cycling properties of the BDHS sample. Source: Reproduced with permission from Zhou et al. [138]. ©2019, Royal Society of Chemistry.

has a highly porous skeleton for rapid ion transport but also provides more active sites for electrochemical reactions. As a result, the BDHSs achieved the highest capacity retention, retaining 96.5% and 95.2% of the capacity after three hundred cycles at 0.5 and 5 $A\,g^{-1}$, respectively (Figure 3.37i, j).

Despite the progress that has been made on the cathode materials of the ZIBs, the construction of special vanadium compounds nanostructure to achieve electrochemically enhanced mass transfer remains a challenge. Lin reported a novel hollow $V_6O_{13-\delta}$/C nanoscroll, which has enlarged interlayer spaces, highly porous carbon-based shells, and defect-rich crystals [99h]. By simultaneously enlarging the distance between adjacent shells and engineering the defective structure for oxide crystal, the $V_6O_{13-\delta}$/C nanoscrolls achieve fast ion transport kinetics, high cycling durability, and superior energy storage capability.

3.2.3.4 Composite Material

Composite vanadium-based materials with other materials can also realize the improvement of electrochemical performance. Combining with conductive materials (e.g. direct mechanical mixing with graphene/carbon nanotubes [85e, 140], in-situ growth, surface coating [141], etc.) is one of the most common means to improve the electronic conductivity of electrode materials. This composite structure not only improves the electronic conductivity but also helps to prevent the dissolution of the active material, thereby improving its structural stability. According to

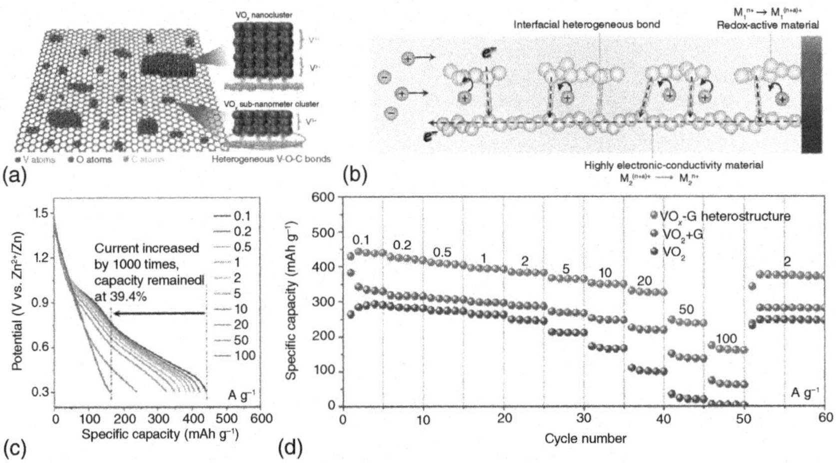

Figure 3.38 (a) Schematic diagram of the VO_x-G heterostructure, (b) schematic of the interface pseudocapacitance. (c) Galvanostatic discharge curves of the VO_x–G heterostructure varying from 0.1 to 100 A g^{-1}. (d) Rate capability of the VO_x–G heterostructure, VO_2+G, and VO_2. Source: Reproduced with permission from Dai et al. [140c]. Copyright 2021, John Wiley and Sons.

their elemental composition, composite materials can be divided into inorganic and organic phases. The inorganic phase mainly refers to the metal compound, and the organic phase refers to the conductive carbon or conductive polymer. Here, several representative examples of composite structures will be introduced.

Highly conductive carbon is usually indispensable, such as carbon cloth, graphene, CNTs, and rGO. Pang et al. synthesized $H_2V_3O_8$/graphene composites by solvothermal method. The desirable structure enabled a uniform mixture and intimate contact between the $H_2V_3O_8$ NWs and graphene surfaces, which significantly improved the charge–transfer kinetics and stability of the composite electrode [85e].

In addition, Dai et al. constructed a VO_x sub-nanometer cluster/rGO cathode material composed of interfacial V—O—C bonds (Figure 3.38) [140c]. Therein, a new mechanism was revealed, where Zn^{2+} ions are predominantly stored at the interface between VO_x and rGO, which causes anomalous valence changes and exploits the storage ability of nonenergy-storing active yet highly conductive rGO. Further, this interface-dominated storage triggers decoupled transport of electrons/Zn^{2+} ions, and the reversible destruction/reconstruction allow the interface to store more ions than the bulk.

As a highly conductive and extremely stable polymer, PEDOT has been reported to not only effectively improve the electronic conductivity of metal oxides, but also serve as a protective interface to restrain the nanostructure from collapsing. The V_2O_5@PEDOT core/shell nanosheet arrays have been grown on a flexible carbon cloth substrate through a three-step process [142]. As compared to the V_2O_5/CC, the V_2O_5@PEDOT nanosheets become thicker and rougher, in which the V_2O_5 nanosheets are intimately wrapped by a uniform PEDOT shell ($t \approx 5$ nm). The synergistic effects between the nanosheet array structure on the conductive carbon cloth and the PEDOT shell protection contribute to the high-rate capability and long cycling performance. Specifically, the V_2O_5@PEDOT/CC cathode shows a superior rate capability (232 mAh g^{-1} at a large current density of 20 A g^{-1}) and superior cycling performance (86.1% retention after 100 cycles at 0.2 A g^{-1}).

Except for the organic coating, inorganic oxide coating is another promising strategy to effectively improve the electrochemical performance of cathode materials. The difference between them is mainly reflected in the optimization of interface transport kinetics. Inorganic coating of cathode materials is expected to improve the rate capability by boosting the interface ion transport kinetics [114, 143]. Alshareef and coworkers constructed HfO_2-coated $Zn_3V_2O_7(OH)_2 \cdot 2H_2O$(ZVO) by atomic layer deposition (ALD) (Figure 3.39a) [139]. The inactive hafnium (IV) oxide (HfO_2) film not only decreases byproduct formation on the surface of ZVO but also suppresses ZVO cathode dissolution in the electrolyte. Meanwhile, the Zn^{2+} could be transported through the interface quickly (Figure 3.39b,c). Based on this, the

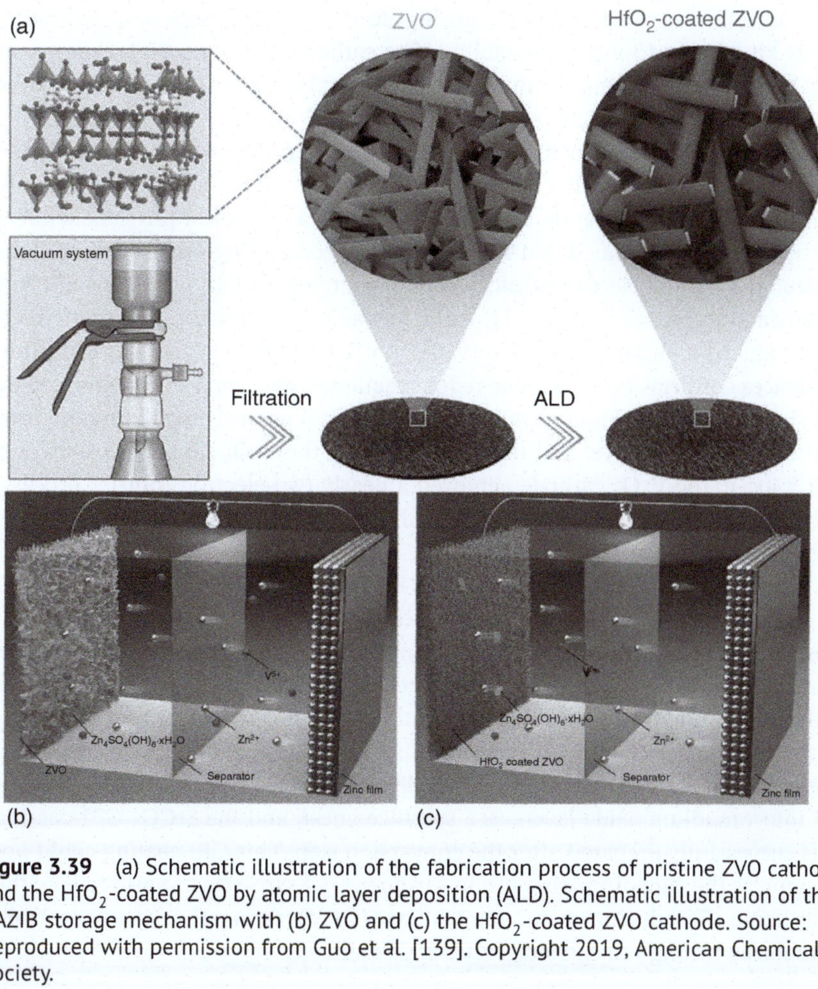

Figure 3.39 (a) Schematic illustration of the fabrication process of pristine ZVO cathode and the HfO_2-coated ZVO by atomic layer deposition (ALD). Schematic illustration of the RAZIB storage mechanism with (b) ZVO and (c) the HfO_2-coated ZVO cathode. Source: Reproduced with permission from Guo et al. [139]. Copyright 2019, American Chemical Society.

obtained HfO_2-coated ZVO cathodes delivered higher capacity and better cycle life compared with pristine ZVO at both low and high current rates.

3.2.3.5 Electrochemical Activation

The electrochemical activation is usually concomitant with a structural transition that can make inherently electrochemically inactive materials effective Zn-ion storage hosts, even leading to the activation of a two-electron reaction. Unexpectedly, the stoichiometric rock salt $VN_{0.9}O_{0.15}$ was found to undergo a conversion reaction upon initial charging, where high-valent anion nitrogen (N^{3-}) was partially substituted by low-valent oxygen (O^{2-}), giving rise to anion-disordered $VN_{0.2}O_{2.1}$ rock salt with abundant vacancies/defects [113g]. Subsequently, such an electrochemical activation strategy was used to design high-performance cathode materials. Wang et al. designed hierarchically porous $Zn_{0.3}V_2O_5 \cdot 1.5H_2O$ (ZVO) nanoflowers by a spontaneous phase transition from VOOH hollow nanospheres during electrochemical cycling [103]. The hierarchically porous structure provides abundant contact with electrolytes, shortens ion diffusion path, and provides a cushion for relieving strain generated during electrochemical processes, facilitating both fast kinetics and long-term stability. The cathode delivers a high capacity of 426 mAh g^{-1} at 0.2 A g^{-1}, and an unprecedented ultralong cycle life is achieved over 20 000 cycles with 96% capacity retention at 10 A g^{-1}.

Moreover, an in situ electrochemical activation strategy has also been extended to other unsuitable materials. For instance, low-valent V_2O_3 could not be effectively discharged due to its inherently unsuitable structure and inferior physicochemical properties. Luo demonstrated that V_2O_3 can be utilized as a high-performance cathode material by in situ electrochemical oxidation strategy [135b]. The ultra-efficient anodic oxidation process of the V_2O_3 cathode was achieved within the first charging and induced a phase transition to $V_2O_{5-x} \cdot nH_2O$ (Figure 3.40a). The evolution of V valences confirms the reversible redox reaction, which can be substantiated by the ex-situ XPS spectra (Figure 3.40b). Through simultaneously regulating the concentration of the electrolyte and the morphology of the V_2O_3 sample, as shown in Figure 3.40c–e, the V_2O_3 cathode achieved a nearly two-electron transfer process, enabling high discharge capacity of 625 mAh g^{-1} at 0.1 A g^{-1} and high-rate performance (486 mA h g^{-1} at 20 A g^{-1}).

It is widely reported that the intercalation metal ions would leach out from vanadium oxide layers during the electrochemical process [144], leading to structural collapse and capacity deterioration [145]. Inspired by this phenomenon, it would be practical that the leached metal ions directly precipitate upon getting out of the vanadium-based cathode to in situ form a cathode electrolyte interphase (CEI) coating layer on the surface [146]. In view of this, an in situ artificial CEI strategy was proposed to kinetically suppress vanadium dissolution [147]. Strontium ion is introduced into vanadium oxide layers as a sacrifice guest, and the $SrCO_3$ CEI-coating layer is successfully prepared after the conversion step. This CEI coating could not only suppress the vanadium dissolution in aqueous electrolytes but also reduce the self-discharge effect during open-circuit voltage (OCV) resting; thus, better cyclability is achieved through this in situ artificial CEI strategy.

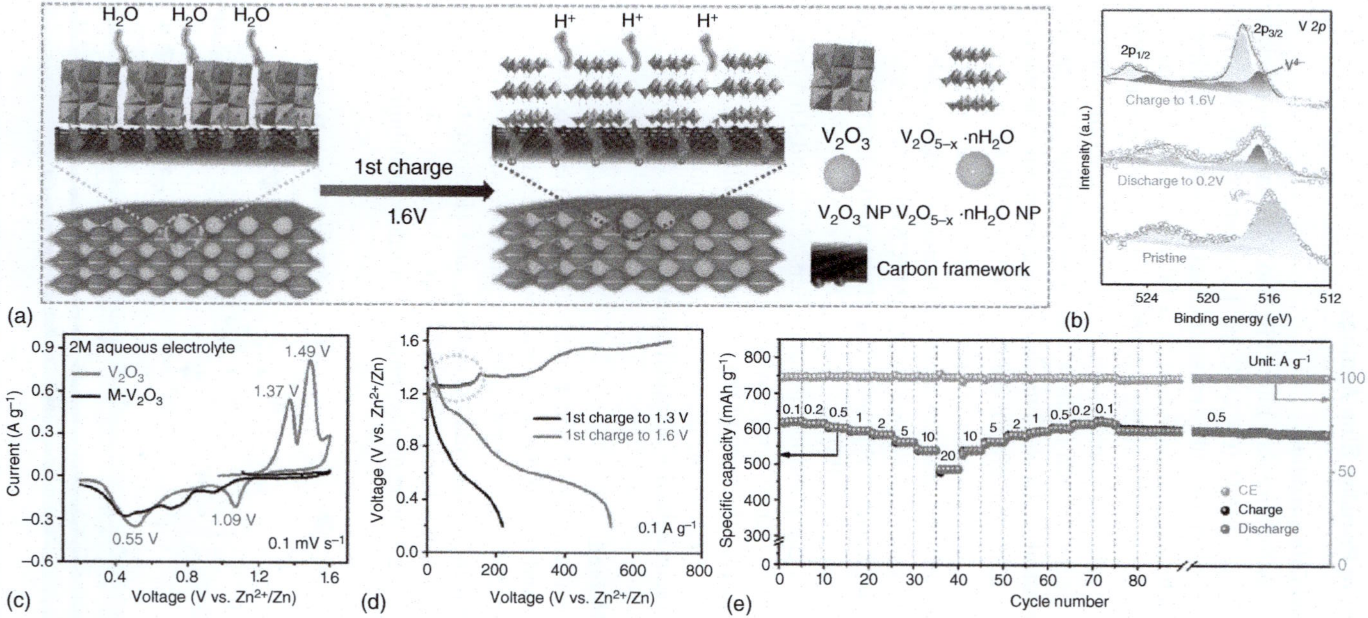

Figure 3.40 (a) Schematic illustration of the oxidation of V_2O_3 during the first full charging. (b) XPS spectra of V 2p regions at pristine, fully discharged, and charged states in the 20th cycle. (c) Initial CV curves of M-V_2O_3 and V_2O_3 at 0.1 mV s^{-1}, (d) galvanostatic charge/discharge curves, and rate performance (e) of V_2O_3. Reproduced with permission from Luo et al. [135b]. Copyright 2020, American Chemical Society.

3.3 Prussian Blue Analogs

3.3.1 Introduction to Prussian Blue Analogs

3.3.1.1 Structure and Categorization

The chemical formulas for Prussian blue (PB) and Prussian blue analogs (PBA) are $A_xM[M'(CN)_6]_{1-y} \cdot \square_y \cdot nH_2O$ $(0<x<2,\ y<1)$, where A, M, M', □ represent the embedded/removed cation, the transition metal attached to nitrogen, the transition metal attached to carbon (when M = M' = Fe, it is PB (Standard PB)), and the $[M'(CN)_6]$ vacancy, respectively. PB/PBAs, which are typically FCC crystal structures, are 3D network structures in which transition metals are typically linked by CN ligands (Figure 3.41). Each of these unit cells contains eight gap sites that can hold not only neutral molecules but also transition metal ions with balanced charges. Interstitial water is usually present in the 3D network structure of PB/PBA. The PB/PBA usually has two types of interstitial water: zeolite water and bound water, where zeolite water occupies the octahedral center of the sub-cell and bound water is coordinated with M ions, due to the $[M'(CN)_6]$ vacancy located in the center of the unit cell (Figure 3.41) [148]. The 3D network structure, as the most prominent structural feature of PB/PBA, has an open channel along the direction of [100] and provides a gap position with diameters of 3.2 and 4.6 Å, respectively. Such an open structure leaves large cavities extending in three directions and tunnels in three directions throughout the porous lattice, allowing various cations to diffuse rapidly [149]. In this case, the electrochemical properties of PB/PBAs can be attributed to the redox behavior of transition metal ions.

3.3.1.2 Synthesis Method

PB/PBAs cathode material has a wider application prospect because of its low cost, easy synthesis, and environmental friendliness [150]. At present, its synthesis methods mainly include the co-precipitation method, single-source method, electrochemical deposition, ball-milling method, and electrostatic spray-assisted coprecipitation [151]. Considering the differences between PB/PBA synthesis methods in terms of yield, crystallinity, production safety, feasibility of large-scale

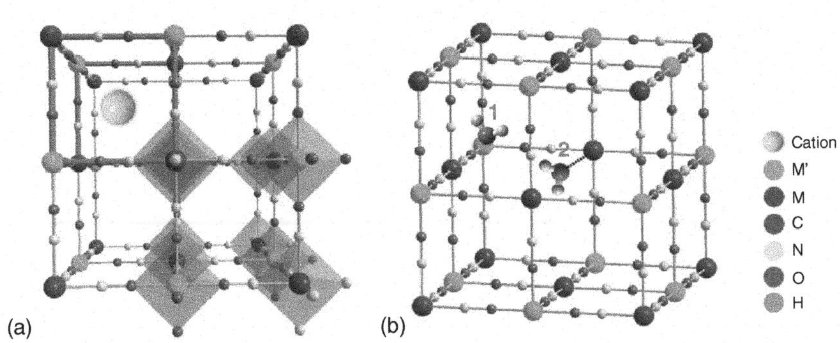

Figure 3.41 Crystal structure of Prussian blue analogs. Source: Zhang et al. [148]. Copyright 2017, John Wiley and Sons.

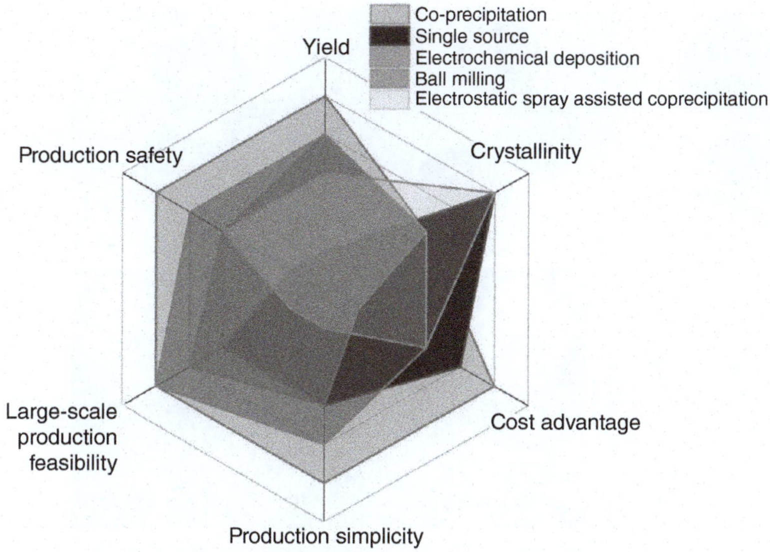

Figure 3.42 Comparison of the characteristics of common synthesis methods for PB/PBA. Source: Yang et al. [151]. Copyright 2022, Elsevier.

production, simplicity of production, and cost, the characteristics of these methods are shown in Figure 3.42. Among them, the advantages of co-precipitation and ball milling in each index are in line with the requirements of large-scale production. The co-precipitation method has significant advantages in terms of yield, production safety, feasibility of mass production, production simplicity, and cost. Although PB/PBA prepared by the single-source method has high crystallinity, toxic CN^- may be produced during the production process, which limits its large-scale production. Other methods, such as electrochemical deposition and electrostatic spray-assisted coprecipitation, are more complex and expensive for large-scale production.

3.3.2 Strategies

The following factors, such as few sites for redox, dissolution of metal ions, structural distortion, defects in the lattice, and crystalline water, limit the practical application of PB/PBA as an anode for zinc ion batteries with high voltage, high specific capacity, and long cycle life. Recently, various inspiring strategies (metal substitution, crystal structure design, surface modification, etc.) have been developed to address the above issues, and much progress has been made. Several modification strategies for PB/PBAs are briefly described here [152].

Metal-substitution and morphology design: The active site and crystal structure of PBA can be changed by atomic substitution, thereby affecting its electrochemical performance. For example, Lou and coworkers [153] prepared copper-substituted manganese-PBA double-shell nano-boxes (CuMn-PBA DSNBs) by tannic acid etching and cation exchange for efficient storage of zinc ions (Figure 3.43). The unique hollow structure exposes a rich number of active sites and mitigates volume

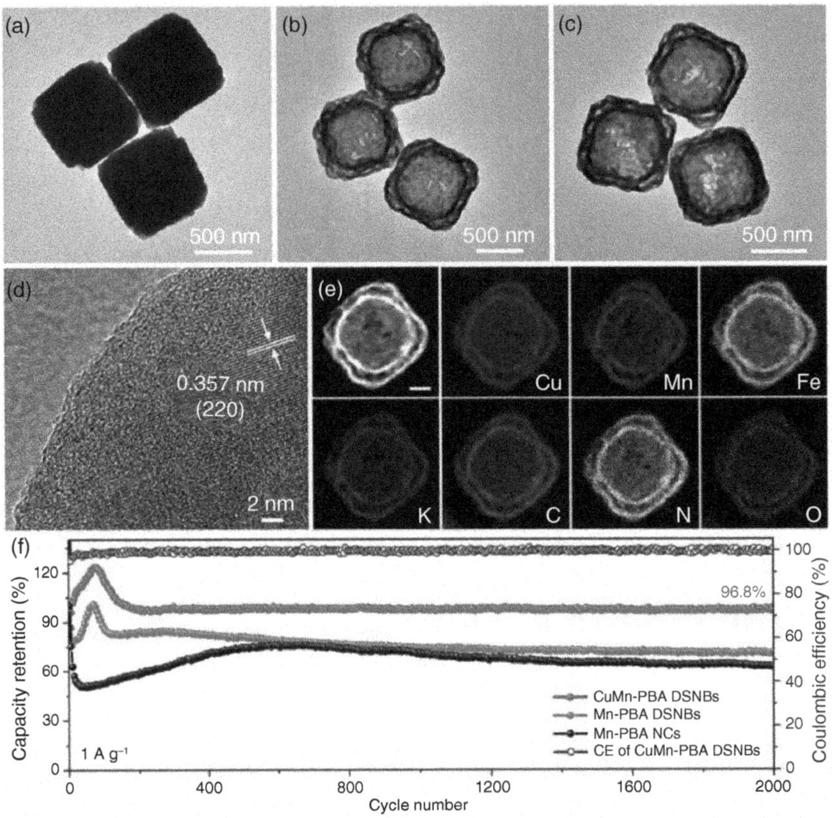

Figure 3.43 TEM images of (a) Mn-PBA NCs, (b) Mn-PBA DSNBs, and (c) CuMn-PBA DSNBs. (d) HRTEM image, (e) HAADF-STEM and elemental mapping images of CuMn-PBA DSNBs. (f) Cycling performance tested at 1 A g^{-1} of the MnPBA NCs, Mn-PBA DSNBs, and CuMn-PBA DSNBs electrodes. Scale bar in (e): 200 nm. Source: Zeng et al. [153]. © 2022, John Wiley and Sons.

variations during cyclic testing. In addition, partial Cu substitution and induced Mn vacancies may inhibit J–T twisting of the Mn-N_6 octahedron, thereby prolonging its lifetime. In-situ characterization shows that the charge storage mechanism of CuMn-PBA DSNBs mainly involved reversible redox reactions of transition metals and Zn^{2+} ion insertion/removal. Based on this substitution design, CuMn-PBA DSNB exhibits high reversible capacity, good rate performance, and excellent cycle stability in 2000 cycles. Zeng et al. [154] constructed co-substituted manganese-rich PBA hollow spheres (denoted as CoMn–PBA HSs) using a simple anion exchange process. The hollow structure can not only provide a large surface area and sufficient active sites but also accommodate large volume changes during repetitive charging and discharging. In addition, co substitution will further stabilize the crystal structure of CoMn–PBA HSs and improve cycle life. As expected, when used as the positive electrode of AZIBs, the as-prepared CoMn–PBA HSs exhibit high reversible capacity, good rate performance, and long cycle stability.

Crystal structure design: Crystalline water and defects are present in normal KMHCF ($K_{1.6}Mn[Fe(CN)_6]_{0.94} \cdot 0.63H_2O$) crystals, which leads to electrodes with low discharge capacity and poor cycling stability. Therefore, achieving a PBA_S structure with fewer defects and water is beneficial to the overall performance of PBAs. For example, Xia and coworkers [155] controlled the crystallization process of KMHCF by adding polyvinylpyrrolidone (PVP), resulting in water-less, low-defect KMHCF material. The KMHCF as a cathode material for zinc ion batteries exhibited excellent electrochemical performance and provided a specific capacity of 140 mAh g^{-1}. The results also show that KMHCF undergoes a phase transition during cell activation, forming a new stable phase, KZMHCF ($K_{0.25}Zn_{0.25}MnFe(CN)_6$) (phase transition completed after two cycles). Subsequently, zinc ions with stable structures continue to be inserted/extracted in the lattice gap in the new reaction stage.

In common electrolytes, PB/PBA requires high charging voltages during the charging process, and the water decomposition that occurs at high voltages can exacerbate side reactions at the cathode interface, so regulating the properties of the electrolyte is an effective strategy to achieve structural stability of PB/PBAs. For example, Deng et al. [156] effectively enhanced the performance of Zn//KMnHCF hybrid cells by using high voltage electrolysis (30 m KFSI + 1 m $Zn(CF_3SO_3)_2$). Characterization of the phase structure of the cathode during cycling showed that Zn^{2+} was embedded in the cathode together with K^+ and that KMnHCF gradually transformed into a stable KZnHCF phase as the number of cycles increased (as shown in Figure 3.44, KMnHCF is transformed into KZnHCF after 100 cycles). The transformation results in a stable electrochemical performance of the cathode.

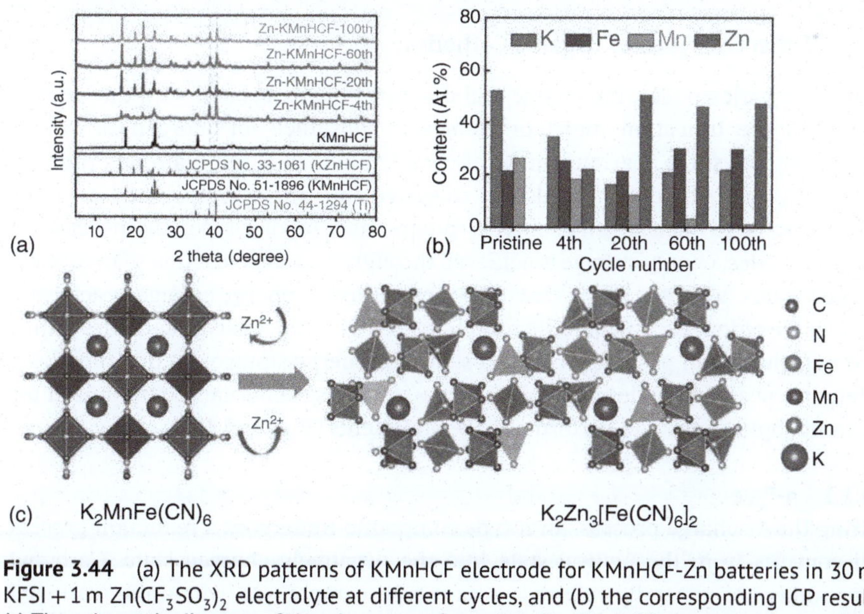

Figure 3.44 (a) The XRD patterns of KMnHCF electrode for KMnHCF-Zn batteries in 30 m KFSI + 1 m $Zn(CF_3SO_3)_2$ electrolyte at different cycles, and (b) the corresponding ICP results. (c) The schematic diagram of the phase transformation for KMnHCF electrode. Source: Deng et al. [156]. Copyright 2021, John Wiley and Sons.

Surface modification: The surface coating is considered to be one of the most effective surface modification methods to improve the electrochemical properties of PBAs by enhancing their electronic conductivity and structural stability, as well as inhibiting side reactions between the electrode material and the electrolyte. For example, Liu et al. [157] achieved effective improvement in ZnHCF ($Zn_3[Fe(CN)_6]_2$) capacity and cycle stability by coating PANI on ferricyanide (ZnHCF). The PANI coating adds a discharge stage to the composite electrode, which significantly increases the capacity of ZnHCF. At the same time, it also alleviates the dissolution problem of ZnHCF during cycling and enhances its cycle stability. This strategy makes the final flexible zinc-ion battery present two long and flat discharge platforms and a high capacity of 150 mAh g^{-1}, which provides a new idea for the development of PBA in zinc-ion batteries.

3.4 Organic Materials

Organic materials are regarded as appealing candidates as cathodes for AZIBs and have been widely studied and explored for decades due to their significant advantages. They are cost-effective because they consist of light and inexpensive elements such as C, H, O, N, and S, and can be synthesized controllably or obtained from artificial synthesis or abundant biomass resources which have environmental compatibility [158]. In addition, they have the advantages of low toxicity, high theoretical capacity, and good structural diversity. Herein, a brief systematic introduction to organic cathode materials for AZIBs is provided in this section.

3.4.1 Different Types of Organic Cathodes

As for inorganic anodes, the charge and discharge are associated with the valence change of the transition metal or elemental substance for inorganics. But in organic cathodes, the rearrangement of chemical bonds is the main determining factor. Organic cathode materials are classified into three types based on their charge/discharge mechanisms: n-type, p-type, and bipolar-type. In the case of n-type cathodes, energy storage is achieved through the redox reaction of negatively charged states, whereas in the case of p-type cathodes, energy storage is achieved through the redox reaction of positively charged states. Bipolar-type cathodes utilize both negatively and positively charged states for energy storage. By utilizing these different types of cathode materials, energy–storage systems can be optimized for different applications and performance requirements (Figure 3.45).

3.4.1.1 n-Type

During the discharge process, an n-type compound undergoes a reduction process and converts from the neutral state into the negatively charged state. Carbonyl compounds are generally referred to as n-type compounds, containing carbonyl groups (C=O), which can reversibly store/release Zn^{2+} by switching between C=O and C—O$^-$ [159]. Carbonyl compounds mainly include ketones and quinones,

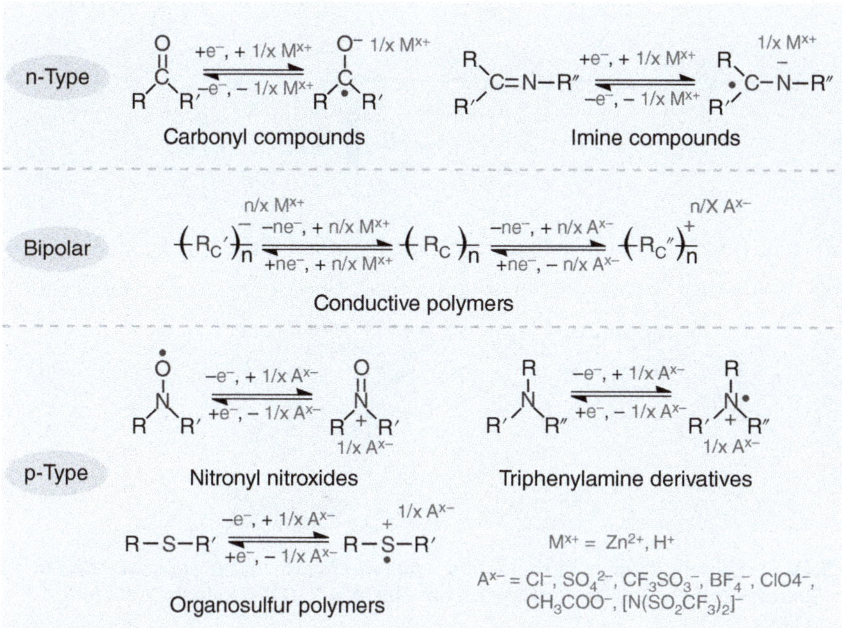

Figure 3.45 The charge/discharge mechanisms of three types of organic cathode materials in aqueous ZIBs. Source: Reproduced with permission from Lu et al. [161]. Copyright 2018 Elsevier.

which are attractive, renewable, and environmentally friendly cathode materials due to their abundance in nature.

Ketones are kinds of very common carbonyl compounds, taking up a large part of the item. Guo et al. [160] reported pyrene-4,5,9,10-tetraone (PTO) as an environmentally friendly cathode of AIZBs (Figure 3.46). The PTO cathode shows a specific capacity of 336 mAh g^{-1} at 0.04 A g^{-1} and delivers capacity retention of approximately 70% after 1000 cycles at 3 A g^{-1}, along with Coulombic efficiency approaching 100%.

Quinones, which contain conjugated carbonyl groups, have been widely studied due to their higher electrochemical activity and specific capacity. Kundu et al. [162] reported on a novel organic cathode, tetrachloro-1,4-benzoquinone (p-chloranil), whose inherently soft crystal structure could provide reversible and efficient Zn^{2+} storage. The organic materials facilitate obtaining a high specific capacity of 207 mAh g^{-1} at 1.1 V with a voltage polarization of ~50 mV for Zn/2e$^-$ reaction. However, the dissolution of p-chloranil in the electrolyte resulted in capacity decay and inferior cycle stability. To improve the problem mentioned above, Chen and coworkers [163] first selected various quinone compounds as the cathode materials of aqueous ZIBs, including 1,2-NQ, 1,4-NQ, 9,10-AQ, 9,10-PQ and C4Q, and used the Nafion membranes as the separators to prevent anions from crossing and forming a by-product on the anode side. C4Q displays a high capacity of 335 mAh g^{-1}, a low charge/discharge potential gap of 70 mV at 20 mA g^{-1}, and a much more

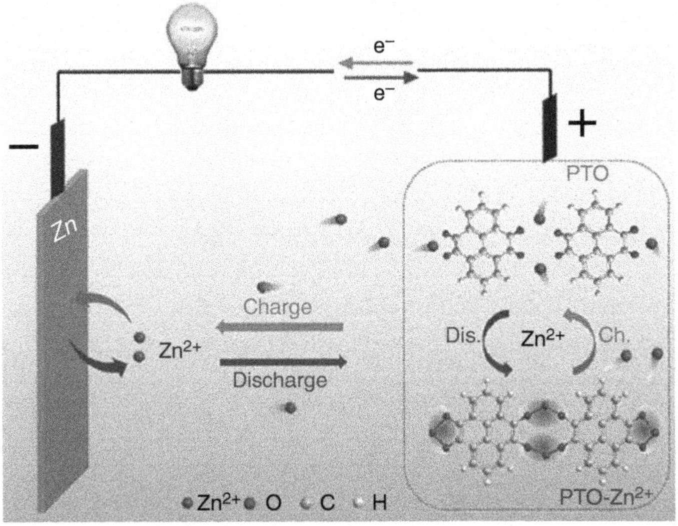

Figure 3.46 Schematic diagram of charge/discharge mechanism of the aqueous PTO//Zn battery. Source: Reproduced with permission from Guo et al. [160]. Copyright 2018 John Wiley and Sons.

stable cycling performance (87% capacity retention over 1000 cycles at 500 mA g^{-1}). The two carbonyls on the top and four carbonyls on the bottom uptake three Zn ions, so one reduction peak corresponds to the reaction from C4Q to Zn_3C4Q, as Figure 3.47d shows that one oxidation peak corresponds to the reaction from Zn_3C4Q to C4Q, respectively.

Beyond that, imine compounds are normal n-type compounds. Tie et al. [164] developed aqueous ZIBs based on diquinoxalino [2,3-a: 2' ,3'-c] phenazine (HATN), which contains six imine active sites experiencing an H^+ uptake/removal, showing a high discharge capacity of 405 mAh g^{-1} at 100 mA g^{-1}. The Zn–HATN batteries have demonstrated outstanding capacity retention of 93.3% even after 5000 cycles at 5 A g^{-1}. This remarkable performance is attributed to the π-conjugated aromatic structure of HATN, which strengthens π–π intermolecular interactions and effectively mitigates the dissolution of HATN in aqueous electrolyte.

3.4.1.2 p-Type

p-type compounds go through an oxidation process and transform into cations, as shown in Figure 3.44 [165]. Amine compounds are typical p-type compounds, including PANI, triphenylamine, PPy, and polyindole. PANI is widely reported to be a promising cathode material for AZIBs [166]. During the charging process, the chemical bonds between carbon and nitrogen (C—N) are converted into C—N$^+$ to interact with anions. The resulting C—N$^+$ has a strong electrophilicity, which can react with the water molecules in the aqueous $ZnSO_4$ electrolyte, ultimately leading to significant capacity degradation over time [167].

As shown in Figure 3.48, 1,4-bis(diphenylamino)benzene (BDB) is used as a cathode by Glatz et al. [168], which undergoes an oxidation process accompanied by

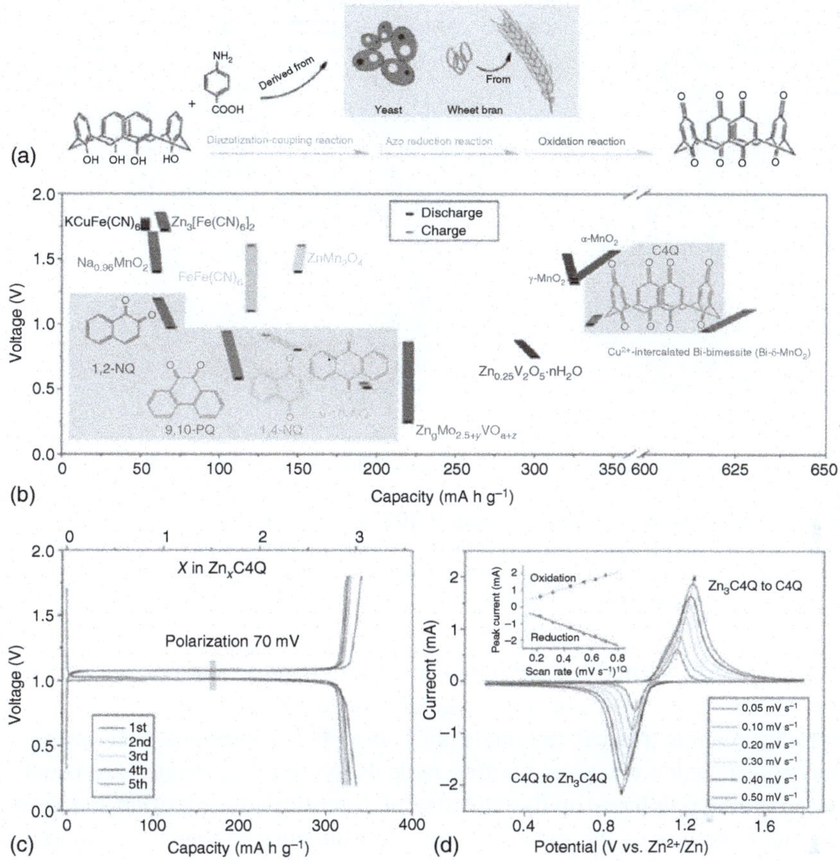

Figure 3.47 (a) Schematic diagram for preparing C4Q. (b) Discharge/charge voltages and capacities of a series of quinone compounds and typically reported inorganic electrodes in aqueous ZBs are also listed for comparison. (c) Galvanostatic discharge/charge curves of Zn–C4Q battery at the current density of 20 mA g^{-1}. (d) CV curves of Zn-C4Q batteries at some sweeping rates. Source: Reproduced with permission from Zhao et al. [163]. Copyright 2018 John Wiley and Sons.

the insertion/release of anions from the electrolyte anion. The BDB cathode delivered 112 mAh g^{-1} with a capacity retention of 82% after 500 cycles at a 3 C rate and 1000 cycles with 75% capacity retention at a 6 C rate, at nearly 100% Coulombic efficiency.

PPy is a typical p-type cathode for AZIB. It possesses a conjugated structure with nitrogen, resulting in electrochemical activity (Figure 3.49) [169]. PPy is synthesized by the polymerization of pyrrole monomer, as in equation [170]: $n(\mathrm{Py}) + ny\mathrm{Cl}^- \rightarrow [\mathrm{PPy}_y{}^+(\mathrm{Cl}^-)_y]n + \mathrm{nye}$.

Indole can be seen as pyrrole with a benzene ring, whose mechanism of charge/discharge is similar to pyrrole. Thus, polyindole obtained by polymerization of indole is a p-type material (Figure 3.53b). The specific capacity of the battery

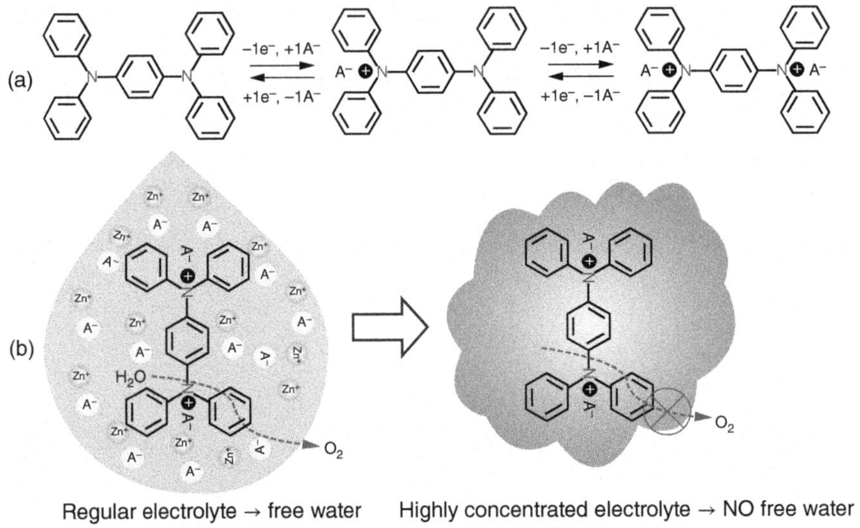

Figure 3.48 (a) Oxidation/reduction process of BDB accompanied by anion uptake/expulsion; A^-: OTf^-/$TFSI^-$. (b) Illustration of the catalytic oxidation of free H2O by BDB^{2+} species in regular electrolytes. Source: Reproduced with permission from Glatz et al. [168]. Copyright 2019 American Chemical Society.

using a polyindole cathode is saturated after 30 cycles and keeps a discharge capacity of 70 mAh g^{-1} after the 200 charge/discharge cycles [171].

Generally, organic radical polymers are open-shell molecules that undergo electrochemical redox reactions via free radicals. Nitronyl nitroxides are usually p-type active materials based on the oxidation of the radical toward cation and rarely pass by the reduction of the radical to the anion because of their lower stability and reversibility [172]. It features fast electrode kinetics, reactant recyclability, and high redox capacity [173]. Compared to other organic materials that store and transport charges by chain skeleton and rearrangement of electrons, nitroxides gain or lose electrons directly during charge/discharge. As shown in Figure 3.50a, Poly(2,2,6,6-tetramethylpiperidinyloxy-4-yl vinylether) (PTVE) was utilized as an active material for the cathode in ZIB. The PTVE half-cell exhibited an ultrafast full charging of 3 mC cm^{-2} in as short as three seconds due to an ultrafast one-electron oxidation capability [174]. As a p-type material, PTVE also has a high-discharge voltage plateau (1.73 V), which accelerates side reactions, including oxygen evolution reaction (OER) and decomposition of some ions. To solve this problem, aqueous electrolytes with high stable potential window need to be developed. It is worth noting that nitronyl nitroxides are more difficult to synthesize compared with carbonyl compounds, imine compounds, and conductive polymers.

In addition, a poly(acetylene)-based 9,10-di(1,3-dithiol-2-ylidene)-9,10-dihydroanthracene (exTTF) can be developed as a p-type cathode for AZIBs due to reversible two-electron oxidation (shown in Figure 3.50b) [175]. This material exhibited a specific capacity of 111 mAh g^{-1} and even could be cycled at 120 C rate with good cycling stability.

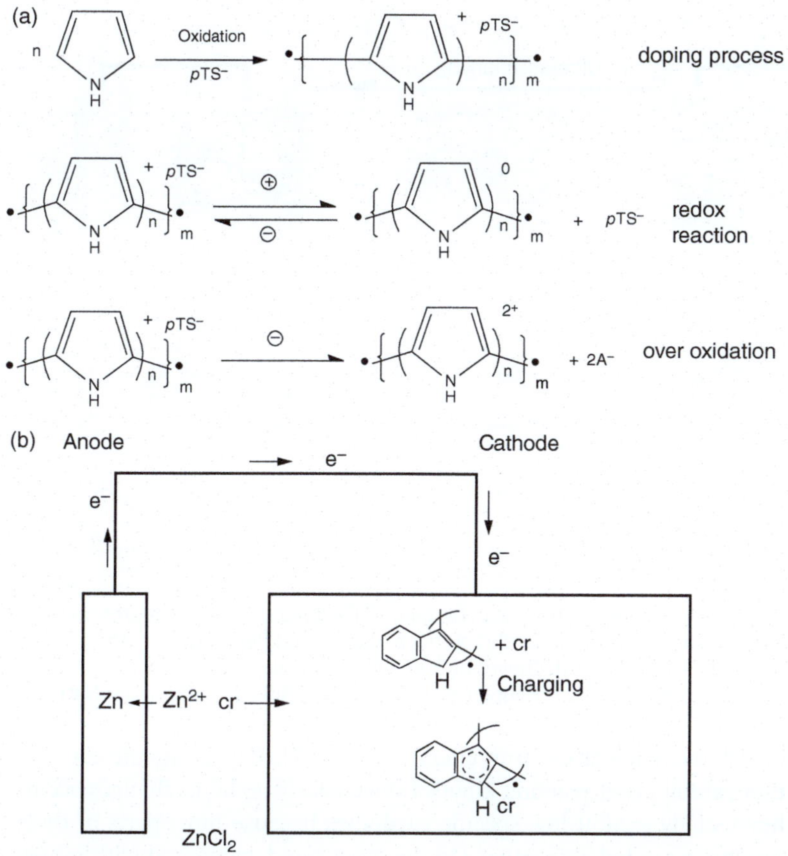

Figure 3.49 (a) Reduction and oxidation reactions of the PPy. (b) The charge/discharge mechanism of zinc/polyindole battery. Source: (a) Reproduced with permission from Li et al. [169]. Copyright 2013 Elsevier. (b) Reproduced with permission from Zhijiang et al. [171]. Copyright 2011 Elsevier.

3.4.1.3 Bipolar-Type

A bipolar-type compound combining the features of both n-type and p-type can transform from the neutral state either to a positively charged state through an oxidation process or to a negatively charged state through a reduction process. Aqueous $Zn(CF_3SO_3)_2$ electrolyte was used to enhance the cycling performance of PANI, and the batteries exhibited a capacity of 95 mAh g^{-1} with a high capacity retention of 92% after 3000 cycles at 5 A g^{-1} [176]. Interestingly, the half-oxidation state of PANI shows a bipolar-type mechanism (Figure 3.51a). The half-oxidation PANI includes doped $=NH^+$— and undoped =N— groups, reduced to —NH— and $—N^-$— during the initial discharge process, respectively. During the following charge process, the —NH— and $—N^-$— groups can be reversibly oxidized to =NH— and =N— groups.

It should be noted that p-type organic materials generally present higher redox potential than n-type and bipolar-type materials (Figure 3.51b). It is because p-type materials pass through an electrochemical redox reaction by losing electrons,

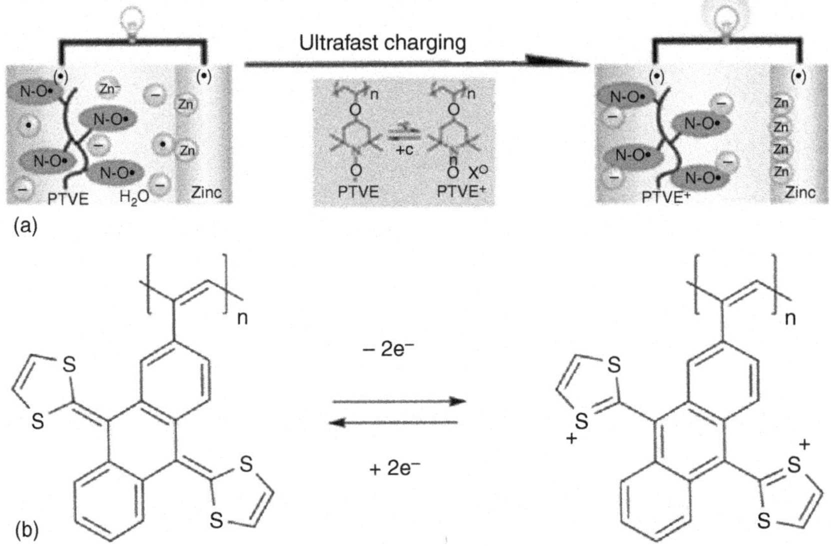

Figure 3.50 (a) Redox couple and ultrafast charging of PTVE. (b) Schematic representation of the redox reaction of poly(2-ethinylanthraquinone). Source: (a) Reproduced with permission from Koshika et al. [174]. Copyright 2009 Royal Society of Chemistry. (b) Reproduced with permission from Häupler et al. [175]. Copyright 2016 Springer Nature.

resulting in a relatively lower electron energy level [177]. P-type organic cathode materials often exhibit faster reaction kinetics since nearly no bond changes occur. However, they usually exhibit low specific capacities because most parts of their molecular structures are redox-inactive. On the other hand, compared with p-type cathode materials, n-type materials generally possess a higher specific capacity since they have more active sites per weight in them. Moreover, bipolar-type conductive polymers show higher electronic conductivity than p-type and n-type materials, improving electron transfer and rate performance. Unfortunately, all of them often deliver limited specific capacity due to the low level of available stimulants.

3.4.2 Main Challenges Faced by Organic Cathode Materials

Although organic cathode materials have lots of advantages, we have to face serious problems in practice, mainly poor electrical conductivity, low energy density, and poor cycling stability. To develop high-performance organic cathodes, these challenges are carefully analyzed to look for strategies.

3.4.2.1 Poor Electrical Conductivity

Organic cathode materials have low electrical conductivity in most cases, even when conductive additives are added. The low electrical conductivity not only impedes charge transfer, increases polarization, and reduces the utilization of active materials, but also decreases active sites due to the inefficient electron transfer during charge/discharge. The poor electrical conductivity will form a weak electric field

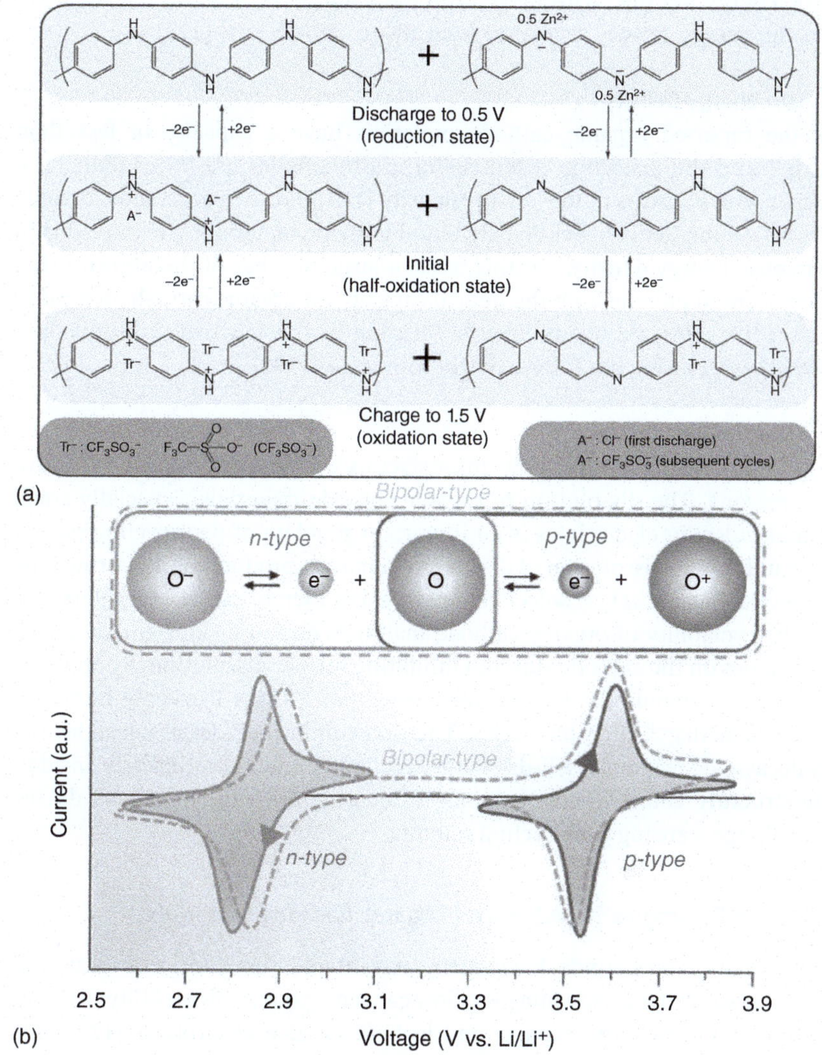

Figure 3.51 (a) The redox mechanism of PANI/CFs. (b) Schematic diagram of redox reaction and representative CV curves of three types: n-type, p-type, and bipolar-type. (O: organic molecule). Source: (a) Reproduced with permission from Alfaruqi et al. [21]. Copyright 2018 Elsevier. (b) Reproduced with permission from Lee et al. [177]. Copyright 2018 John Wiley and Sons.

inside cathode materials, which results in sluggish ionic diffusion and slow reaction kinetics. PANI is widely reported to combine with conductive materials as cathode materials even though it is conductive; for example, the highly conductive and 3D-structured CF brought PANI cathode with fast electron transfer and ultrafast rate capability [178]. Although the addition of conductive materials will increase electrical conductivity, it will decrease the energy density of organic cathodes, particularly volumetric energy density. With decreasing the loading of PANI on the same CF substrate, the energy density decreased from 389 to 159 Wh kg^{-1} [178].

3.4.2.2 Low Energy Density

Most of the reported organic cathode materials have a capacity of less than 200 mAh g^{-1} and the discharge potentials of most cathodes are lower than 1.0 V vs. Zn/Zn^{2+}, which results in low energy density [179]. Furthermore, most organic materials have a markedly lower density than inorganic materials, which results in low volumetric energy density. On this basis, a large number of theoretical redox active sites seem to provide high theoretical capacity for organic cathode materials, but many active sites are not utilized by burring in the inactive skeletons, and multi-active centers were not fully active in some cases [180].

3.4.2.3 Poor Cycling Stability

Most organic cathode materials show unsatisfactory cycling performance in AZIBs, for many reasons. The dissolution of organic cathode materials, especially small organic molecules, in electrolytes is an important problem of poor cycling performance, including the dissolution of discharged/charged products and intermediate states in electrolytes [181]. A selectively permeable membrane was developed to enhance the cyclability of organic cathode materials due to locking the dissolved active molecules on the cathode side and inhibiting further dissolution. By applying a membrane to assemble the battery, the capacity of PT was markedly improved from 2.4 to 73 mAh g^{-1} after 300 cycles. The dissolution problem will also lead to capacity decay, side reactions, self-discharge, and low Coulombic efficiency. In addition, the structure could be easily damaged after storage/release of ions during discharge/charge, causing poor cycling stability.

3.4.3 Design Strategies for Advanced Organic Cathode Materials

Although organic compounds have many advantages, most organic materials reported so far show the disadvantages of low capacity, poor cyclic stability, sluggish ion diffusion, low discharge voltage, and lack of suitable electrolytes [178, 182]. When optimizing the performance of organic cathodes, the key parameters should be considered comprehensively, including capacity, voltage, rate, cycle number, capacity retention, and preparation complexity [165]. Fortunately, organic compounds are easy to synthesize and have tunable structures to optimize their electrochemical performances. Here are some strategies for optimizing performance for the challenges.

3.4.3.1 Enhancing Electrical Conductivity

Improving the electronic conductivity of organic molecules can be achieved by enlarging π-electron conjugated structure and introducing highly active redox groups such as free radicals, which are intrinsic methods to the molecular conductivity from the controllable molecular structure design. On the other hand, we can increase specific surface area and hybridize with suitable carbon materials, including graphite, graphene, and carbon nanotubes. This can increase the conductivity by extrinsic methods, enhancing the conductivity through more sufficient utilization of the active sites. The addition of CNT to polyimides (PI@CNT) ensures ultrafast reaction kinetics and also improves the poor cycle life of multivalent battery chemistries. Full cells assembled with PI@CNT cathode and lithium metal anode exhibit long cycle life (>10 000 cycles), and fast kinetics (>20 C) [183]. Moreover, the extension of the π-conjugated system is an efficient strategy to improve rate performances because of the enhanced intermolecular interactions and the improved charge transfer [184].

3.4.3.2 Increasing Energy Density

The energy density is determined by the capacity and output voltage of batteries. Where the capacity is determined by the number of transferred electrons in the reaction process, and the output voltage is determined by the $\Delta G/n$ of the reaction process or potentials of the cathode and anode. Their relationship can be described as following equations [161]:

$$E_d = E \times Q \tag{3.27}$$

$$Q = \frac{nF}{3.6M_w} \tag{3.28}$$

$$E = -\frac{\Delta G}{nF} = E_+ - E_- \tag{3.29}$$

where E_d is the energy density, E is the output voltage, Q is the specific capacity ($mAh\,g^{-1}$), n is the number of transferred electrons in the reaction process, F is the Faraday constant, and M_w is the molecular weight.

The capacity is the amount of transferred electricity after the completion of an electrochemical reaction, so it is a physical quantity of width related to the chemical reaction progress and the amount of electricity transferred per unit of reaction progress. To measure the capacity more conveniently, the mass-specific capacity is introduced, which is a physical quantity of strength property only related to the material and its electrochemical reaction. The general method to increase the specific capacity is to increase the proportion of active groups in compounds, including increasing the number of active groups and decreasing the number and mass of inactive atoms. However, it is important to note that the active site is not fully utilized, and measures should be taken to improve its utilization. To date, macrocyclic compounds, oligomers, and fused ring polymers have been reported to be great multiple active center materials [161]. In addition, the low electrical conductivity of organic materials usually results in limited utilization of active materials. Besides optimizing the intrinsic electroconductibility of materials, we can add some conductive materials such as carbon fiber fabric, carbon fiber felt [185], and graphene [186]. Furthermore, the amorphous materials are more beneficial for the diffusion of multivalent metal ions in the cathode, thus improving the utilization of the active site. Walter et al. found that amorphous polypyrene is significantly greater than crystalline pyrene under the same electrochemical testing conditions, showing a larger initial capacity of ~92 mAh g^{-1} at 200 mA g^{-1} with ~76% capacity retention after 300 cycles [187].

Energy density, a very important performance indicator of the battery, is the theoretical discharge energy contained in a unit volume or mass. The energy density is not only directly related to the theoretical capacity but also to the discharge voltage of the battery. Organic materials usually exhibit low discharge voltage plateau, which will result in low energy densities. Therefore, increasing the voltage of the battery is also an important way to improve the energy density of the battery. The introduction of appropriate electron-drawing groups such as —F, —Cl, and —C≡N into the organic compounds can raise the LUMO energy and thus increase the voltage of the battery [188]. For example, Fu and coworkers [184b] reported that the dichlone@CNT composite cathode with electron-withdrawing halogen (2Cl atoms) maintained an average discharge voltage of 0.84 V, which is higher than the unsubstituted naphthoquinone. The p-type organic materials with nitroxide free radicals or conjugated electron-rich groups are promising candidates with higher redox potentials, which can significantly increase energy density [189].

3.4.3.3 Improving Cycling Stability

Many organic materials for AZIBs show limited cycle life because of their unstable or soluble nature [190]. Thus, reducing and suppressing the solubility of organic compounds in electrolytes will greatly increase cyclic stability. Small organic compounds should be changed into polymers that have lower solubility in water [191]. In addition, organic molecules can be combined with polymer frameworks to inhibit dissolution. For example, the phenanthroline covalent organic framework (Figure 3.52a) shows an average capacity decay of only 0.38% per cycle after 10 000 cycles at 1.0 A g^{-1} (Figure 3.52b) [192]. The formation of coordination sites with

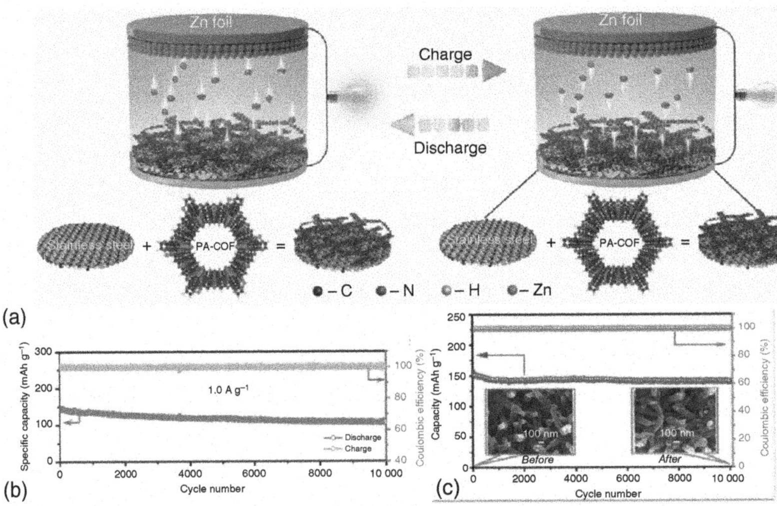

Figure 3.52 (a) Schematic configuration of PA-COF||Zn. (b) Long-term cycling stability of PA-COF. (c) Cycling stability of poly(1,5-NAPD) 10 A g^{-1}. Source: (b) Reproduced with permission from Wang et al. [192]. Copyright 2020 American Chemical Society. (c) Reproduced with permission from Li et al. [185]. Copyright 2020 Elsevier.

electrochemically active compounds can also enhance cycling performance [193]. Furthermore, decreasing the entangled molecular structure and increasing the space of diffusion is a good approach for long cycling stability during the charging/discharging processes. The poly(1,5-NAPD) supported by carbon felt shows excellent cycling stability with capacity retention of 100% after 10 000 cycles at $10\,A\,g^{-1}$ due to ample space for ion insertion or extraction (Figure 3.52c) [185].

The selection of appropriate electrolytes has been extensively investigated to improve cyclic stability, including hydrated eutectic electrolytes [194] and organic electrolytes [195]. The Zn^{2+}-based dual-ion battery with the super-concentrated electrolyte of 30 M $ZnCl_2$ shows a high energy density ($\sim$160 Wh kg^{-1}) and remarkable stability after 38 000 cycles at $3\,A\,g^{-1}$ [196]. In addition, adding additives to electrolytes is one of the most convenient and economical methods. DMSO is an electrolyte additive that can provide higher adsorption energy to Zn(002) plane, which will induce Zn deposition to form the Zn(002) texture with no dendrite formation [50]. The Zn/Zn cell provides ultra-stable cycling performance for more than 2100 h at $1\,mA\,cm^{-1}$ by using the electrolyte with 20% DMSO. The larger π-conjugated system of organic materials can improve rate performance as well as suppress the dissolution and improve the stability of the active material. For example, the triangular PQ–Δ exhibits higher capacity and cycling performance compared to the monomer PQ in rechargeable aluminum organic batteries (Figure 3.53a) [51]. The rich multielectron–active centers also deliver good stability because coordination with multivalent metal ions can improve the stability of the materials. As shown in Figure 3.53b, HATNQ shows an ultrahigh capacity (482.5 mAh g^{-1} at $0.2\,A\,g^{-1}$) and long cycle life (upon 11 000 cycles at $5\,A\,g^{-1}$) due to the chelation of O···Zn···N from adjacent C=O and C=N groups [161].

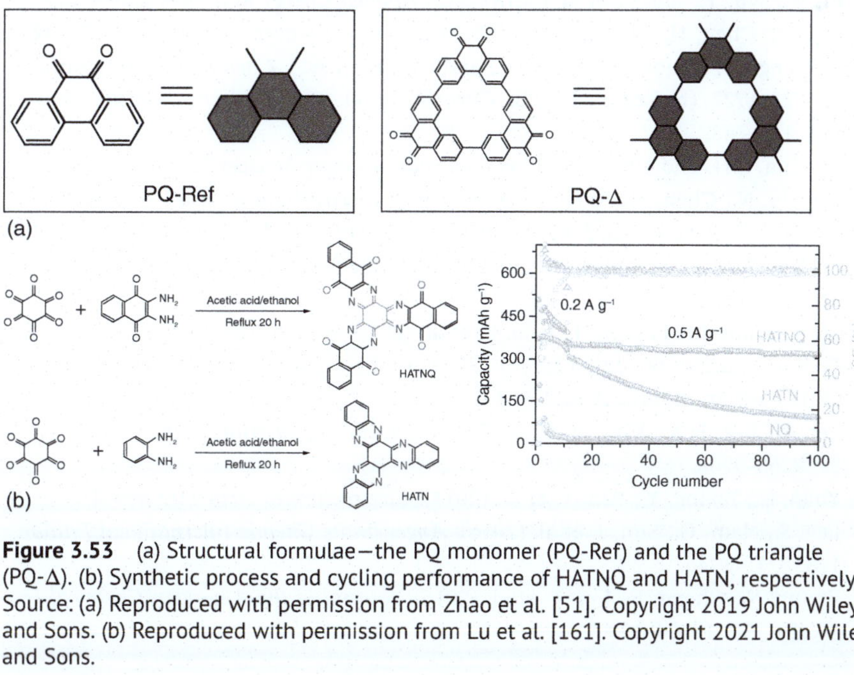

Figure 3.53 (a) Structural formulae—the PQ monomer (PQ-Ref) and the PQ triangle (PQ-Δ). (b) Synthetic process and cycling performance of HATNQ and HATN, respectively. Source: (a) Reproduced with permission from Zhao et al. [51]. Copyright 2019 John Wiley and Sons. (b) Reproduced with permission from Lu et al. [161]. Copyright 2021 John Wiley and Sons.

References

1 Jia, X., Liu, C., Neale, Z. et al. (2020). *Chemical Reviews* 120 (15): 7795–7866.
2 Selvakumaran, D., Pan, A., Liang, S., and Cao, G. (2019). *Journal of Materials Chemistry A* 7: 18209–18236.
3 Blanc, L.E., Kundu, D., and Nazar, L.F. (2020). *Joule* 4: 771–799.
4 Han, M., Qin, L., Liu, Z. et al. (2021). *Materials Today Energy* 20: 100626.
5 Feng, Q., Yanagisawa, K., and Yamasaki, N. (1998). *Journal of Porous Materials* 5: 153–162.
6 Xu, C., Du, H., Li, B. et al. (2009). *Electrochemical and Solid-State Letters* 12: A61.
7 Alfaruqi, M.H., Gim, J., Kim, S. et al. (2015). *Journal of Power Sources* 288: 320–327.
8 Lee, B., Yoon, C.S., Lee, H.R. et al. (2014). *Scientific Reports* 4: 1–8.
9 Lee, B., Seo, H.R., Lee, H.R. et al. (2016). *ChemSusChem* 9: 2948–2956.
10 Pan, H., Shao, Y., Yan, P. et al. (2016). *Nature Energy* 1: 16039.
11 Munichandraiah, N. (2008). *Journal of Physical Chemistry C* 112: 4406–4417.
12 Islam, S., Alfaruqi, M.H., Mathew, V. et al. (2017). *Journal of Materials Chemistry A* 5: 23299–23309.
13 Zhang, N., Cheng, F., Liu, J. et al. (2017). *Nature Communications* 8: 1–9.
14 (a) Massé, R.C., Uchaker, E., and Cao, G. (2015). *Science China Materials* 58: 715–766. (b) Goodenough, J.B. and Kim, Y. (2010). *Chemistry of Materials* 22: 587–603.
15 Alfaruqi, M.H., Mathew, V., Gim, J. et al. (2015). *Chemistry of Materials* 27: 3609–3620.
16 Bi, W., Wu, Y., Liu, C. et al. (2018). *ACS Applied Energy Materials* 2: 668–677.
17 Sun, Z., Chen, H., Shu, D. et al. (2012). *Journal of Power Sources* 203: 233–242.
18 (a) Feng, Q., Kanoh, H., Miyai, Y., and Ooi, K. (1995). *Chemistry of Materials* 7: 1722–1727. (b) Lee, J., Ju, J.B., Cho, W.I. et al. (2013). *Electrochimica Acta* 112: 138–143.
19 (a) Kim, S.H. and Oh, S.M. (1998). *Journal of Power Sources* 72: 150–158. (b) Qiu, N., Chen, H., Yang, Z. et al. (2018). *Electrochimica Acta* 272: 154–160. (c) Guo, C., Liu, H., Li, J. et al. (2019). *Electrochimica Acta* 304: 370–377.
20 Li, Y., Wang, S., Salvador, J.R. et al. (2019). *Chemistry of Materials* 31: 2036–2047.
21 (a) Alfaruqi, M.H., Gim, J., Kim, S. et al. (2015). *Electrochemistry Communications* 60: 121–125. (b) Alfaruqi, M.H., Islam, S., Putro, D.Y. et al. (2018). *Electrochimica Acta* 276: 1–11.
22 Lee, B., Lee, H.R., Kim, H. et al. (2015). *Chemical Communications* 51: 9265–9268.
23 Yuan, C., Zhang, Y., Pan, Y. et al. (2014). *Electrochimica Acta* 116: 404–412.
24 Lee, S., Nam, G., Sun, J. et al. (2016). *Angewandte Chemie International Edition* 128: 8741–8746.
25 Zhang, N., Cheng, F., Liu, Y. et al. (2016). *Journal of the American Chemical Society* 138: 12894–12901.

26 Knight, J.C., Therese, S., and Manthiram, A. (2015). *Journal of Materials Chemistry A* 3: 21077–21082.

27 Hao, J., Mou, J., Zhang, J. et al. (2018). *Electrochimica Acta* 259: 170–178.

28 Jiang, H. and Ji, X. (2020). *Carbon Energy* 2: 437–442.

29 Wu, Y., Fee, J., Tobin, Z. et al. (2020). *ACS Applied Energy Materials* 3: 1627–1633.

30 (a) Liu, C., Zhang, C., Song, H. et al. (2016). *Journal of Materials Chemistry A* 4: 3362–3370. (b) Liu, C., Fu, H., Pei, Y. et al. (2019). *Journal of Materials Chemistry A* 7: 7831–7842. (c) Liu, C., Zhang, C., Song, H. et al. (2016). *Nano Energy* 22: 290–300.

31 Wang, J., Wang, J.-G., Liu, H. et al. (2019). *Journal of Power Sources* 438: 226951.

32 Zhu, C., Fang, G., Liang, S. et al. (2020). *Energy Storage Materials* 24: 394–401.

33 Cockayne, E., Levin, I., Wu, H., and Llobet, A. (2013). *Physical Review B* 87: 184413.

34 Feng, D., Gao, T.-N., Zhang, L. et al. (2020). *Nano-Micro Letters* 12: 1–13.

35 Liu, C., Chi, X., Han, Q., and Liu, Y. (2020). *Advanced Energy Materials* 10: 1903589.

36 Jiang, B., Xu, C., Wu, C. et al. (2017). *Electrochimica Acta* 229: 422–428.

37 Liu, Y., Zhou, X., Liu, R. et al. (2019). *ACS Applied Materials and Interfaces* 11: 19191–19199.

38 Zhang, Z., Li, W., Shen, Y. et al. (2022). *Journal of Energy Storage* 45: 103729.

39 (a) Kim, J.-S., Kim, K., Cho, W. et al. (2012). *Nano Letters* 12: 6358–6365. (b) Pang, W.K., Peterson, V.K., Sharma, N. et al. (2014). *Journal of Physical Chemistry C* 118: 3976–3983.

40 Yuan, Y., Nie, A., Odegard, G.M. et al. (2015). *Nano Letters* 15: 2998–3007.

41 Mizuno, Y., Okubo, M., Hosono, E. et al. (2013). *Journal of Physical Chemistry C* 117: 10877–10882.

42 (a) Gallaway, J.W., Hertzberg, B.J., Zhong, Z. et al. (2016). *Journal of Power Sources* 321: 135–142. (b) Bailey, M.R. and Donne, S.W. (2012). *Journal of the Electrochemical Society* 159: A2010.

43 Shannon, R.D. (1976). *Acta Crystallographica Section A: Crystal Physics, Diffraction, Theoretical and General Crystallography* 32: 751–767.

44 (a) Li, X., Xu, Y., and Wang, C. (2009). *Journal of Alloys and Compounds* 479: 310–313. (b) Shin, E.M., Senthurchelvan, R., Munoz, J. et al. (1996). *Journal of the Electrochemical Society* 143: 1562.

45 (a) Sun, X., Duffort, V., Mehdi, B.L. et al. (2016). *Chemistry of Materials* 28: 534–542. (b) Su, D., Ahn, H.-J., and Wang, G. (2013). *NPG Asia Materials* 5: e70.

46 Chen, L., Yang, Z., Qin, H. et al. (2019). *Journal of Power Sources* 425: 162–169.

47 Preisler, E. (1976). *Journal of Applied Electrochemistry* 6: 311–320.

48 (a) Chung, I., Song, J.-H., Im, J. et al. (2012). *Journal of the American Chemical Society* 134: 8579–8587. (b) Zhou, W. and Fu, H. (2018). *Inorganic Chemistry Frontiers* 5: 1240–1254.

49 Chen, L., Yang, Z., Qin, H. et al. (2019). *Electrochimica Acta* 317: 155–163.

50 Jiang, L., Wu, Z., Wang, Y. et al. (2019). *ACS Nano* 13: 10376–10385.

51 Zhao, Q., Chen, X., Wang, Z. et al. (2019). *Small* 15: 1904545.

52 Wang, Y., Wu, Z., Jiang, L. et al. (2019). *Nanoscale Advances* 1: 4365–4372.

53 (a) Liu, Y., Chi, X., Han, Q. et al. (2019). *Journal of Power Sources* 443: 227244. (b) Wang, J., Wang, J.-G., Liu, H. et al. (2019). *Journal of Materials Chemistry A* 7: 13727–13735. (c) Zhao, T., Zhang, G., Zhou, F. et al. (2018). *Small* 14: 1802320. (d) Shi, M., Xiao, P., Yang, C. et al. (2020). *Journal of Power Sources* 463: 228209. (e) Kim, S.H., Kim, J.M., Ahn, D.B., and Lee, S.Y. (2020). *Small* 16: 2002837.

54 (a) Zhu, C., Fang, G., Zhou, J. et al. (2018). *Journal of Materials Chemistry A* 6: 9677–9683. (b) Wang, J., Zhang, Z., Yan, X. et al. (2020). *Nano-Micro Letters* 12: 1–14. (c) Guo, X., Li, J., Jin, X. et al. (2018). *Nanomaterials* 8: 301. (d) Fu, Y., Wei, Q., Zhang, G. et al. (2018). *Advanced Energy Materials* 8: 1801445. (e) Moon, H., Ha, K.H., Park, Y. et al. (2021). *Advancement of Science* 8: 2003714. (f) Wu, X., Xiang, Y., Peng, Q. et al. (2017). *Journal of Materials Chemistry A* 5: 17990–17997.

55 (a) Zhang, Y., Deng, S., Li, Y. et al. (2020). *Energy Storage Materials* 29: 52–59. (b) Tang, F., Gao, J., Ruan, Q. et al. (2020). *Electrochimica Acta* 353: 136570. (c) Yang, C., Han, M., Yan, H. et al. (2020). *Journal of Power Sources* 452: 227826. (d) Zhang, X., Li, J., Ao, H. et al. (2020). *Energy Storage Materials* 30: 337–345. (e) Liu, C., Li, Q., Sun, H. et al. (2020). *Journal of Materials Chemistry A* 8: 24031–24039. (f) Chen, L., Yang, Z., Cui, F. et al. (2020). *Materials Chemistry Frontiers* 4: 213–221. (g) Wu, B., Zhang, G., Yan, M. et al. (2018). *Small* 14: 1703850.

56 (a) Li, Z., Huang, Y., Zhang, J. et al. (2020). *Nanoscale* 12: 4150–4158. (b) Zang, X., Wang, X., Liu, H. et al. (2020). *ACS Applied Materials & Interfaces* 12: 9347–9354.

57 Huang, J., Tang, X., Liu, K. et al. (2020). *Materials Today Energy* 17: 100475.

58 Zhu, X., Cao, Z., Wang, W. et al. (2021). *ACS Nano* 15: 2971–2983.

59 Nam, K.W., Kim, H., Choi, J.H., and Choi, J.W. (2019). *Energy & Environmental Science* 12: 1999–2009.

60 Jiang, Y., Ba, D., Li, Y., and Liu, J. (2020). *Advancement of Science* 7: 1902795.

61 Huang, J., Wang, Z., Hou, M. et al. (2018). *Nature Communications* 9: 2906.

62 Guo, S., Liang, S., Zhang, B. et al. (2019). *ACS Nano* 13: 13456–13464.

63 Zhang, Y., Deng, S., Pan, G. et al. (2020). *Small Methods* 4: 1900828.

64 (a) Zhang, Y., Tao, L., Xie, C. et al. (2020). *Advanced Materials* 32: 1905923. (b) Xie, J., Yang, X., and Xie, Y. (2020). *Nanoscale* 12: 4283–4294. (c) Xu, W., Sun, C., Zhao, K. et al. (2019). *Energy Storage Materials* 16: 527–534.

65 Xiong, T., Yu, Z.G., Wu, H. et al. (2019). *Advanced Energy Materials* 9: 1803815.

66 Tan, Q., Li, X., Zhang, B. et al. (2020). *Advanced Energy Materials* 10: 2001050.

67 Fang, G., Zhu, C., Chen, M. et al. (2019). *Advanced Functional Materials* 29: 1808375.

68 Zhang, H., Wang, J., Liu, Q. et al. (2019). *Energy Storage Materials* 21: 154–161.

69 (a) Lu, F., Chen, Q., Geng, S. et al. (2018). *Journal of Materials Chemistry A* 6: 24232–24244. (b) Sun, C., Wang, Y.-J., Gu, H. et al. (2020). *Nano Energy* 77: 105092.

70 Zhang, D., Cao, J., Zhang, X. et al. (2021). *Advanced Functional Materials* 31: 2009412.

71 Tao, Y., Li, Z., Tang, L. et al. (2020). *Electrochimica Acta* 331: 135296.

72 Koketsu, T., Ma, J., Morgan, B.J. et al. (2020). *Energy Storage Materials* 25: 154–163.

73 (a) Yi, Z., Chen, G., Hou, F. et al. (2021). *Advanced Energy Materials* 11: 2003065. (b) Chamoun, M., Brant, W., Tai, C. et al. (2018). *Energy Storage Materials* 15: 351–360.

74 Li, W., Gao, X., Chen, Z. et al. (2020). *Chemical Engineering Journal* 402: 125509.

75 (a) Ding, J., Gao, H., Ji, D. et al. (2021). *Journal of Materials Chemistry A* 9: 5258–5275. (b) Xu, X., Xiong, F., Meng, J. et al. (2020). *Advanced Functional Materials* 30: 1904398. (c) Song, M., Tan, H., Chao, D., and Fan, H.J. (2018). *Advanced Functional Materials* 28: 1802564.

76 (a) Selvakumaran, D., Pan, A., Liang, S., and Cao, G. (2019). *Journal of Materials Chemistry A* 7: 18209–18236. (b) Zhang, S., Tan, H., Rui, X., and Yu, Y. (2020). *Accounts of Chemical Research* 53: 1660–1671.

77 (a) Mallick, S. and Raj, C.R. (2021). *ChemSusChem* 14: 1987–2022. (b) Liu, S., Kang, L., Kim, J.M. et al. (2020). *Advanced Energy Materials* 10: 2000477.

78 Zhang, Y., Ang, E.H., Dinh, K.N. et al. (2021). *Materials Chemistry Frontiers* 5: 744–762.

79 Zhang, N., Dong, Y., Jia, M. et al. (2018). *ACS Energy Letters* 3: 1366–1372.

80 Yan, M., He, P., Chen, Y. et al. (2017). *Advanced Materials* 30: 1703725.

81 (a) Islam, S., Alfaruqi, M.H., Putro, D.Y. et al. (2019). *Journal of Materials Chemistry A* 7: 20335–20347. (b) Yang, Y., Tang, Y., Fang, G. et al. (2018). *Energy & Environmental Science* 11: 3157–3162.

82 Sai Gautam, G., Canepa, P., Richards, W.D. et al. (2016). *Nano Letters* 16: 2426–2431.

83 (a) Zhang, Y., Zhao, L., Chen, A., and Sun, J. (2021). *Fundamental Research* 1: 418–424. (b) Wu, F., Wang, Y., Ruan, P. et al. (2021). *Materials Today Energy* 21: 100842.

84 Tang, B., Fang, G., Zhou, J. et al. (2018). *Nano Energy* 51: 579–587.

85 (a) Kundu, D., Hosseini Vajargah, S., Wan, L. et al. (2018). *Energy & Environmental Science* 11: 881–892. (b) Cao, Z., Chu, H., Zhang, H. et al. (2019). *Journal of Materials Chemistry A* 7: 25262–25267. (c) Shen, C., Li, X., Li, N. et al. (2018). *ACS Applied Materials & Interfaces* 10: 25446–25453. (d) Rastgoo-Deylami, M., Chae, M.S., and Hong, S.-T. (2018). *Chemistry of Materials* 30: 7464–7472. (e) Pang, Q., Sun, C., Yu, Y. et al. (2018). *Advanced Energy Materials* 8: 1800144. (f) Rastgoo-Deylami, M., Heo, J.W., and Hong, S.T. (2019). *ChemistrySelect* 4: 11711–11717. (g) He, P., Quan, Y., Xu, X. et al. (2017). *Small* 13: 1702551.

86 Alfaruqi, M.H., Mathew, V., Song, J. et al. (2017). *Chemistry of Materials* 29: 1684–1694.

87 (a) Cai, Y., Liu, F., Luo, Z. et al. (2018). *Energy Storage Materials* 13: 168–174. (b) Hu, P., Zhu, T., Wang, X. et al. (2018). *Nano Letters* 18: 1758–1763. (c) Soundharrajan, V., Sambandam, B., Kim, S. et al. (2018). *Nano Letters* 18: 2402–2410.

88 (a) Chen, S., Zhang, Y., Geng, H. et al. (2019). *Journal of Power Sources* 441: 227192. (b) Bin, D., Huo, W., Yuan, Y. et al. (2020). *Chem* 6: 968–984.

89 Wu, J., Kuang, Q., Zhang, K. et al. (2021). *Energy Storage Materials* 41: 297–309.

90 (a) Zhang, Y., Chen, A., and Sun, J. (2021). *Journal of Energy Chemistry* 54: 655–667. (b) Mathew, V., Sambandam, B., Kim, S. et al. (2020). *ACS Energy Letters* 5: 2376–2400. (c) Tang, B., Shan, L., Liang, S., and Zhou, J. (2019). *Energy & Environmental Science* 12: 3288–3304. (d) Liu, N., Li, B., He, Z. et al. (2021). *Journal of Energy Chemistry* 59: 134–159.

91 (a) Fang, G., Zhou, J., Pan, A., and Liang, S. (2018). *ACS Energy Letters* 3: 2480–2501. (b) Yong, B., Ma, D., Wang, Y. et al. (2020). *Advanced Energy Materials* 10: 2002354.

92 (a) Shi, H.Y., Song, Y., Qin, Z. et al. (2019). *Angewandte Chemie International Edition* 58: 16057–16061. (b) Wan, F., Zhang, Y., Zhang, L. et al. (2019). *Angewandte Chemie International Edition* 58: 7062–7067. (c) Wang, F., Sun, W., Shadike, Z. et al. (2018). *Angewandte Chemie International Edition* 57: 11978–11981.

93 Lai, J., Zhu, H., Zhu, X. et al. (2019). *ACS Applied Energy Materials* 2: 1988–1996.

94 Zhang, N., Jia, M., Dong, Y. et al. (2019). *Advanced Functional Materials* 29: 1807331.

95 (a) He, P., Yan, M., Zhang, G. et al. (2017). *Advanced Energy Materials* 7: 1601920. (b) Yu, D., Wei, Z., Zhang, X. et al. (2020). *Advanced Functional Materials* 31: 2008743.

96 Wu, Z., Lu, C., Wang, Y. et al. (2020). *Small* 16: e2000698.

97 Liu, W., Hao, J., Xu, C. et al. (2017). *Chemical Communications* 53: 6872–6874.

98 (a) Zhang, L., Miao, L., Zhang, B. et al. (2020). *Journal of Materials Chemistry A* 8: 1731–1740. (b) Wei, T., Li, Q., Yang, G., and Wang, C. (2018). *Journal of Materials Chemistry A* 6: 8006–8012. (c) Cao, Z., Wang, L., Zhang, H. et al. (2020). *Advanced Functional Materials* 30: 2000472. (d) Chen, L., Yang, Z., and Huang, Y. (2019). *Nanoscale* 11: 13032–13039. (e) Park, J.-S., Jo, J.H., Aniskevich, Y. et al. (2018). *Chemistry of Materials* 30: 6777–6787. (f) Liu, Y., Hu, P., Liu, H. et al. (2020). *Materials Today Energy* 17: 100431. (g) Chen, L., Ruan, Y., Zhang, G. et al. (2019). *Chemistry of Materials* 31: 699–706. (h) Li, Z., Ren, Y., Mo, L. et al. (2020). *ACS Nano* 14: 5581–5589. (i) Ding, J., Du, Z., Gu, L. et al. (2018). *Advanced Materials* 30: e1800762.

99 (a) Shin, J., Choi, D.S., Lee, H.J. et al. (2019). *Advanced Energy Materials* 9: 1900083. (b) He, P., Liu, J., Zhao, X. et al. (2020). *Journal of Materials Chemistry A* 8: 10370–10376. (c) Lin, Y., Zhou, F., Chen, M. et al. (2020). *Chemical Engineering Journal* 396: 125259. (d) Shan, L., Zhou, J., Zhang, W.

et al. (2019). *Energy Technology* 7: 1900022. (e) Wang, X., Ye, L., Zou, Y. et al. (2021). *Materials Today Energy* 19: 100593. (f) Meng, W., Pigliapochi, R., Bayley, P.M. et al. (2017). *Chemistry of Materials* 29: 5513–5524. (g) Shi, W., Yin, B., Yang, Y. et al. (2021). *ACS Nano* 15: 1273–1281. (h) Lin, Y., Zhou, F., Xie, M. et al. (2020). *ChemSusChem* 13: 3696–3706.

100 Li, Z., Ganapathy, S., Xu, Y. et al. (2019). *Advanced Energy Materials* 9: 1900237.

101 Spencer Braithwaite, J., Richard, C., Catlow, A. et al. (2001). *Physical Chemistry Chemical Physics* 3: 4052–4059.

102 Jo, J.H., Sun, Y.-K., and Myung, S.-T. (2017). *Journal of Materials Chemistry A* 5: 8367–8375.

103 Wang, L., Huang, K.-W., Chen, J., and Zheng, J. (2019). *Science Advances* 5: eaax4279.

104 Guo, S., Fang, G., Liang, S. et al. (2019). *Acta Materialia* 180: 51–59.

105 Kaveevivitchai, W. and Manthiram, A. (2016). *Journal of Materials Chemistry A* 4: 18737–18741.

106 Djerdj, I., Sheptyakov, D., Gozzo, F. et al. (2008). *Journal of the American Chemical Society* 130: 11364–11375.

107 (a) Zheng, C., Zeng, L., Wang, M. et al. (2014). *CrystEngComm* 16: 10309–10313. (b) Butt, F.K., Cao, C., Ahmed, R. et al. (2014). *CrystEngComm* 16: 894–899. (c) De Juan-Corpuz, L.M.Z., Nguyen, M.T., Corpuz, R.D. et al. (2019). *ACS Applied Nano Materials* 2: 4247–4256. (d) Liu, Y., Li, C., Xu, J. et al. (2020). *Nano Energy* 67: 104211.

108 Tang, W., Lan, B., Tang, C. et al. (2020). *ACS Sustainable Chemistry & Engineering* 8: 3681–3688.

109 (a) Zhang, N., Cheng, F., Liu, Y. et al. (2016). *Journal of the American Chemical Society* 138: 12894–12901. (b) Lu, J.S., Maggay, I.V.B., and Liu, W.R. (2018). *Chemical Communications* 54: 3094–3097. (c) Tan, Q., Li, X., Zhang, B. et al. (2020). *Advanced Energy Materials* 10: 2001050. (d) Jiang, L., Wu, Z., Wang, Y. et al. (2019). *ACS Nano* 13: 10376–10385.

110 Wang, F., Hu, E., Sun, W. et al. (2018). *Energy & Environmental Science* 11: 3168–3175.

111 (a) Hu, P., Zou, Z., Sun, X. et al. (2020). *Advanced Materials* 32: e1907526. (b) Zhang, X., Rui, X., Chen, D. et al. (2019). *Nanoscale* 11: 2556–2576. (c) Li, G., Yang, Z., Jiang, Y. et al. (2016). *Nano Energy* 25: 211–217. (d) Hu, P., Zhu, T., Wang, X. et al. (2019). *Nano Energy* 58: 492–498.

112 (a) Li, W., Wang, K., Cheng, S., and Jiang, K. (2018). *Energy Storage Materials* 15: 14–21. (b) Ni, Q., Jiang, H., Sandstrom, S. et al. (2020). *Advanced Functional Materials* 30: 2003511. (c) Song, W., Wu, Z., Chen, J. et al. (2014). *Electrochimica Acta* 146: 142–150.

113 (a) Wang, L., Sun, J., Song, R. et al. (2016). *Advanced Energy Materials* 6: 1502067. (b) Wang, Y., Jiang, M., Yang, Y., and Ran, F. (2016). *Electrochimica Acta* 222: 1914–1921. (c) Sun, Z., Zhang, J., Yin, L. et al. (2017). *Nature Communications* 8: 14627. (d) Djire, A., Pande, P., Deb, A. et al. (2019). *Nano Energy* 60: 72–81. (e) Lu, X., Yu, M., Zhai, T. et al. (2013). *Nano*

Letters 13: 2628–2633. (f) Wang, R., Lang, J., Zhang, P. et al. (2015). *Advanced Functional Materials* 25: 2270–2278. (g) Ding, J., Du, Z., Li, B. et al. (2019). *Advanced Materials* 31: e1904369. (h) Li, T., Wang, J., Li, X. et al. (2020). *Advanced Science* 7: 1903276. (i) Fang, G., Liang, S., Chen, Z. et al. (2019). *Advanced Functional Materials* 29: 1905267.

114 Fu, Y., Wei, Q., Zhang, G. et al. (2018). *Advanced Energy Materials* 8: 1801445.

115 (a) Bock, D.C., Marschilok, A.C., Takeuchi, K.J., and Takeuchi, E.S. (2013). *Journal of Power Sources* 231: 219–225. (b) Xia, C., Guo, J., Lei, Y. et al. (2018). *Advanced Materials* 30: 1705580.

116 Wan, F. and Niu, Z. (2019). *Angewandte Chemie International Edition* 58: 16358–16367.

117 Dong, Y., Miao, L., Ma, G. et al. (2021). *Chemical Science* 12: 5843–5852.

118 Lu, Y., Zhu, T., van den Bergh, W. et al. (2020). *Angewandte Chemie International Edition* 59: 17004–17011.

119 Wan, F., Zhang, L., Dai, X. et al. (2018). *Nature Communications* 9: 1656.

120 Tang, B., Zhou, J., Fang, G. et al. (2019). *Journal of Materials Chemistry A* 7: 940–945.

121 Guo, X., Fang, G., Zhang, W. et al. (2018). *Advanced Energy Materials* 8: 1801819.

122 Wu, T., Zhu, K., Qin, C., and Huang, K. (2019). *Journal of Materials Chemistry A* 7: 5612–5620.

123 (a) Cai, Y., Chua, R., Kou, Z. et al. (2020). *ACS Applied Materials & Interfaces* 12: 36110–36118. (b) Chen, Z., Hu, J., Liu, S. et al. (2021). *Chemical Engineering Journal* 404: 126536. (c) Du, Y., Wang, X., and Sun, J. (2020). *Nano Research* 14: 754–761. (d) He, D., Peng, Y., Ding, Y. et al. (2021). *Journal of Power Sources* 484: 229284. (e) Luo, H., Wang, B., Wang, C. et al. (2020). *Energy Storage Materials* 33: 390–398.

124 (a) Xiong, T., Zhang, Y., Lee, W.S.V., and Xue, J. (2020). *Advanced Energy Materials* 10: 2001769. (b) Zhang, Y., Tao, L., Xie, C. et al. (2020). *Advanced Materials* 32: e1905923. (c) Zhao, D., Zhu, Q., Li, X. et al. (2022). *Batteries & Supercaps* 5: e202100341. (d) Zong, Q., Wang, Q., Liu, C. et al. (2022). *ACS Nano* 16: 4588–4598. (e) Zu, D., Wang, H., Lin, S. et al. (2019). *Nano Research* 12: 2150–2163.

125 Liao, M., Wang, J., Ye, L. et al. (2020). *Angewandte Chemie International Edition* 59: 2273–2278.

126 Yang, W., Dong, L., Yang, W. et al. (2019). *Small Methods* 4: 1900670.

127 Xiao, L., Zhao, Y., Yang, Y. et al. (2008). *Electrochimica Acta* 54: 545–550.

128 (a) Yang, Y., Tang, Y., Liang, S. et al. (2019). *Nano Energy* 61: 617–625. (b) Xia, C., Guo, J., Li, P. et al. (2018). *Angewandte Chemie International Edition* 57: 3943–3948. (c) Tian, M., Liu, C., Zheng, J. et al. (2020). *Energy Storage Materials* 29: 9–16. (d) Kundu, D., Adams, B.D., Duffort, V. et al. (2016). *Nature Energy* 1: 16119.

129 Meng, J., Liu, Z., Niu, C. et al. (2016). *Journal of Materials Chemistry A* 4: 4893–4899.

130 Zhao, Y., Han, C., Yang, J. et al. (2015). *Nano Letters* 15: 2180–2185.

131 (a) Yaghoobnejad Asl, H., Sharma, S., and Manthiram, A. (2020). *Journal of Materials Chemistry A* 8: 8262–8267. (b) Jiang, H., Zhang, Y., Pan, Z. et al. (2020). *Materials Chemistry Frontiers* 4: 1434–1443. (c) Xue, D.J., Liu, S.C., Dai, C.M. et al. (2017). *Journal of the American Chemical Society* 139: 958–965.

132 (a) Hou, Z., Dong, M., Xiong, Y. et al. (2020). *Small* 16: e2001228. (b) Li, Y., Chen, J., Chen, Y. et al. (2019). *The Journal of Physical Chemistry C* 123: 3048–3057. (c) An, Q., Li, Y., Deog Yoo, H. et al. (2015). *Nano Energy* 18: 265–272.

133 Liu, N., Wu, X., Fan, L. et al. (2020). *Advanced Materials* 32: e1908420.

134 (a) Liu, S., Zhu, H., Zhang, B. et al. (2020). *Advanced Materials* 32: e2001113. (b) Liu, X., Ni, W., Wang, Y. et al. (2022). *Small* 18: e2105796. (c) Ma, X., Cao, X., Yao, M. et al. (2022). *Advanced Materials* 34: e2105452.

135 (a) Liu, P., Liu, W., Huang, Y. et al. (2020). *Energy Storage Materials* 25: 858–865. (b) Luo, H., Wang, B., Wang, F. et al. (2020). *ACS Nano* 14: 7328–7337.

136 (a) Dai, X., Wan, F., Zhang, L. et al. (2019). *Energy Storage Materials* 17: 143–150. (b) Tian, Y., An, Y., Wei, H. et al. (2020). *Chemistry of Materials* 32: 4054–4064.

137 (a) Chen, D., Rui, X., Zhang, Q. et al. (2019). *Nano Energy* 60: 171–178. (b) Hu, P., Zhu, T., Ma, J. et al. (2019). *Chemical Communications* 55: 8486–8489. (c) Li, Y., Huang, Z., Kalambate, P.K. et al. (2019). *Nano Energy* 60: 752–759.

138 Zhou, F., Lin, Y., Li, T. et al. (2019). *Journal of Materials Chemistry A* 7: 10589–10600.

139 Guo, J., Ming, J., Lei, Y. et al. (2019). *ACS Energy Letters* 4: 2776–2781.

140 (a) Qin, H., Yang, Z., Chen, L. et al. (2018). *Journal of Materials Chemistry A* 6: 23757–23765. (b) Wei, T., Li, Q., Yang, G., and Wang, C. (2019). *Advanced Energy Materials* 9: 1901480. (c) Dai, Y., Liao, X., Yu, R. et al. (2021). *Advanced Materials* 33: e2100359.

141 (a) Ren, H.-Z., Zhang, J., Wang, B. et al. (2022). *Rare Metals* 41: 1605–1615. (b) Jiang, H., Zhang, Y., Liu, Y. et al. (2020). *Journal of Materials Chemistry A* 8: 15130–15139.

142 Xu, D., Wang, H., Li, F. et al. (2019). *Advanced Materials Interfaces* 6: 1801506.

143 Liu, Y., Wang, J., Zeng, Y. et al. (2020). *Small* 16: e1907458.

144 Zhang, X., Li, D., Ruan, Q. et al. (2023). *Materials Today Energy* 32: 101232.

145 Peng, X., Zhang, S., Zou, Z. et al. (2023). *Journal of Electronic Materials* 52: 4413–4433.

146 Lv, T., Peng, Y., Zhang, G. et al. (2023). *Advanced Science* 10: 2206907.

147 Zhang, L., Zhang, B., Hu, J. et al. (2021). *Small Methods* 5: e2100094.

148 Zhang, C., Xu, Y., Zhou, M. et al. (2017). *Advanced Functional Materials* 27: 1604307.

149 Zuo, S., Xu, X., Ji, S. et al. (2021). *Chemistry—A European Journal* 27: 830–860.

150 H. W. Lee, C. Yi, 2016. Prussian blue analogue materials for sodium-ion batteries. ECS Meeting Abstracts, MA2016-02:666.

151 Yang, Y., Zhou, J., Wang, L. et al. (2022). *Nano Energy* 99: 107424.

152 (a) Zampardi, G. and La Mantia, F. (2020). *Current Opinion in Electrochemistry* 21: 84–92. (b) Li, Z., Liu, T., Meng, R. et al. (2021). *Energy & Environmental Materials* 4: 111–116.

153 Zeng, Y., Xu, J., Wang, Y. et al. (2022). *Angewandte Chemie International Edition* 61: e202212031.

154 Zeng, Y., Lu, X.F., Zhang, S.L. et al. (2021). *Angewandte Chemie International Edition* 60: 22189–22194.

155 Cao, T., Zhang, F., Chen, M. et al. (2021). *ACS Applied Materials & Interfaces* 13: 26924–26935.

156 Deng, W., Li, Z., Ye, Y. et al. (2021). *Advanced Energy Materials* 11: 2003639.

157 Liu, Q., Ma, Z., Chen, Z. et al. (2022). *Chemical Communications* 58: 8226–8229.

158 Cui, J., Guo, Z., Yi, J. et al. (2020). *ChemSusChem* 13: 2160–2185.

159 Yang, H., Lee, J., Cheong, J.Y. et al. (2021). *Energy & Environmental Science* 14: 4228–4267.

160 Guo, Z., Ma, Y., Dong, X. et al. (2018). *Angewandte Chemie International Edition* 57: 11737–11741.

161 Lu, Y., Zhang, Q., Li, L. et al. (2018). *Chem* 4: 2786–2813.

162 Kundu, D., Oberholzer, P., Glaros, C. et al. (2018). *Chemistry of Materials* 30: 3874–3881.

163 Zhao, Q., Huang, W., Luo, Z. et al. (2018). *Science Advances* 4: eaao1761.

164 Tie, Z., Liu, L., Deng, S. et al. (2020). *Angewandte Chemie (International Ed. in English)* 59: 4920–4924.

165 Tie, Z. and Niu, Z. (2020). *Angewandte Chemie International Edition* 59: 21293–21303.

166 (a) Fu, X., Zhang, W., Lan, B. et al. (2020). *ACS Applied Energy Materials* 3: 12360–12367. (b) Shi, H.-Y., Ye, Y.-J., Liu, K. et al. (2018). *Angewandte Chemie International Edition* 57: 16359–16363. (c) Yu, H., Liu, G., Wang, M. et al. (2020). *ACS Applied Materials & Interfaces* 12: 5820–5830.

167 Jiménez, P., Levillain, E., Alévêque, O. et al. (2017). *Angewandte Chemie International Edition* 56: 1553–1556.

168 Glatz, H., Lizundia, E., Pacifico, F., and Kundu, D. (2019). *ACS Applied Energy Materials* 2: 1288–1294.

169 Li, S., Sultana, I., Guo, Z. et al. (2013). *Electrochimica Acta* 95: 212–217.

170 Grgur, B.N., Gvozdenović, M.M., Stevanović, J. et al. (2008). *Electrochimica Acta* 53: 4627–4632.

171 Zhijiang, C. and Chengwei, H. (2011). *Journal of Power Sources* 196: 10731–10736.

172 Friebe, C. and Schubert, U.S. (2017). *Topics in Current Chemistry* 375: 19.

173 Nishide, H. and Oyaizu, K. (2008). *Science* 319: 737–738.

174 Koshika, K., Sano, N., Oyaizu, K., and Nishide, H. (2009). *Chemical Communications* 7: 836–838.

175 Häupler, B., Rössel, C., Schwenke, A.M. et al. (2016). *NPG Asia Materials* 8: e283.

176 Wan, F., Zhang, L., Wang, X. et al. (2018). *Advanced Functional Materials* 28: 1804975.

177 Lee, S., Kwon, G., Ku, K. et al. (2018). *Advanced Materials* 30: 1704682.

178 Kim, C., Ahn, B.Y., Wei, T.-S. et al. (2018). *ACS Nano* 12: 11838–11846.

179 Qin, K., Huang, J., Holguin, K., and Luo, C. (2020). *Energy & Environmental Science* 13: 3950–3992.

180 Lu, Y., Hou, X., Miao, L. et al. (2019). *Angewandte Chemie International Edition* 58: 7020–7024.

181 Wang, C.L., Jiang, C., Xu, Y. et al. (2016). *Advanced Materials* 28: 9182.

182 (a) Yu, X. and Manthiram, A. (2018). *ACS Applied Energy Materials* 1: 273–277. (b) Chen, Y., Fan, K., Gao, Y., and Wang, C. (2022). *Advanced Materials* 34: e2200662.

183 Fan, X., Wang, F., Ji, X. et al. (2018). *Angewandte Chemie International Edition* 57: 7146–7150.

184 (a) Mirle, C., Medabalmi, V., and Ramanujam, K. (2021). *ACS Applied Energy Materials* 4: 1218–1227. (b) Kumankuma-Sarpong, J., Tang, S., Guo, W., and Fu, Y. (2021). *ACS Applied Materials & Interfaces* 13: 4084–4092.

185 (a) Li, Y., Liu, L., Lu, Y. et al. (2021). *Advanced Functional Materials* 31: 2102063. (b) Nam, K.W., Kim, H., Beldjoudi, Y. et al. (2020). *Journal of the American Chemical Society* 142: 2541–2548. (c) Chen, Y., Li, J., Zhu, Q. et al. (2022). *Angewandte Chemie International Edition* 61: e202116289.

186 Zhao, Y., Wang, Y.N., Zhao, Z.M. et al. (2020). *Energy Storage Materials* 28: 64–72.

187 Zhang, S.Q., Zhao, W.T., Li, H., and Xu, Q. (2020). *ChemSusChem* 13: 188–195.

188 Walter, M., Kravchyk, K.V., Böfer, C. et al. (2018). *Advanced Materials* 30: 1705644.

189 Zhao, Q., Zhu, Z., and Chen, J. (2017). *Advanced Materials* 29: 1607007.

190 Wang, Y.R., Wang, C.X., Ni, Z.G. et al. (2020). *Advanced Materials* 32: 2000338.

191 Haüpler, B., Rössel, C., Schwenke, A.M. et al. (2016). *NPG Asia Materials* 8: e283.

192 Wang, W.X., Kale, V.S., Cao, Z. et al. (2020). *ACS Energy Letters* 5: 2256–2264.

193 Yao, H., Li, Q.J., Zhang, M.S. et al. (2020). *Chemical Engineering Journal* 392: 123653.

194 Yang, W., Du, X., Zhao, J. et al. (2020). *Joule* 4: 1557–1574.

195 Zhang, J., Chang, Z., Zhang, Z. et al. (2021). *ACS Nano* 15: 15594–15624.

196 Luo, Y., Zheng, F., Liu, L. et al. (2020). *ChemSusChem* 13: 2239–2244.

4

Anode Materials for Aqueous Zinc-Ion Batteries

Zinc metal, one of the most widely used anode materials, is not only widely used in aqueous zinc-ion batteries (AZIBs) but also in other zinc-based battery systems due to its thermodynamic stability, such as Zn–air, Ni–Zn, and Zn–MnO_2 batteries [1–3]. Zinc-based batteries have attracted the attention of researchers owing to their relatively high theoretical gravimetric capacity (820 mAh g^{-1}) and volumetric capacity with two valence electrons (5851 mAh cm^{-3}). Meanwhile, zinc metal is more stable in aqueous electrolytes than lithium, sodium, and potassium metals, and inhibits the occurrence of hydrogen evolution reactions due to its high redox potential (−0.76 V compared with standard hydrogen electrode [SHE]) [4, 5]. Moreover, combining the advantages of high abundance and good safety, zinc metal has attracted huge interest in the aqueous battery industry, which is believed to have promising practical applications for large-scale energy storage systems.

Although zinc metal anodes exhibit great potential in AZIBs for energy storage, their practical large-scale application is greatly hindered due to dendrite formation and low plating/stripping efficiency. In particular, zinc metal tends to dissolve in alkaline electrolytes, leading to the formation of supersaturated zincate solutions **($Zn(OH)_4^{2-}$)** [6]. When the concentration of **$Zn(OH)_4^{2-}$** is high enough, the electrochemically inert by-products, such as zinc oxide (ZnO) and hydroxide ($Zn(OH)_2$), will be precipitated to induce the formation of interface passivation layers [7, 8]. These insulating ZnO precipitates cannot be reconverted into zinc metal during charging processes in AZIBs, reducing the rechargeability and utilization of the zinc anodes. In contrast, such dissolution–precipitation processes will be greatly suppressed in weakly acidic electrolytes such as zinc sulfate ($ZnSO_4$) or zinc trifluoromethanesulfonte ($Zn(CF_3SO_3)_2$) aqueous solutions (based on the zinc Pourbaix diagram, Figure 4.1) [9, 10]. Meanwhile, zinc dendrite growth is still unavoidable during repeated cycling in both electrolytes. Uncontrolled dendritic growth can pierce the separator and cause severe battery short circuits. Besides, issues like corrosion and hydrogen evolution become more severe due to the increased specific surface area of anodes induced by zinc dendrites, which further lowers Coulombic efficiency (CE). Therefore, the development of zinc anodes with high stability in weakly acidic electrolytes and no dendrite growth during operation is the key to realizing the commercial application of AZIBs.

Aqueous Zinc Ion Batteries: Fundamentals, Materials, and Design, First Edition.
Haiyan Wang, Qi Zhang, Yixin Li, and Yougen Tang.

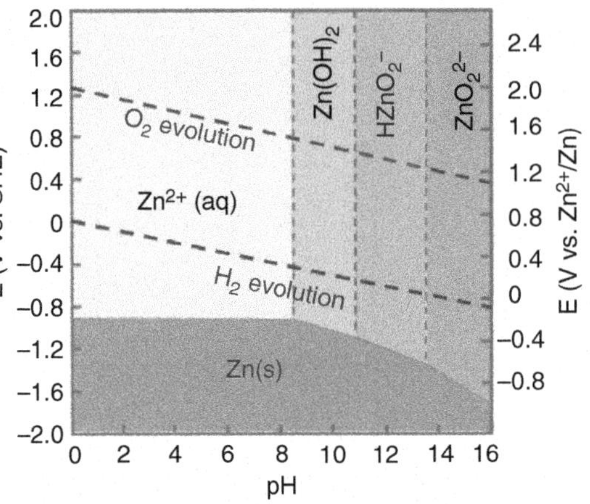

Figure 4.1 Pourbaix diagram of zinc speciation in the aqueous solution. When the pH value is lower than 8.5, by-products such as ZnO and $Zn(OH)_2$ are not formed, indicating that dendrite formation in neutral or weakly acidic electrolytes is not as severe as in alkaline electrolytes. Source: Reproduced with permission from Jia et al. [9]. Copyright 2020, American Chemical Society.

At present, there are two main research directions for the modification strategies of dendrite-free zinc metal anodes. The first one focuses on bulk structure design, such as the construction of three-dimensional (3D) zinc architectures [11], zinc alloys [12], and zinc-plated hierarchical anodes [13]. The second strategy is interface modification engineering, including the host-zinc interface and zinc-electrolyte interface. The former corresponds to the process of preparing the zinc anode by electrodeposition on the current collector material, while the latter is the electrodeposition process that occurs on the surface of the zinc bulk.

4.1 Structural Design

4.1.1 3D Zinc Anodes

Designing 3D porous architectures to enhance the electrolyte–electrode contact area has become a generally accepted approach to alleviate the zinc anode dendrite issues [14–16]. The high surface area of the 3D zinc architecture can reduce the local current density, resulting in low polarization and accelerated battery kinetics. Also, the electrodeposition of zinc preferentially occurs in the pores and limits the growth of zinc dendrites through the 3D spatial hierarchical network, significantly increasing the cyclic stability of the battery (Figure 4.2a) [17]. In addition, other advantages can be presented when the planar zinc anode is transformed into a 3D structure, as follows: (i) the 3D zinc electrode can enhance the wettability and contact area with the electrolyte and increase the number of nucleation sites; (ii) the internally connected 3D network of electrodes possesses good electronic conductivity, which can guarantee fast interfacial electron exchange during charging and discharging; (iii) the 3D design helps to mitigate volume change of zinc anode during zinc deposition/dissolution. Based on the 3D architectural features, designing a zinc anode with a porous, monolithic, 3D aperiodic structure can achieve 90% zinc utilization without visible dendrites after electrochemical cycles, which opens a new avenue for improving Zn utilization [19]. An inexpensive strategy to fabricate the well-developed 3D hierarchical pore zinc architecture was exploited by using organic acid etching [11]. After chemical corrosion, the 3D zinc

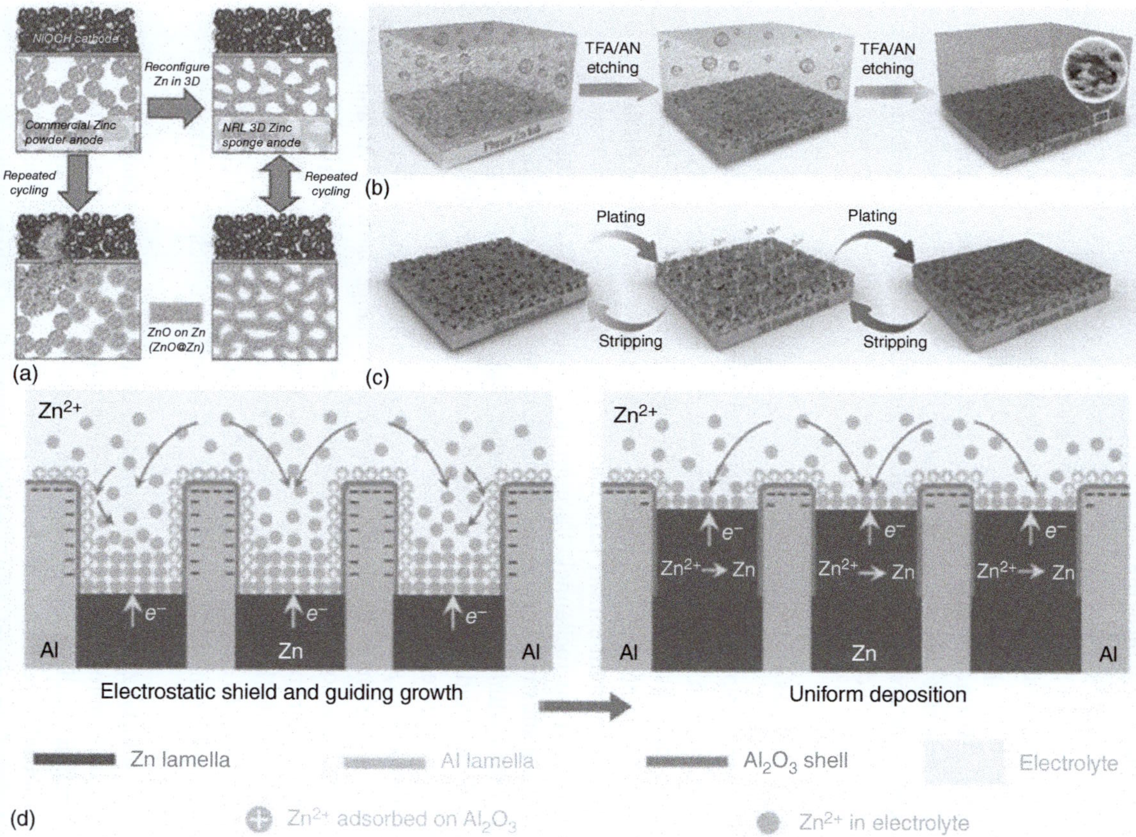

Figure 4.2 (a) Schematic illustration of rechargeable Zn–Ni batteries with commercial zinc powder and 3D zinc sponge anodes. (b) Schematic diagram of fabrication of 3D porous zinc foil by organic acid etching. (c) Schematic illustration of the plating/stripping process using the 3D zinc anode. (d) Schematic illustration of eutectic strategy for dendrite suppression. Source: (a) Reproduced with permission from Parker Joseph et al. [17]. Copyright 2017 American Association for the Advancement of Science; (b, c) Wang et al. [11]/Reproduced with permission from John Wiley & Sons; (d) Wang et al. [18]/Springer Nature/Licensed under CC BY 4.0.

anode exhibits a pore-enriched and clean surface appearance without a passivation layer (Figure 4.2b). Accordingly, the electrodeposition of zinc is confined within the cavity with suppressed dendrite growth (Figure 4.2c), which is attributed to the abundant nucleation sites provided by well-developed hierarchical pores and voids and improved electric field distribution.

The 3D zinc architecture reduces the volumetric capacity and energy density of the battery due to its porous character. Also, it should be kept in mind that the 3D zinc anode has a higher active surface area than the planar one, indicating that more active zinc is contacted with the electrolyte. However, metallic zinc is thermodynamically unstable in an aqueous solution [20]. In this case, the corrosion reaction on the surface of the 3D zinc anode is worsened, which accelerates the consumption of active zinc and degrades the overall performance of the battery [21]. Therefore, these adverse effects induced by the structural design of 3D zinc electrodes should be considered in future studies to construct high-performance zinc metal anodes.

4.1.2 Zinc Alloy Anodes

Optimizing the composition of zinc electrodes and creating different structures are other effective ways to build high-performance zinc anodes. Generally, metal deposition quality, such as size and location, is closely related to the current density, confirming the effectiveness of introducing a high-specific-area electroplating matrix to reduce local current density [22]. For example, porous current collectors with vertically aligned microchannels have greater current densities within the channels than at the substrate surface, which leads to preferential nucleation and growth of metal ions within the channel openings [23]. Along this line, eutectic Zn/Al alloys with alternating Zn and Al lamella nanostructures are designed as a dendrite-free anode material for high-performance AZIBs [18]. In the charging process, Al lamella acts as a 2D supporting framework, and it generates insulating Al_2O_3 shells at the periphery to block the electron exchange between Al and Zn^{2+} ions. As a result, a positive electrostatic shielding field around the Al/Al_2O_3 lamellas is formed, inducing the deposition of zinc ions along the zinc precursor sites between the Al/Al_2O_3 lamellas (Figure 4.2d). This Zn–Al eutectic alloy strategy with alternating Al and Zn sheet interfaces benefits from interfacial matrix current density principles and symbiotic nonprecious Al lamellas, which is worthy of further development.

In addition, the alloyed zinc anode can provide a better interface for nucleation and simultaneously suppress the formation of by-products. When zinc is soaked in a common mildly acidic $ZnSO_4$ electrolyte, corrosion reactions, and hydrogen evolution inevitably occur during rest or electrochemical cycling due to the chemical reactivity of zinc metal, resulting in the formation of a sluggish interface with nonconducting and insoluble zinc hydroxide sulfate (ZHS) and bubbles [12]. The increased electrochemical impedance caused by the inert interface further deteriorates the overall performance of the cell. Metallic copper possesses good anticorrosion capability in aqueous electrolytes due to positive electrode potential (+0.34 V vs. SHE) and can be readily alloyed with zinc. Cu–Zn alloy/Zn composites were developed to alleviate these corrosion problems of zinc metal anodes in mildly acidic electrolytes. The Cu/Zn composite was synthesized through a displacement reaction between zinc and Cu^{2+} ions in an ethanol solution before electrochemical in situ transformation into a Cu–Zn/Zn electrode. Electrochemical corrosion performance tests revealed that Cu/Zn electrodes exhibited more positive corrosion potential and lower corrosion current than pristine Zn electrodes, suggesting improved chemical stability and lower corrosion rate for Cu-integrated Zn electrodes. Furthermore, a 3D nanoporous Cu–Zn alloy anode was reported to improve the limited capacity and cycling stability of the anode [24]. Specifically, this unique structural design endows porous alloy anodes with numerous advantages, such as fast ion transport, shortened ion transport paths, a high electrolyte–electrode contact area, and high electronic conductivity. High-voltage aqueous zinc-bromine batteries based on 3D Cu–Zn alloy anodes were fabricated and delivered a high areal capacity (1.56 mAh cm^{-2}) comparable to Li-ion batteries (1–3 mAh cm^{-2}).

4.1.3 Zinc-Plated Hierarchical Anodes

Apart from the design of 3D zinc bulk structures, preparing zinc-plated hierarchical electrodes is also the most common strategy to construct 3D composite zinc anodes. Given the importance of gravitational energy density in practical applications of zinc batteries, the selection of current collectors has important implications for the deposition of zinc electrodes. Conductive materials with 3D porous structures are widely used as hosts in the field of battery research [25]. The 3D skeleton structure can uniformize the electric field distribution and disperse current density at the current collector interface, avoiding encounters with the critical current density for dendrite formation. Carbon-based (fibers, foams, felts, and aerogels) [26–30], metallic (meshes and foams) [13, 31], and metal-organic framework (MOF)-based materials have been widely used as hosts for zinc ions due to their chemical stability in neutral electrolytes, robust mechanical performance, and adaptive electronic conductivity [32].

4.1.3.1 3D Carbon-Based Hosts

Functional carbon material is one of the most advanced energy materials, and advantages lie in adjustable structure and satisfactory performance. Tuning the dimensions of carbon nanomaterials can expand their application fields and enhance their physicochemical properties. 3D carbon fibers woven by 1D carbon nanowires were directly used as a zinc metal deposition substrate to prepare 3D zinc–carbon fiber (Zn@CFs) composite anodes [30]. The Zn@CFs anode exhibits improved electrochemical energy and relieves internal pressure and stress during the zinc deposition/stripping process compared to the zinc anode plated on stainless steel mesh. However, such pristine 3D carbon fibers cannot meet the practical application requirements of zinc metal batteries due to fragmented nucleation sites. To fully exploit the high surface area properties of 3D carbon nanotube (CNT) architectures, a flexible carbon cloth-supported random CNT array was reported for zinc plating/stripping substrate [27]. This porous CNT framework has a higher specific surface area of 24.2 m^2 g^{-1} than that of bare carbon cloth (1.2 m^2 g^{-1}). As a result, the Zn/CNT anode exhibits excellent electrochemical performance with low voltage hysteresis (84 vs. 192 mV) and cycling stability with a depth of discharge (DOD) of 28%, which can be attributed to the uniform electric field distribution provided by CNT arrays on the interface (Figure 4.3a). In contrast, the charges are concentrated in the vicinity of isolated zinc nucleation sites on the bare carbon cloth, leading to preferential deposition of zinc ions at independent nucleation centers to induce dendrite growth (Figure 4.3b,c). In addition, the DOD coefficient of an electrode is important to improve the utilization of active materials and the energy density of batteries. The CNT-assembled, highly conductive 3D carbon foam with internal interconnection can mechanically confine the deposited zinc nanosheets within the internal framework and avoid the formation of zinc dendrites [29]. Quite different from individual CNT and carbon fiber substrates, both the internal sites and surfaces of 3D carbon foams can act as accessible centers for zinc deposition. Also, the electrically interconnected CNT architectures within the foam can effectively accelerate the transport of electrons in the bulk and ions in the electrolyte. Symmetric cells based on CNT foam zinc anodes are capable of maintaining capacity retention of 100% with a DOD of 81% after 100 cycles, while planar zinc foil anodes only sustain several cycles. These works accelerate the research of CNT in high-performance metal anodes for AZIBs.

Reducing the weight of the current collector to load an appropriate amount of active electrode material is another feasible strategy for developing high-energy-density batteries. Graphene, among carbon materials, is one of the most ideal lightweight host materials due to its ultrahigh specific surface area (2675 m^2 g^{-1}), low density, and excellent electrical conductivity (0.5–100 S m^{-1}) [34, 35]. However, as-prepared graphene or graphene oxide nanosheets tend to be re-stacked during subsequent processing and eventually aggregate into inactive carbon films. To inherit the excellent properties of 2D graphene, 3D graphene architectures, and composites extending along the vertical direction of the 2D graphene sheets are highly desired. Graphene-like carbon films (GCFs) assembled from raw graphite paper by simple electrochemical processing are reported as anode and cathode substrates for AZIBs [26]. The full cell assembled with electrolyzed $NaMnO_2$ nanosheets on GCF and electrodeposited Zn on GCF as cathode and anode, respectively, exhibits outstanding electrochemical performance with a record energy density of 511.9 Wh kg^{-1}. Although the 3D framework substrate enables uniform electric field distribution and confines the deposited zinc in the interior space, no consideration for the ordered structure design and nucleation site guidance may lead to the preferential deposition of metals at the top of the 3D host (i.e. top growth mode), especially at high current densities [36–38]. Inspired by surface and structural alterations, a flexible 3D Mxene/graphene aerogel (MGA) with a highly zincophilic framework for zinc encapsulation was designed [28]. The morphologies of bare and MGA-coated copper foils were snapshotted during electroplating by in situ optical microscopy observations at a current density of 3 mA cm^{-2}. The MGA thickness increased from 60 to 92 μm with the increase in plating time from 15 to 30 minutes (Figure 4.3d), and no zinc deposition was observed at the interface

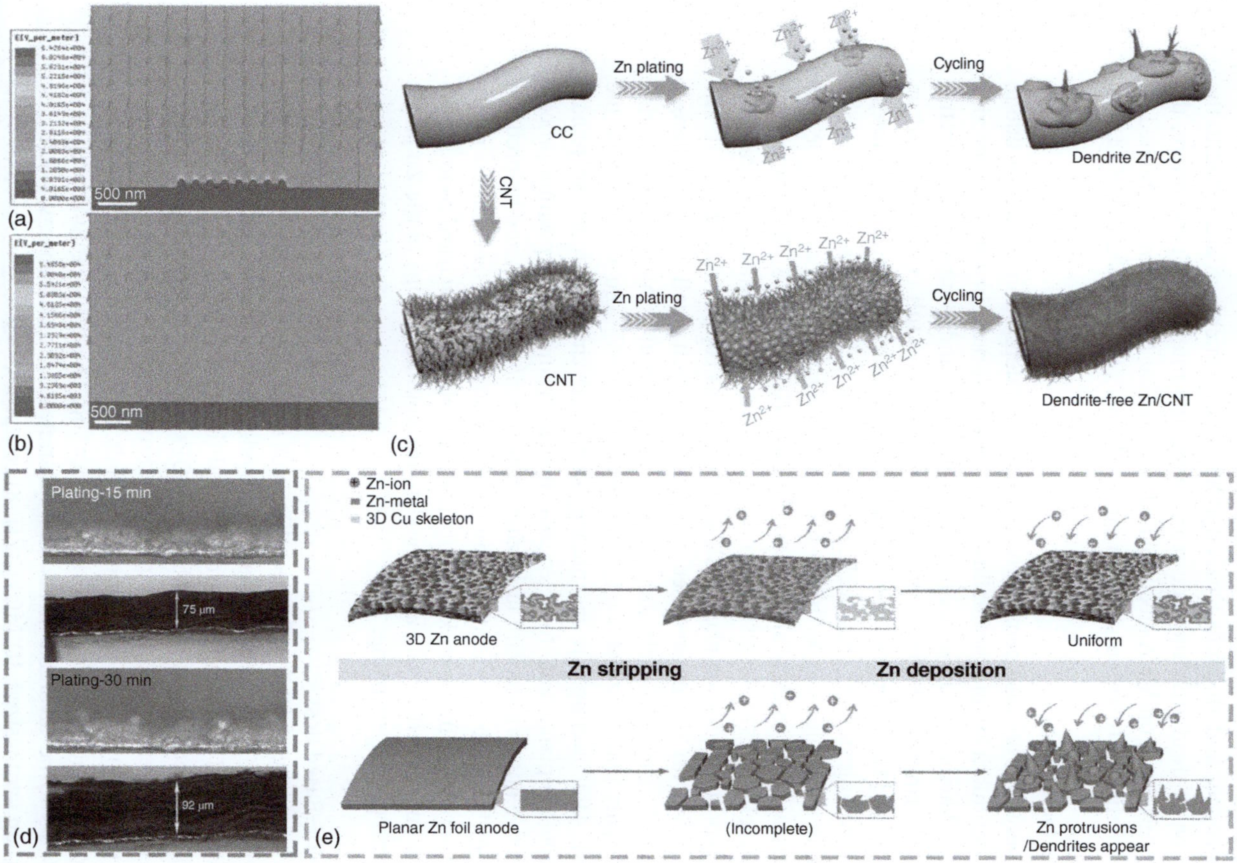

Figure 4.3 Simulations of electric field distribution on (a) carbon cloth and (b) CNT arrays. (c) The schematic illustrations of zinc deposition on carbon cloth and CNT electrodes. (d) In situ optical microscopy visualization of zinc plating on bare Cu and MGA-coated Cu. (e) Schematic illustrations of zinc stripping/plating on 3D zinc and planar zinc anodes. Source: (a–c) Zheng et al. [27]/Reproduced with permission from John Wiley & Sons. (d) Adapted with permission from Zhou et al. [28]/© 2022/John Wiley & Sons. (e) Reproduced with permission from Kang et al. [33]. Copyright 2019 American Chemical Society.

between the MGA and the electrolyte, indicating that the metal zinc was successfully encapsulated inside the 3D framework. The half-cell directly assembled with MGA and Zn foil could stably operate for 600 cycles with a CE of 99.67% at a current density of 10 mA cm^{-2} and a set capacity of 1 mAh cm^{-2}. However, the half-cell employing copper and zinc foils is failed at 86 cycles, which was attributed to the accumulation of dead zinc and side reactions at the interface. Furthermore, the symmetric cell based on MGA electroplated with zinc at 5 mAh cm^{-2} exhibited an ultra-long lifetime of 1050 hours and an extremely low voltage hysteresis of 64 mV.

4.1.3.2 3D Metallic Host

3D metal networks are also commonly used as metal deposition substrates due to their high conductivity and availability. Zhou and coworkers investigated the zinc deposition behavior on copper foam and nickel foam and compared the electrochemical performance of symmetric cells, half cells, and full cells assembled on these substrates [31]. Copper foams were found to exhibit lower voltage hysteresis and longer cycling stability than nickel foams, planar copper foils, and zinc foils, suggesting that metallic copper-based substrates have potential as hosts for zinc metal deposition due to their high redox potentials (0.342 V vs. SHE) compared to metallic nickel (−0.257 V vs. SHE). 3D copper-based zinc anodes were fabricated by electrodepositing zinc on chemically etched porous copper networks [33]. The symmetric cell with 3D copper-based zinc anode shows a small voltage hysteresis of about 40 mV and long cycle stability of 350 hours, better than the planar zinc foil counterpart with a larger voltage hysteresis (100 mV) and shorter cycle life (110 hours). After cycling, the morphology of the 3D zinc electrode can remain intact, while the planar zinc foil electrode shows an unstable deposition process and a defective appearance with large zinc protrusions/dendrites, and by-products (Figure 4.3e). This enhancement mechanism is attributed to the fact that the multifunctional 3D copper framework can not only maintain high electrical conductivity at all times but also effectively accommodate the volume change of the Zn anode during electroplating/stripping.

4.1.3.3 MOF-Based Host

MOFs are attractive candidates to meet the needs of next-generation energy-storage technologies [32]. Their modular nature provides great synthetic tunability and fine chemical and structural control. In addition, numerous multidisciplinary characterization techniques applicable to MOFs allow extensive experiments to elucidate the underlying determinants of performance and provide key insights for the development of next-generation materials. MOFs can be easily transformed into 3D porous conductive frameworks by annealing. These outstanding properties enable MOFs to be favorable hosts and substrates for metal-energy electrodes. Wang's group synthesized 3D MOFs (ZIF-8) composed of ZnN_4 tetrahedral units bridged by imidazolate linkers [39]. After heat treatment in an inert atmosphere, the Zn^{2+} in the ZIF-8 framework was transformed into a trace amount of ZnO with an even distribution in bulk by thermal reduction, and the inherent porous structure was well preserved. Annealed ZIF-8 (ZIF-8-500) as a zinc metal deposition substrate exhibits high CE (98.6–99.8%) over a wide range of current density (1–30 mA cm^{-2}),

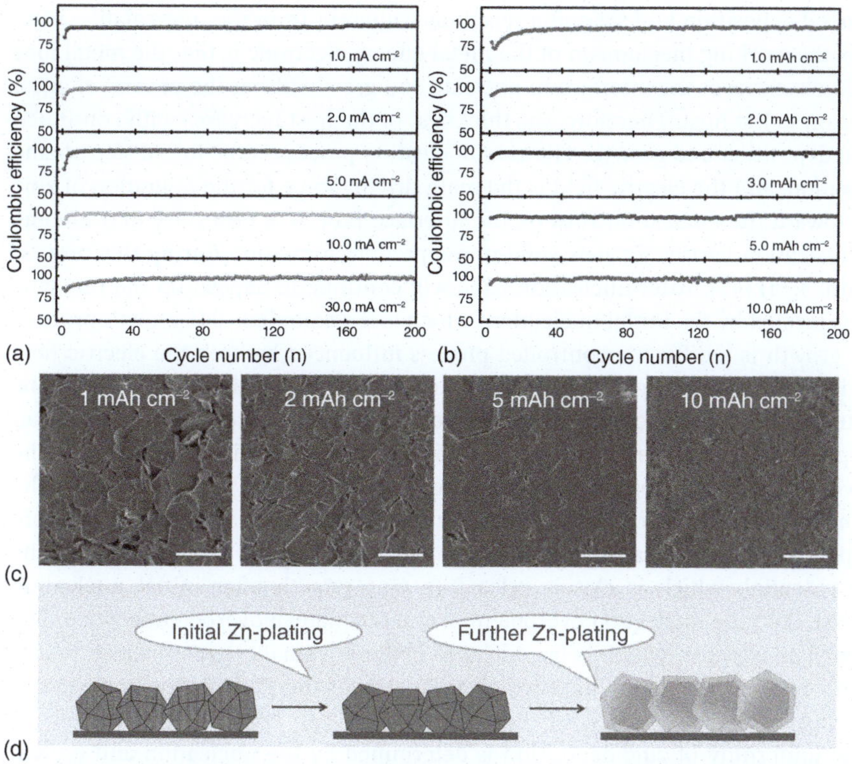

Figure 4.4 (a) CE of zinc plating on ZIF-8-500 at different current densities. (b) CE and (c) SEM images of zinc plating on ZIF-8-500 at different capacities. (d) Schematic diagrams of zinc plating on ZIF-8-500. Source: (c) Reproduced with permission from Wang et al. [39], © 2019/Elsevier.

deposition capacity (1–10 mAh cm^{-2}), and cycling stability (Figure 4.4a,b). There are always smooth surfaces with uniform morphology on ZIF-8-500 at different plating capacities (Figure 4.4c), which indicates that ZIF-8-500 electrodes are promising for the development of high-capacity zinc anodes. Such excellent plating/stripping performance results from the unique structure of ZIF-8-500. During the initial stage of electroplating, zinc tends to deposit in and/or on the ZIF-8-500 particles in the electrode because ZnO formed by annealing is a favorable site for nucleation. With continued electroplating, the zinc-plated ZIF-8-500 particles can serve as a deposition substrate with uniform zinc nuclei, which results in uniform and compact zinc deposition morphology (Figure 4.4d). When the ZIF-8-500 anode after zinc plating is coupled with an activated carbon or iodine (I_2) cathode, both hybrid capacitors and full cells exhibit enhanced electrochemical performance.

4.2 Surface Modifications

As mentioned above, zinc metal deposition involves two interfaces: the zinc–electrolyte interface and the host–zinc interface. Both interfaces are the main sites for metal deposition and stripping reactions. Different from the intercalation electrode, the working mechanism of the metal energy electrode is that the metal ions obtain electrons at the interface to nucleate and grow without the process of guest diffusion in the host. Therefore, the thickness of the host increases with continued metal deposition and growth. The zinc nucleation process is the initial step of zinc deposition, and the electric field is the main driving force for the migration of zinc ions toward the nucleation sites on the interface [40]. It is necessary to overcome the nucleation barrier for new solid-phase product formation during this process (Figure 4.5a) [41]. After nucleation, zinc will continue to deposit on the fine zinc grains formed by the initial nucleation, which is the complete zinc growth process. Zinc growth is a diffusion-controlled process influenced by both the electric field and ion concentration gradient [22, 39], and the quality of the deposition can be evaluated with two important parameters: the nucleation overpotential and plateau overpotential. The nucleation overpotential is defined as the difference between the tip potential and the subsequently stabilized plateau potential, which represents the thermodynamic cost of critical cluster formation at the interface during nucleation; the plateau overpotential is the difference between the plateau potential and the zero potential, which is closely related to zinc growth after initial nucleation [44, 45]. Both the nucleation overpotential and the plateau overpotential should be as small as possible, which is beneficial to obtaining smaller zinc nucleation and growth resistance, thereby improving the quality of the zinc plating coating and reducing the energy loss.

The uniformity of zinc deposition is determined by the nucleation and growth behavior of zinc, and the zinc nucleation process plays an important role in the subsequent zinc growth process. During the nucleation process, the random distribution of zinc ions affects the relatively balanced electric field at the interface, and the disturbed electric field affects the ion distribution in reverse [30, 46]. It can be seen from the theoretical analysis simulation (Figure 4.5b,c) that the nonuniform zinc seed particles formed during the nucleation process will strongly interfere with the electric field and ion distribution, and larger zinc grains can induce more severe inhomogeneous electric field and ion distribution. Electrons and ions tend to accumulate at the tips of the zinc seed particles, which accelerates the growth of zinc dendrites since the priority of zinc growth depends on the electric field strength and ion concentration (Figure 4.5d). Therefore, a flatter surface can minimize the inhomogeneity of zinc nucleation and growth [27, 42, 43].

It can be deduced that electrons and ions are the two most critical factors affecting the quality of zinc deposition on the reactive interface. Also, there are two main approaches to achieving dendrite-free zinc anodes: homogenizing the interfacial electric field and uniformizing the interfacial ion distribution. A uniform electric field facilitates the uniform nucleation process of zinc, which is crucial for the compactness of the interface between the current collector and zinc metal. In addition, it has been demonstrated in previous works that current density and deposition-specific capacity also affect the quality of zinc plating coatings (Figure 4.5e) [42]. Under the condition of low current density, there are fewer active sites on the surface for zinc nucleation, and zinc tends to grow at a dispersed

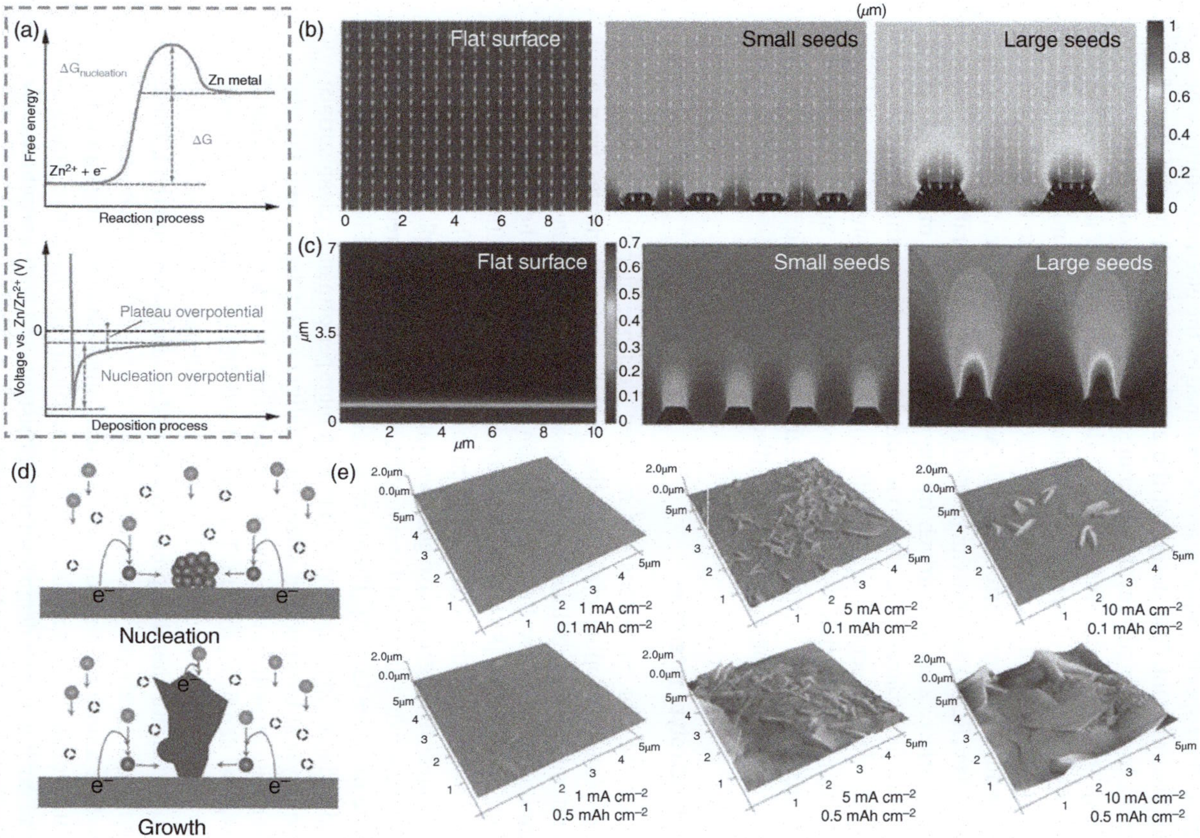

Figure 4.5 (a) Nucleation barrier and voltage curve of zinc deposition. Simulations of (b) electric field and (c) ion distribution on zinc anode surface under different conditions. (d) Schematic illustration of zinc dendrites. (e) AFM images of zinc foil after plating at different current densities and capacities. Source: (a, b) Adapted from Zhang et al. [41] with permission. Copyright 2017 American Chemical Society. (c) Reproduced from Yang et al. [42] with permission. Copyright 2019 Wiley-VCH. (d) Yin et al. [43]/with permission. Copyright 2019 Wiley-VCH/John Wiley & Sons. (e) Reproduced from Yang et al. [42] with permission. Copyright 2019 Wiley-VCH.

and large size. Zinc ions with sufficient energy at large current densities can be deposited at arbitrary sites along the entire interface. However, the higher deposition-specific capacity brings about the adverse effect of selective growth, which induces the formation of zinc dendrites. It should be noted that the current density and deposition-specific capacity are determined by the electrical equipment and operating environment. The design of the zinc anode should be capable of steady operation in a wide current density range and a high deposition-specific capacity [47, 48]. Based on such considerations, the design principle for stabilizing the zinc anode by tuning the interfacial electronic behavior is not only to homogenize the electric field during zinc deposition and stripping but also to find some approaches to reduce the local current density. As for ion regulation, the classical electric double-layer (EDL) structure can be used to describe the migration behavior of ions at the interface. Zinc ions driven by the electric field first migrate to the diffusion layer on the surface of the electrode and then transfer to the low-energy barrier sites in the dense layer through 2D diffusion to obtain electrons for the zinc nucleation reaction [49]. In this section, the specific modification strategies and intrinsic mechanisms for the current collector–zinc metal interface and the zinc metal–electrolyte interface engineering will be discussed in detail.

4.2.1 Zinc–Electrolyte Interface

It is widely accepted that the porous and ramified electrode morphology formed after repeated plating/stripping cycles of zinc metal anodes originates from a variety of fundamental instabilities, either physical (e.g. mass transport limitations and nonuniform electric field distribution), chemical (e.g. heterogeneous solid-electrolyte interphases [SEIs]), or the aforementioned interactions [50, 51]. Furthermore, zinc deposition is highly reactive and very unstable in aqueous electrolytes because hydrogen or oxygen generated from water splitting cannot precipitate on the anode surface to form an SEI as a mediator for reconciling the zinc–electrolyte interface [52, 53]. Therefore, other strategies should be considered to modulate zinc deposition. The first is to build a high-performance, clean surface to remove the physical barriers that affect the quality of zinc deposition, thereby preventing dendrite growth in the separator or guiding the uniform deposition of zinc to alleviate the dendrite problem. The second is building an artificial interface layer that is similar in function to SEI. There are two main methods for constructing the interface layer: in situ chemical pretreatment and ex situ coating method. These additional protective layers can alleviate the direct contact of water with the zinc anode and induce a uniform ion flow, thereby inhibiting surface hydrogen evolution, chemical corrosion, and dendrite growth. The principles of these two strategies to improve the quality of zinc deposition are consistent, and both are achieved by tuning the electronic and ionic behavior at the zinc–electrolyte interface.

4.2.1.1 Design of High-Performance Surface

Physical barriers to uniform zinc deposition exist on commercial zinc foils, such as native passivation layers consisting of oxides and carbonates, manufacturing-induced surface defects (scratches, cracks, and crease lines) from rolling and coiling, and inhomogeneous interfacial charge distribution on the polycrystalline surface [54–56]. Note that a problem usually corresponds to multiple solutions, and a strategy can also solve multiple problems. Different questions and approaches are interrelated. In this section, high-performance surface engineering and its intrinsic mechanisms will be classified and analyzed in detail.

4.2.1.2 Electrochemical Protocol to Uniformize Surface

It is known that 2D nucleation and the growth of deposits require a sufficiently large overpotential to form a high nucleation density on the electrode surface since the nucleation density is strongly dependent on the overpotential. Achieving 2D nucleation in conventional electrolytes is also a strategy to address 3D nucleation-induced dendrite growth. For instance, a 10 ms-long potentiostatic deposition pulse was applied during the first nucleation process in lithium metal batteries using a 0.02 M $LiPF_6$ electrolyte to obtain 2D nucleation and growth over the entire surface [57]. In this case, a large number of homogeneously distributed nuclei can be obtained, driven by high potentials, which allow lithium to grow uniformly under normal galvanostatic conditions. This electrochemical pulse protocol to stabilize metal deposition is also applicable to redox reactions in alkaline zinc metal anodes. During the reduction of ZnO to Zn, the concentration of $Zn(OH)_4^{2-}$ near the electrode surface is depleted, which accelerates the formation of dendrites. When pulsed direct current or alternating current charging is introduced after a brief $Zn(OH)_4^{2-}$ reduction period, the $Zn(OH)_4^{2-}$ concentration at the electrode surface can be re-established during this relaxation time, thereby eliminating competition from the adjacent site for a limited amount of $Zn(OH)_4^{2-}$, and achieving uniform growth on the entire electrode surface [58].

In weakly acidic electrolytes, zinc deposition stability also depends on the applied current density since a space-charge model was built to describe the linkage of dendrite formation time (Sand's time, τ) to the applied current density (J) [59, 60]. τ is inversely proportional to J^2 ($\tau \propto 1/J^2$), which implies that zinc dendrites are easier to form in a short time at high J. However, high J can accelerate the nucleation rate (νn) thermodynamically, leading to uniform deposition in the initial stage [22, 61]. The theoretical relationships of τ–J, and νn–J are shown in Figure 4.6a. Hou et al. [62] applied short galvanostatic pulses of high J during the initial zinc deposition process to generate sufficient nuclei, inducing uniform metal deposition at standard J during the subsequent deposition process. It is well known that the stripping step of the Zn anode always takes precedence over deposition in ZIBs matched to zinc-free cathodes. Therefore, it is important to control the first stripping step, as the stripping step creates a surface on which subsequent deposition steps will take place. At low current densities, zinc tends to strip at specific sites with high activity due to the limited electric driving force, resulting in a rough surface with dispersed cavities. A fundamental principle of energy barrier-dominated surfaces is proposed to describe the heterogeneous stripping phenomenon [63]. Under the limited driving force induced by the low current density, zinc is preferentially stripped in

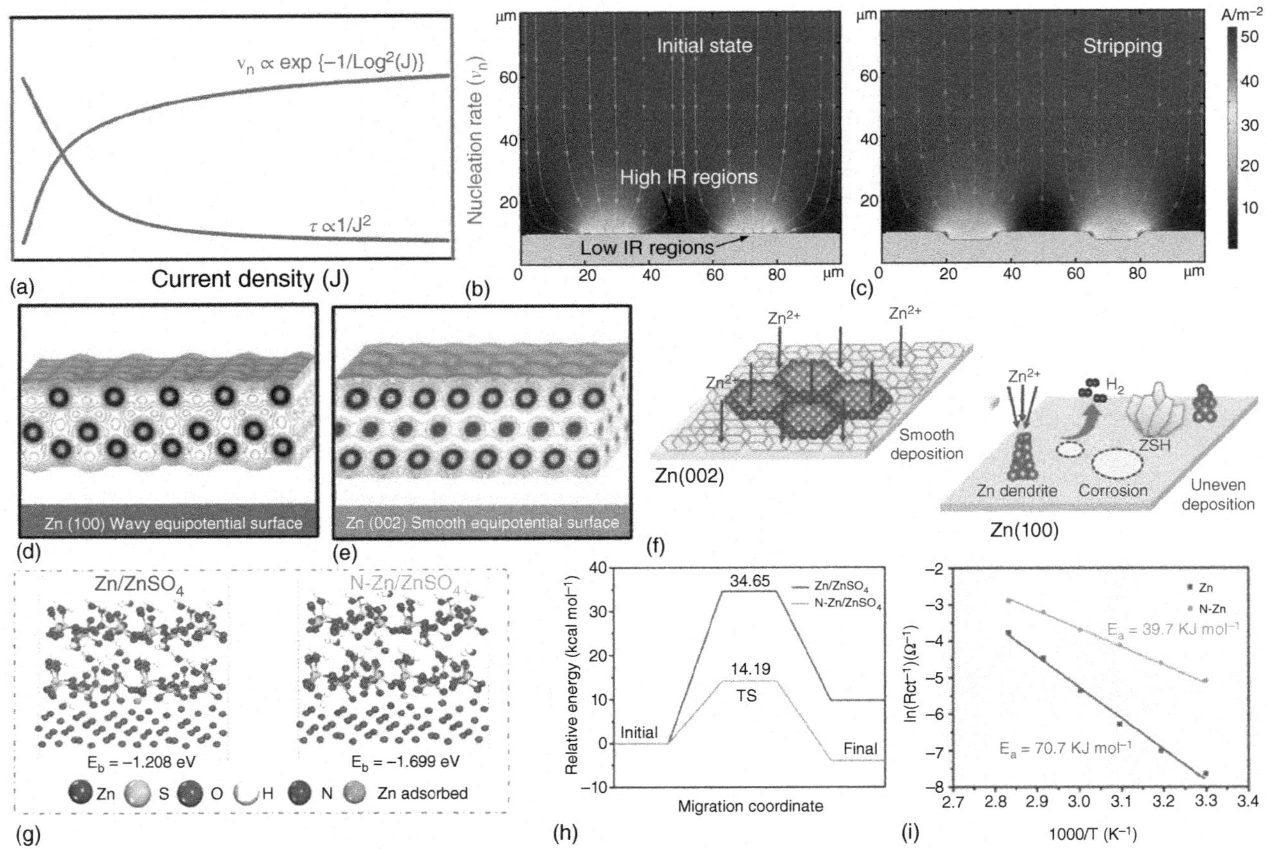

Figure 4.6 (a) Sand's time and nucleation rate are impacted by applied current densities. Simulation of an electric field on a zinc surface with different interface resistances at (b) initial and (c) stripped states. Surface atomic arrangement and electron equipotential plane of (d) zinc (100) and (e) zinc (002). (f) Schematic diagrams of zinc deposition on (002) and (100) planes of zinc. (g) Theoretical calculation of binding energy of zinc ions on bare zinc and N-doped zinc surface. (h) The migration barrier energy of zinc ions from bare zinc and N-doped zinc surfaces to electrolytes. (i) The activation energy of bare zinc and N-doped zinc electrodes. Source: (a) Reproduced from Hou et al. [62] with permission. Copyright 2021 Wiley-VCH Verlag; (b, c) Reproduced from Yang et al. [63] with permission. Copyright 2022 Elsevier; (d–f) Zhou et al. [55]/Reproduced with permission from John Wiley & Sons; (g–i) [63]/John Wiley & Sons/Licensed under CC BY 4.0.

the low-energy barrier region at the interface. However, when the applied overpotential is higher than the highest energy barrier over the region on the interface, an overall uniform stripping can be achieved (Figure 4.6b,c). In our previous work, the zinc deposition/stripping behavior of commercial zinc foils was systematically evaluated at a wide range of current density. A homogeneous inner surface with high activity is obtained by electro-stripping at a current density of 100 mA cm^{-2} or higher. More homogeneous surfaces with more exposed inner surface areas can be achieved at higher stripping current densities, but it comes at the expense of increased energy consumption and the risk of water decomposition. Based on the above consideration, a suitable stripping current density of 150 mA cm^{-2} is selected for the homogeneous interface reconstruction of the commercial zinc foils. The reconstituted zinc surface exhibits overall zinc deposition and highly reversible cycling stability. The electrochemical stripping activation protocol was directly applied to zinc-ion hybrid capacitors through the first large current discharge for the fresh capacitors since the capacitors can operate at wide current densities, leading to significant improvements in cycling stability.

4.2.1.3 Physically and Chemically Polished Surface

The manufacturing process of zinc foil involves rolling and coiling because zinc is a relatively soft metal (Mohs hardness 2.5; Brinell hardness 327–412 MPa). Zinc foil deformation is inevitable during rolling, such as scratches, microcracks, and fold lines, and the surface becomes rougher during continuous grain deformation. Note that the local current density can be enhanced on rough surfaces, which is favorable for the rapid formation of zinc protrusions. Furthermore, a cutting step is required to obtain zinc electrodes of specific shape and size, resulting in exposure of more fresh edges with high activity. These intrinsic surface defects of zinc foils can be eliminated by facile physical polishing [54]. The polished zinc foil exhibits a flat topography, which is conducive to more uniform electric field distribution at the electrode and electrolyte interface and more uniform nucleation and growth. Furthermore, zinc foil tends to react with oxygen in the air to form a chemically inert passivation layer during storage due to the high chemical reactivity of zinc metal. However, these passivation layers increase the nonuniform surface electrochemical impedance, resulting in nonuniform zinc plating/stripping [65]. The chemical etching method is another strategy to remove the uneven passivation layer from surfaces [66]. Remarkably, the zinc foil exposes a surface with numerous grain boundaries and defects after brief acid etching. Such surfaces are more reactive than bulk zinc structures. Deep etching is further performed to construct a 3D ridge-dominated surface. As a result, the 3D ridged zinc electrode with a high exposed active area not only increases the accessible surface area of the electrode to the electrolyte to facilitate zinc dissolution and deposition but also homogenizes the ion flow at the interface to reduce the possibility of dendrite growth. In addition to these surface topographical defects and inert layers, metal processing may also result in an uneven distribution of residual stress on the surface, which could induce localized whisker growth [67]. It is necessary to consider the effect of surface stress distribution on zinc deposition when developing high-performance zinc metal electrodes.

4.2.1.4 The Textured Surface

Zinc metal acts as a hostless electrode, indicating that full deep discharge is impossible. However, to truly compete with state-of-the-art rechargeable batteries, the pre-stored metal should be utilized as much as possible during each discharge and charge cycle. This means that the road to practical zinc metal batteries requires the fabrication of extremely thin electrodes and the preservation of full electrochemical activity over the lifetime of the battery. However, most commercial metal foils exhibit crystal heterogeneity with a widespread distribution of grain sizes and orientations. The morphology and electrochemical instability of the zinc foil are also factors that affect the failure of the battery during cycling. The dissolution rate of the metal depends on the exposed crystal facets [68, 69]. In other words, the orientation heterogeneity of the grains will be converted into morphological heterogeneity upon battery discharge [70]. The crystal facet with high activity can induce the preferential deposition of zinc and accelerate the formation of zinc dendrites. In addition, the atomic arrangement on the (100) facet of zinc crystals is wavy, which makes the (100) plane exhibit nonuniform interfacial charge density (Figure 4.6d). Conversely, there is a uniform charge density on the smooth zinc (002) facet due to the special atomic arrangement, which favors uniform zinc ion deposition (Figure 4.6e). A (002)-preferred zinc foil anode shows highly reversible and stable zinc plating/stripping behavior, as reported by Zhou et al. [55]. Combined with theoretical calculations, fewer side reactions occur on the (002)-preferred zinc anode surface owing to the higher hydrogen adsorption free energy of the (002) facet than the (100) facet (Figure 4.6f). Besides, the zinc metal spontaneously reacts with the zinc sulfate electrolyte to form ZHS through the consumption of the electrolyte and active zinc. It is found that the interfacial energy between the (001) face of ZHS and the (002) face of zinc is lower than that of other zinc facets, indicating the high affinity between the (001)-face ZHS and the (002)-face zinc [71]. In this way, when the (002)-preferred zinc metal foil is immersed in the zinc sulfate electrolyte, a uniform and dense ZHS layer dominated by the (001) facet spontaneously forms through an epitaxial growth mechanism. The (001)-preferred ZHS layer acts as an artificial SEI with high ionic conductivity, which enables zinc anodes with significantly improved corrosion inhibition and reduced hydrogen evolution.

4.2.1.5 The Plasma-Treated Surface

The plasma-processing technique is considered an efficient tool for material preparation or surface modification due to its low energy consumption and environmentally friendly features [72]. It is reasonable to optimize the zinc foil surface to change the original surface properties by plasma treatment. Some active groups in the electrolyte additives play important roles in the zinc deposition process, suggesting that the direct introduction of free radicals on the surface of the zinc foil may affect the zinc plating performance [73]. It is proven that zinc ions have stronger interactions with nitrogen atoms than with zinc atoms (Figure 4.6g). This strong interaction can reduce the zinc nucleation overpotential. Deep nitrogen doping on the zinc foil surface can be achieved by bombardment with high-energy ionized nitrogen. The doped zinc anode surface provides uniform N-active sites, which induce uniform Zn-ion deposition [64]. Furthermore, there is a lower ion diffusion barrier and improved desolvation on N-doped Zn surfaces compared to bare Zn anodes (Figure 4.6h,i).

4.2.1.6 Introduction of Interface Layer

The purpose of constructing the surface coating is to achieve uniform zinc nucleation and flat zinc deposition by regulating the Zn^{2+} ion flux, thereby improving the interfacial stability and cycle life of the zinc anode. In general, an effective interfacial layer must be designed to provide the following characteristics: (i) chemically and electrochemically inert, avoiding reactions with electrolytes; (ii) mechanically strong enough to resist volume change and dendrite growth; and (iii) good electrolyte wettability to facilitate ion diffusion. According to previous reports, various materials have been used as interfacial layers to achieve high-performance zinc anodes [39, 74–76]. Based on their conductive properties, these materials can be divided into insulating layers, electron-oriented layers, ion-oriented layers, and composite layers. The research progress on these interface coating materials is described in detail below.

4.2.1.7 Insulating Layer

Inorganic materials with neither electrical nor ionic interaction with zinc are suitable as insulating layers at the zinc–electrolyte interface, such as $CaCO_3$, TiO_2, and Al_2O_3 [75, 77, 78]. Insulating coatings with high porosity can be easily penetrated by aqueous electrolytes, leading to a relatively uniform electrolyte flux and zinc plating rate across the entire zinc foil surface. Furthermore, there is a large potential change in the vertical direction of the coating due to its high-insulating properties. As a result, zinc ions can be deposited on the surface of the zinc foil with a low potential, which induces a bottom-up deposition process rather than preferential deposition on the protrusion tips (Figure 4.7a) [84]. Besides the intrinsic insulating properties of inorganic materials, the metal affinity of the coating also plays an important role in guiding the metal deposition process. Commercial spherical TiO_2 powder without facet orientation is used as the zinc anode coating, resulting in cell failure after several cycles because the deposited zinc flakes pierce the coating [79]. Theoretical calculations reveal the differences in the zinc metal affinity of different facets in commercial TiO_2 (Figure 4.7b,c), which may be a key factor in inducing cell failure due to the inhomogeneous ion flux induced by the zincophilic facet. Accordingly, in our previous work, zincophobic TiO_2 with highly exposed (001) facets was prepared by hydrofluoric acid-assisted hydrothermal method and applied as a zinc anode protective layer [79]. This (001) TiO_2 layer can effectively prevent the vertical growth of zinc dendrites and stabilize the interface between the anode and the electrolyte.

The slow ion transport kinetics in the interfacial double layer are also responsible for the low plating/stripping efficiency. Bivalent zinc ions tend to form stronger and tighter solvation sheaths with water molecules on the host surface in contrast to monovalent lithium or sodium ions [84, 85]. It has been demonstrated that improved battery stability and prolonged battery life are realized by tuning the inner

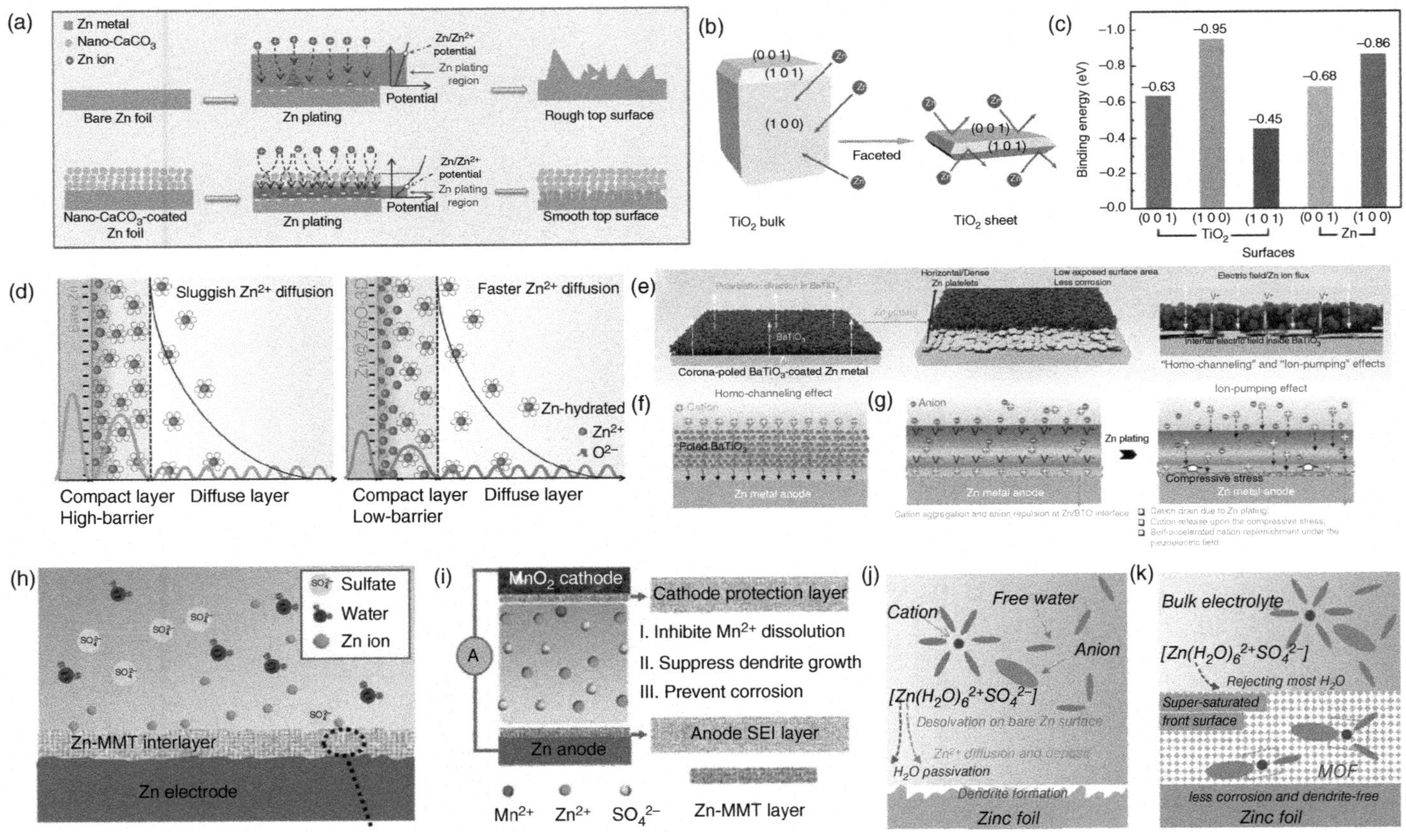

Figure 4.7 (a) Schematic illustrations of morphology evolution for bare and surface-coated zinc foils during plating/stripping. (b) Schematic illustrations of the interaction between zinc and different TiO_2 crystal facets. (c) Calculated binding energy of zinc atoms on different facets of zinc and TiO_2. (d) Electric double-layer structure in contact with the bare zinc anode and interfacial modified zinc anode. (e) Schematic illustration of zinc deposition morphology on zinc surface-coated polarized ferroelectric coatings. (f) Schematic illustration of ion diffusion pathways in ferroelectric coatings. (g) Schematic diagrams for fast cation replenishment at anode–electrolyte interphase by "ion-pumping effect" during the metal plating. (h) Schematic diagrams of the Zn-based MMT interface for inhibiting side reactions and dendrite formation. (i) Schematic illustrations of the full cells assembled by MMT-protected cathode and anode. Schematic illustrations of surface evolution for (j) bare and (k) MOF-coated zinc foils in aqueous electrolyte. Source: (a) Reproduced from Kang et al. [75] with permission. Copyright 2018 Wiley-VCH Verlag; (b, c) Reproduced from Zhang et al. [79] with permission. Copyright 2020 Springer Nature; (d) Reproduced from Xie et al. [80] with permission. Copyright 2019 The Royal Society of Chemistry; (e–g) Adapted from Zou et al. [81] with permission. Copyright 2021 Wiley-VCH Verlag; (h, i) Reproduced from Yan et al. [82] with permission. Copyright 2021 Wiley-VCH Verlag; (j, k) Reproduced from Yang et al. [83] with permission. Copyright 2020 John Wiley and Sons Ltd.

Helmholtz surface in EDLs through the spontaneous formation of thin interfacial layers by foreign molecules in non-aqueous/aqueous lithium-ion batteries [86–88]. Accordingly, 3D nanoporous ZnO structures with a thickness of ~5 μm were grown in situ on the surface of zinc foil by liquid-phase synthesis method [80]. This novel artificial interface-protected zinc anode exhibits fast zinc ion deposition kinetics and low desolvation energy at the EDL interface, improving zinc plating/stripping reversibility and suppressing hydrogen evolution (Figure 4.7d). When zinc ions are deposited on pristine zinc surfaces with protuberances, electroplating is more likely to occur at the protuberances according to the lightning rod theory and Ohm's law [89, 90]. However, if the uneven protrusions on the surface are transformed into array-like protrusions, the local current density on the surface can be significantly weakened [91]. A functionalized ZnO passivation layer was coated on the zinc surface with an array of zinc hexagonal pyramids, where the thickness of the top of the zinc pyramid is much larger than that of the sides and bottom, resulting in zinc deposition on the sides and bottom due to relatively low interfacial resistance. Such interface resistance gradient design plays an important role in eliminating tip deposition.

Polarization effects in dielectric materials can be exploited to tune zinc ion deposition behavior. Typically, ZrO_2 with a high dielectric constant as a zinc anode coating is capable of separating the charge at the interface over a considerable distance through Maxwell–Wagner polarization (known as "space charge polarization" or "interface polarization"), thereby achieving interfacial charge redistribution [92, 93]. To achieve electric field regulation at the interface of the zinc anode, a dielectric coating with spontaneous polarization was introduced into the zinc anode. Under the external electric field, the polarized dipoles of the coating build up a uniform electric field, which enables the orderly migration of zinc ions during the zinc plating/stripping process [94]. However, the rapid supply of ions is still insufficient due to the limited polarization of the ferroelectric layer along the vertical direction of the anode. Pre-polarized ferroelectric coatings for zinc anodes instead of pristine ferroelectric coatings are expected to maximize the piezoelectric capability of the coating along its polarization direction, which enables more uniform and fast zinc ion migration through homogeneous channel effects and ion-pumping effect (Figure 4.7e–g) [81].

To accelerate interfacial zinc ion transport, a compound coating composed of zinc and nonmetal elements was constructed to change the charge distribution at the interface, such as ZnS, ZnP, and ZnSe [95–97]. There is charge transfer between zinc and these compounds due to the bonding interactions between the non-metallic elements and the zinc atoms. The unbalanced charge distribution not only enhances the diffusion of zinc ions at the interface but also increases the adhesion between the coating and zinc [95]. The additional nonmetallic elements in the coating facilitate zinc ion transport and lower the reaction energy barrier during electroplating/stripping. The zinc anode with a ZnP protective layer can achieve a high DOD (82%) with a capacity of 48 mAh cm^{-2} at 15 mA cm^{-2} for 100 cycles [96].

Considering their low cost and environmentally friendly properties, naturally available nanoscale clay minerals have received particular attention in energy-storage applications over the past few decades. These clay minerals possess suitable nanoporosity and high ionic conductivity, which makes them suitable as coating materials for zinc anodes. For example, a porous kaolin coating with a sieve element function and uniform size can effectively limit the growth of zinc dendrites by restricting 3D diffusion [98]. Furthermore, multifunctional zinc-based montmorillonite (MMT) can simultaneously improve the cathode and anode performance of AZIBs owing to its unique layered structure [82]. The negatively charged MMT sheet surface facilitates uniform zinc plating/stripping, and the corrosion reaction is well suppressed by anchoring the anions through the edge with the positive charge of the sheet (Figure 4.7h). When the MMT layer was introduced into the interface between the MnO_2 cathode and the electrolyte, the dissolution of the cathode was effectively suppressed (Figure 4.7i).

Although a well-established porous coating can guide uniform ion flow to eliminate zinc dendrite growth, the contact of active water in the electrolyte with the zinc anode is unavoidable. Parasitic water splitting occurs because of the inherent thermodynamic instability of zinc under hydrous conditions. Furthermore, the solvation of Zn^{2+} located in the EDL requires overcoming the energy barrier to desolvate and release a large number of electrochemically reactive water molecules. The active water molecules close to zinc foil anodes are more easily decomposed than those in the bulk electrolyte, resulting in hydrogen evolution and local pH change [99]. As a result, the alkaline environment accelerates the corrosion of the zinc surface to form non-reactive by-products. MOF with tiny pores allows zinc ion transport but rejects large-sized charged ion complexes. Therefore, such a MOF coating can not only achieve uniform zinc deposition through the channels but also desolvate the solvent sheath at the electrolyte-coating interface under the action of the field, which enables the MOF channel to become supersaturated with the electrolyte solution, resulting in less corrosion and dendrite-free surfaces (Figure 4.7j,k) [83]. In addition, constructing an organophilic MOF interfacial layer with a hydrophobic organic electrolyte at the zinc anode interface can also effectively hinder water splitting [100].

4.2.1.8 Electron-Oriented Layer

Zinc dendrites originate from nanoscale nuclei, which may contain only tens or hundreds of zinc atoms. However, these tiny nuclei can grow into catastrophic dendrites during the long-term operation of the battery due to inhomogeneous zinc growth. Taking nuclei into consideration, uniform zinc nucleation plays a key role in dendrite suppression. The nucleation energy barrier significantly affects the quality of formed zinc nuclei, and their nucleation radius is inversely proportional to the overpotential [101]. The high energy barrier to nucleation promotes the formation of small zinc nuclei, which alter the roughness of the deposited layer and enhance the electric field around it, both deteriorating uniform zinc deposition [102]. Therefore, a low and constant nucleation energy barrier on the zinc surface is required to ensure uniform deposition. Introducing an artificial conductive protective layer with a nucleation agent can achieve a lower energy barrier and preferential nucleation on it. For example, gold coatings with high curvature can preferentially adsorb zinc ions and homogenize the subsequent nucleation process to produce the growth of zinc flake arrays rather than large and nonuniform zinc dendrites [103]. In addition, $CuZn_5$ and $AgZn_3$ alloy buffer layers with high zinc affinity can be formed during the initial galvanizing process on anodes coated with ultra-thin metallic Cu or Ag interfacial layers, which facilitate uniform nucleation and growth of zinc [104].

Although the metals Au and Ag are excellent zinc-nucleating agents, their high cost limits large-scale applications. Considering the cost issues, a lot of research has been focused on non-precious metals to decorate the surface of zinc anodes. Metals with higher redox potentials than zinc (−0.76 V vs. SHE) are intensively studied since these protective layers can be easily obtained by facile chemical displacement reactions, such as the metals In (−0.338 V vs. SHE) and Sn (−0.14 V vs. SHE) [105, 106]. When the metallic zinc foil is immersed in an $InCl_3$ solution, a spontaneous substitution reaction occurs to form a thin In-protective layer on the zinc surface. The formed In layer can act as both a corrosion inhibitor and a nucleating agent due to its high hydrogen evolution overpotential and zinc metal affinity, which results in dendrite-free and corrosion-resistant zinc anodes [106]. Furthermore, a textured Sn coating was constructed by controlling the displacement reaction time in the Sn salt solution. The high zinc affinity nature of Sn provides a lower deposition energy barrier, which promotes the deposition kinetics of zinc. The entire zinc deposition is achieved on the Sn-textured surface due to the improved wettability of the Sn-texture coating [107]. However, we found a novel mechanism of dendrite formation induced by 2D diffusion of surface atoms in addition to the inhomogeneous ion field and electric field effects [108]. Before electro-crystallization, zinc atoms on rough zinc surfaces with low 2D diffusion barriers tend to locally aggregate and nucleate, which in turn evolve into tiny crystallites [109]. In the subsequent deposition, zinc ions in solution preferentially adsorb on the tips of these crystallites and grow into dendrites to minimize the surface energy. From the surface 2D diffusion perspective, the introduction of a copper nitride interface with a high diffusion energy barrier and strong zinc interaction on the zinc surface can induce the formation of spacious atomic clusters rather than localized atomic clusters, leading to whole zinc nucleation and dense zinc growth [108]. Therefore, the limited 2D atomic diffusion can achieve uniform and complete nucleation in the early stage of electroplating, thereby eliminating the formation of zinc protrusions that accelerate the growth of zinc dendrites.

4.2.1.9 Ion-Oriented Layer

Polymers with polar groups such as C=O and N—H bonds can provide abundant adsorption sites and transfer Zn^{2+} along the polymer chain to the reaction interface [110, 111]. The fast pathways obtained for ion diffusion can tune the uniform distribution of Zn^{2+} ions at the molecular scale [112]. Nonconductive polymers with polar groups can promote uniform zinc deposition between the interfacial layer and the zinc anode. The zinc anode was coated with a polyamide layer by the doctor-blade method to control the zinc deposition behavior. This polymer coating has strong coordination with zinc, which could act as a guiding layer to regulate ion migration and inhibit the contact of the zinc anode with the aqueous electrolyte (Figure 4.8a,b) [49]. As shown in Figure 4.8c, a large number of visible dendrites grow on the bare zinc anode after deposition, and the zinc with a coating presents a flat surface without dendrite formation. This dendrite-free behavior on polyamide-coated zinc can be attributed to the confined 2D surface diffusion demonstrated by chronoamperometry (Figure 4.8d). The Zn^{2+} ion absorbed at bare zinc tends to diffuse laterally along the surface and accumulate at sites with a low nucleation energy barrier. The 2D diffusion of Zn^{2+} is limited by the additional energy barrier of lateral motion since the polyamide layer is tightly bound to zinc [115]. As a result, zinc nucleation sites are increased, and further zinc growth is uniform and dense because Zn^{2+} is forced to nucleate at the initial adsorption sites rather than at low surface energy sites. A similar polymer glue coating composed of polyethylene oxide (PEO) and lithium bis(trifluoromethane sulfonyl)imide can achieve the same lateral diffusion confinement effect [113]. Furthermore, the ether-based oxygen atoms in PEO can accelerate the migration of Zn^{2+} ions, which kinetically favors the reduction and adjacent deposition of Zn^{2+} ions to generate more nucleation sites (Figure 4.8e).

Polar atoms (i.e. N and O) with lone pairs of electrons in the coating can reduce water molecules in the solvent sheath of hydrated zinc ions through coordination, which suppresses random zinc ion diffusion at the zinc–electrolyte interface [116, 117]. A self-healing ion regulator (SIR) with abundant active groups is designed as a protective layer for the zinc anode [114]. The strong interaction between the amide group in the SIR and the zinc ion reduces the number of water molecules desolvated at the interface, which effectively alleviates the water-induced corrosion. The zinc ion diffusion path can be regulated by ion absorption sites on the SIR chain for uniform nucleation, thereby alleviating the formation of dendrites and dead zincs (Figure 4.8f). More importantly, the hydrogen bonds between the SIR chains endow the coating with the dynamic capability of in situ repair of cracks induced by the electrochemical cycling process (Figure 4.8g). The strong coordination effect between zinc ions and strongly polar groups ameliorates the common problem of large overpotentials caused by polymer coatings. In addition, the presence of abundant oxygen-containing functional groups can enhance the interfacial adhesion of the newly added interface between the electrolyte and the polymer layer, further reducing the interfacial resistance.

4.2.1.10 Complex Layer

The abovementioned coatings have their advantages and limitations. Inorganic materials form a porous coating with a small amount of binder, which effectively uniformizes the ion flow. However, it is easy to fracture and detach from the zinc bulk after repeated electroplating/stripping due to its brittleness, thereby losing dendrite suppression ability. Furthermore, most organic polymers are hydrophobic, which raises the nucleation barrier for zinc [49]. Therefore, designing composite coatings to hinder the interfacial dendrite growth and corrosion reactions of zinc anodes for long-life batteries is a novel strategy. Organic Nafion membranes, used as separators with selective cation transport in aqueous batteries [118, 119],

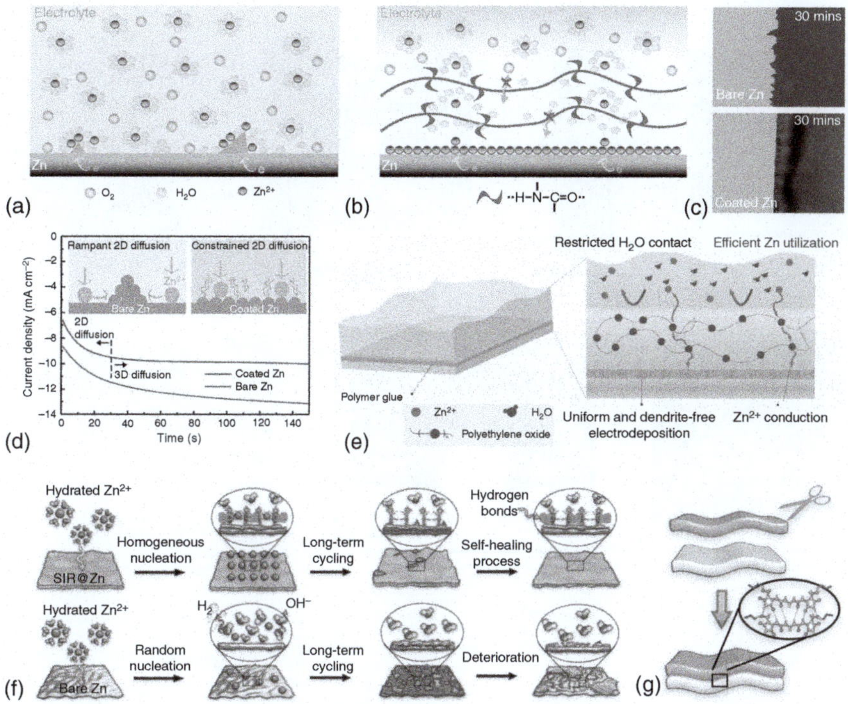

Figure 4.8 Illustration for zinc deposition on a (a) bare zinc anode and (b) polyamide layer-coated zinc anode. (c) Optical morphologies of zinc deposition on zinc anodes with and without coating. (d) Chronoamperograms of bare zinc and coated zinc anodes in symmetric cells. (e) Schematic diagrams of inhibiting corrosion and passivation of zinc foil by polymer glue coating. (f) Schematic diagrams of structural evolution for SIR-coated zinc and bare zinc electrodes during cycling. (g) Schematic illustration of self-healing process for SIR. Source: (a–d) Reproduced from Zhao et al. [49] with permission. Copyright 2019: The Royal Society of Chemistry. (e) Reproduced from Jiao et al. [113] with permission. Copyright 2021: Wiley-VCH Verlag. (f, g) Reproduced from Du et al. [114] with permission. Copyright 2021 John Wiley and Sons Ltd.

consist of hydrophilic and hydrophobic regions, which seem to have the potential for dendrite protection and side reaction inhibition of zinc metal anodes. An organic–inorganic composite protective layer composed of Nafion and Zn-X zeolite was designed to tune the zinc deposition/stripping behavior. Zinc ions in the composite coating bridge the interface between Zn-X zeolite and Nafion to form a dense organic-inorganic interface, which is beneficial to the transport of ion and blocking of water molecules. Generally, the Nafion layer cannot completely hinder the shuttle of small anions and water molecules due to the existence of hydrophilic channels (Figure 4.9a). However, when the Zn-X zeolite is incorporated, the hydrophilic channels are blocked by the zeolite particles, as shown in Figure 4.9b. Zn ions can move freely at the inorganic-organic interface owing to the negatively charged framework and small pore size of the Zn-X zeolite, while other anions and free water are blocked, thereby inhibiting hydrogen evolution and corrosion reactions [120].

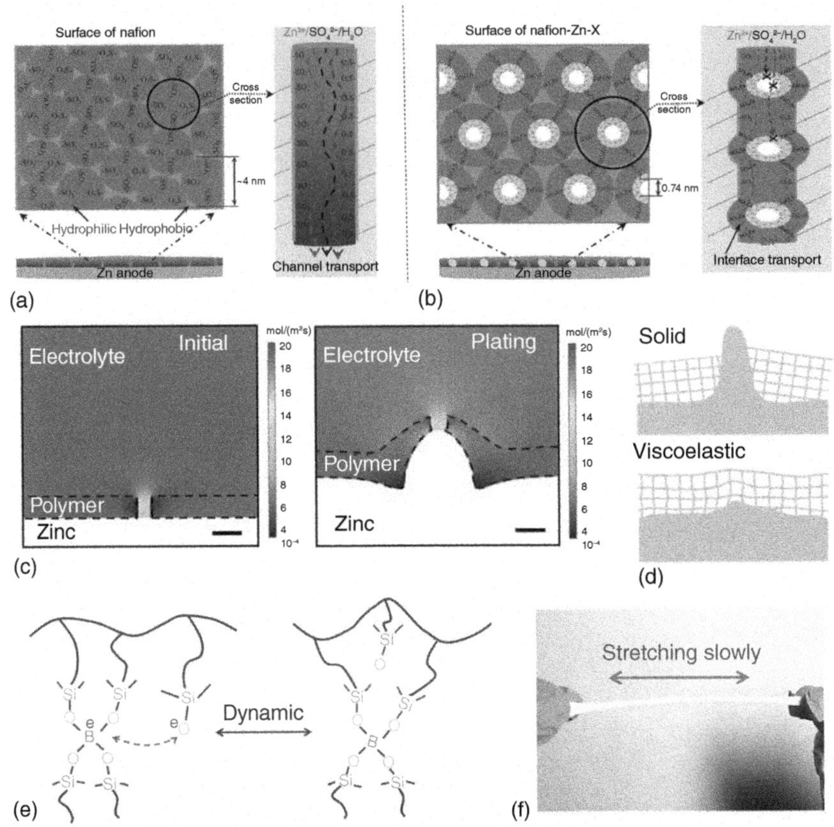

Figure 4.9 Schematic illustrations of ion transport in (a) Nafion and (b) Nafion-Zn-X layer. (c) Simulation of the dendrite evolution on the surface with pinhole. (d) Schematic diagram of action mechanism of viscoelastic and solid coatings. (e) Illustration of dynamic change process for B—O bond in PDMS. (f) Optical photos of PDMS/TiO_{2-x} in stretching process at 0.5 cm s^{-1}. Source: (a, b) Reproduced from Cui et al. [120] with permission. Copyright 2020 John Wiley and Sons Ltd; (c–f) Adapted from Guo et al. [74] with permission. Copyright 2022 Wiley-Blackwell.

The drastic volume change of the zinc anode during charge-discharge cycling can lead to instability of the electrode–electrolyte interface (Figure 4.9c) [74]. However, ordinary rigid interfacial coatings cannot maintain conformal contact with the zinc surface, resulting in irreversible damage to the protective layer. Therefore, constructing a dynamic protective layer to accommodate the continuous volume change in the electrode and maintaining good contact with the electrode at all times can effectively suppress the uncontrollable zinc dendrite growth (Figure 4.9d) [76]. Poly(dimethylsiloxane) (PDMS) is a highly dynamic adaptable polymer due to the broken B—O bonds that can spontaneously reconnect dynamically (Figure 4.9e). When PDMS is combined with inorganic TiO_{2-x} particles with rich oxygen vacancies, the composites (PDMS/TiO_{2-x}) can enhance the fast and uniform transfer of zinc ions (Figure 4.9f). The shape of PDMS/TiO_{2-x} coating on the zinc anode can be adapted to changes in the electrode interface and volume. In other words, if the self-adapting coating perceives the formation of tiny protrusions on the surface, the coating would become hardened to induce uniform zinc deposition under the coating. Therefore, the elastic polymer coating can effectively block the direct contact of active water with the electrode and alleviate the interfacial volume change during electroplating/stripping. The incorporation of inorganic nanoparticles, such as TiO_2 and Si_3N_4, into elastic polymers can enhance the electrochemical performance of the coatings. These particles act as zinc deposition redirectors by providing favorable nucleation sites and modulating local currents, thus ensuring uniform and dense zinc deposition [121, 122].

4.2.2 Host–Zinc Interface

Zinc metal is a host-free anode that is dissolved at stripping and redistributed back as subsequent plating [13]. Pure anodes tend to fail rapidly when using pure zinc foil because the zinc grows randomly and cannot return to its original position, which leads to dendrites in some parts and cracks in others [42]. A host material is necessary to confine zinc growth and serves as a structural framework and conductive network. However, common carbon-based hosts still suffer from irreversible dendrite growth, hydrogen evolution, and side reactions due to their poor electrical conductivity and few nucleation sites, resulting in poor cycle life and low CE. Therefore, some further modifications to the host–zinc interface are required to facilitate the deposition of zinc for high-performance zinc metal anodes.

4.2.2.1 Using Uniform Conductive Host

Copper-based materials have been extensively studied as current collectors [123, 124]. Their high electrical conductivity can maintain a uniform interfacial electric field by rapidly transferring surface charges during zinc deposition. A flat surface is also required to uniformize the electric field distribution, which avoids charge accumulation at the tips of rough surfaces. Pristine copper foil and graphite paper with low surface area usually show poor deposition quality due to excessive local current density. Zinc continues to grow on the previously deposited small zinc seeds and eventually aggregates to form amplified dendrites [125].

The 3D design of the conductive host can increase its surface area to reduce the local current density and homogenize the interfacial charge distribution, thereby suppressing the formation of dendrites and generating a relatively flat plating surface [126]. However, electrodeposited zinc species prefer nucleation and grow into embryos at defect sites of a thermodynamically and kinetically favorable heterogeneous interface. These defects facilitate fast electron transport but deteriorate the inhomogeneous interfacial electric field distribution. Once inhomogeneous nuclei are formed, dendrites rapidly establish growth in these regions. If the surface of the substrate has many dispersive defects, the zinc deposits at the defects will grow as irregular zinc strips. The strip edges with strong curvature become new nucleation sites and subsequently evolve into a hierarchical network with continued plating. However, dendrites are still inevitable due to a limited number of defects. Therefore, designing defect-rich surfaces on conductive 3D frameworks to homogenize the interfacial electric and ionic fields is an effective strategy for solving the dendrite problem. For example, vertical carbon-coated defect-rich V_2O_3 nanosheets grown on 3D stainless steel (SS-V_2O_3@C) meshes were used as zinc hosts to achieve dendrite-free electroplating behavior [127]. During the zinc deposition on bare stainless steel meshes, zinc aggregates into irregular zinc blocks with hundreds of micrometers in size, accelerating the growth of zinc dendrites. In contrast, the carbon-coated defect-rich V_2O_3 nanosheets enhance electron transport and avoid charge accumulation, enabling the SS-V_2O_3@C substrate to maintain 3D morphology after zinc deposition. Therefore, rationally designing conductive scaffolds with rich defects as substrates for zinc deposition is an effective strategy to achieve high-performance anodes.

4.2.2.2 Building Zincophilic Sites

The uneven distribution of zinc ions at the interface also easily induces the formation of zinc dendrites. For the zinc deposition substrate, the uniform zinc ion distribution on substrate surfaces can be realized by enhancing the ability to adsorb and bind zinc (that is, zincophinicity). The high zincophinicity endows the current collector-zinc metal interface with a lower zinc nucleation energy barrier and more nucleation sites. Zinc nucleation on the copper surface is relatively uniform since the Cu—Zn solid solution layer with good zinc affinity is formed in situ during the plating process [13]. Furthermore, the construction of porous polybenzimidazole (PBI) nanofiber networks on the surface of copper substrates can provide uniform nucleation sites. The polar functional groups on the PBI can attract the Zn^{2+} ions in the electrolyte to promote the uniform transport of Zn^{2+} ions in the porous PBI through the electrokinetic effect, which effectively reduces the zinc nucleation overpotential [128]. Contrary to the strong interaction between zinc and copper surfaces, the adsorption effect of zinc ions on carbon surfaces is very weak, and the binding energy between zinc atoms and carbon surfaces is lower than that between zinc atoms and zinc surfaces, which indicates that zinc is more inclined to deposit on nucleated zinc seeds rather than on otherwise exposed carbon surfaces. As shown in Figure 4.10a, the deposited zinc cannot completely cover the surface of ordinary hollow carbon spheres (HCS), which makes subsequent zinc deposition preferentially on the inhomogeneous zinc deposits rather than on the carbon spheres, resulting in more severe dendrite growth. Theoretical calculations demonstrate that the introduction of nitrogen doping can change the interaction between zinc and carbon substrates from zincophobic to zincophilic. It can be seen from the electroplating curves (Figure 4.10b) that the nucleation overpotential of zinc deposition on nitrogen-doped hollow carbon spheres (N-HCS) is lower than that of pristine carbon spheres. The N-HCS substrate provides more spacious sites for zinc nucleation, promoting the formation of zinc clusters on the entire carbon surface [129].

In addition, introducing defects into carbon-based current collectors can also change the electronegativity and local dipoles of the surface sites, which enhance the zincophilicity of a substrate and guide the uniform nucleation of zinc in comparison with pure carbon [131]. Hee-Tak Kim et al. fabricated defect-rich 3D carbon

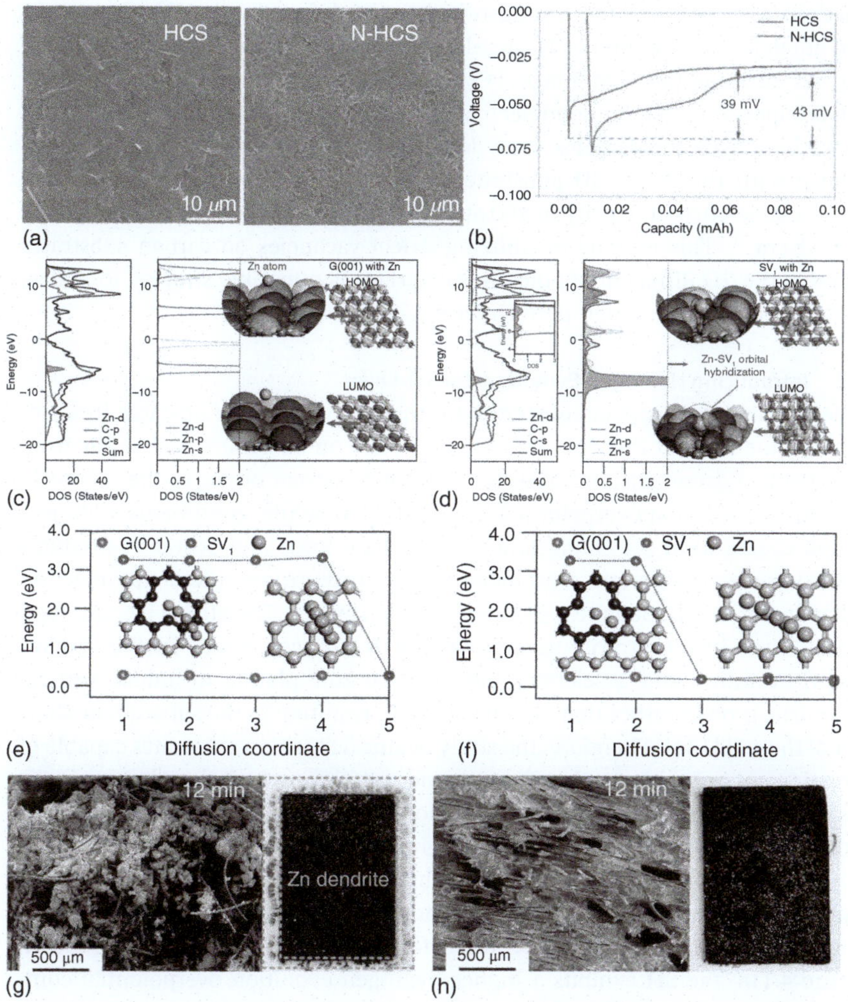

Figure 4.10 (a) SEM images and (b) voltage curves of zinc deposition on HCS and N-HCS. Calculated total density of states and partial density of states of the zinc adsorbed (c) G(001) complex and (d) the single vacancy defect. Changes in the energy profiles of Zn on the single vacancy defect-containing carbon surface with movement of the adsorbed zinc atom crossing (e) the C—C bond and (f) along the C—C bond. SEM images and corresponding photographs of zinc deposition on (g) pristine carbon and (h) defective carbon in zinc–bromine batteries. Source: (a) Reproduced with permission from Xie et al. [129], © 2021/John Wiley & Sons, (b) Reproduced from Xie et al. [129] with permission. Copyright 2021 Wiley-VCH Verlag; (c–f) Adapted from Lee et al. [130] with permission. Copyright 2020 Royal Society of Chemistry, (g–h) Adapted with permission from Lee et al. [130], © 2020/Royal Society of Chemistry.

current collectors by using MOFs as a sacrificial template to directly carbonize MOFs [130]. As shown in Figure 4.10c–f, the adsorption and surface diffusion of zinc atoms on carbon defects are calculated by the density functional theory. The results show that the carbon defect sites have strong orbital hybridization interactions with zinc, and the zinc adsorbed on the defect sites is difficult to move toward nondefective sites in the 2D lateral direction, which prevents the aggregation of zinc. As shown in the SEM images (Figure 4.10g,h), zinc deposition on defective carbon is superior to that on nondefective carbon because this atomically designed defect carbon layer guides dense zinc deposition. The zinc-bromine flow battery assembled with the defect 3D substrate can operate stably for more than 1800 cycles at a large current density of 100 mA cm^{-2} and a high areal-specific capacity of 20 mAh cm^{-2}. This method of utilizing defect vacancies on carbon substrates to suppress the 2D diffusion of zinc on the surface is a feasible strategy to realize reversible zinc anodes for high-performance AZIBs.

4.2.2.3 Introducing Hydrogen Evolution Barrier Layer

Conductive carbon or copper substrates are thermodynamically unstable in aqueous environments, and the zinc deposition process on substrates will suffer from the competitive decomposition of water to form hydrogen owing to the relatively low hydrogen evolution overpotential [95, 132]. Meanwhile, continuous hydrogen evolution causes local pH changes, which generate loose and brittle by-products by depleting electrolytes and zinc metal [133]. Furthermore, these side reaction products increase the roughness of the electrode surface. The strong electric field is concentrated in the protruding area on rough surfaces, which further excites hydrogen evolution and accelerates the selective zinc plating to induce vertical zinc dendrite growth rather than parallel growth, resulting in the subsequent short circuit of the cell [134]. Therefore, the study of zinc deposition substrates capable of suppressing uncontrolled dendrite growth and complex water-induced corrosion is of great significance for the development of long-life AZIBs.

Surface modulation is often required to give the 3D host high hydrogen evolution overpotential and reduce the catalytic ability of the substrate toward hydrogen evolution. Multifunctional metallic tin-decorated carbon felt host (SH) were constructed by magnetron sputtering technique for zinc deposition host [43]. As shown in Figure 4.11a, the SH exhibits a higher hydrogen evolution overpotential compared with the pristine carbon felt host (PH), demonstrating the strong hydrogen evolution resistance at the tin interface. Furthermore, the high metal affinity of zinc to tin reduces the zinc nucleation overpotential and induces a uniform zinc plating/stripping process (Figure 4.11b). To further buffer the volume expansion and enhance the structural stability during zinc deposition, a 3D fibrous network structure composed of adjacent N-doped hollow carbon embedded with tin nanoparticles (Sn@NHCF) was designed as a composite host for zinc metal (Figure 4.11c,d) [135]. Compared with bare zinc foils and conventional carbon substrates, the hierarchical hollow fibers provide sufficient space for zinc deposition and prevent volume expansion from inducing structural collapse. Meanwhile, zincophilic tin nanoparticles and N-doped hollow carbon with a low nucleation barrier and strong interaction with Zn^{2+} can ensure uniform deposition on the inner and outer surfaces of hollow fibers at low overpotentials (Figure 4.11e). Importantly, tin with weak catalytic properties for hydrogen evolution prevents the generation of hydrogen bubbles in the cell (Figure 4.11f,g), suggesting that building a hydrogen evolution barrier

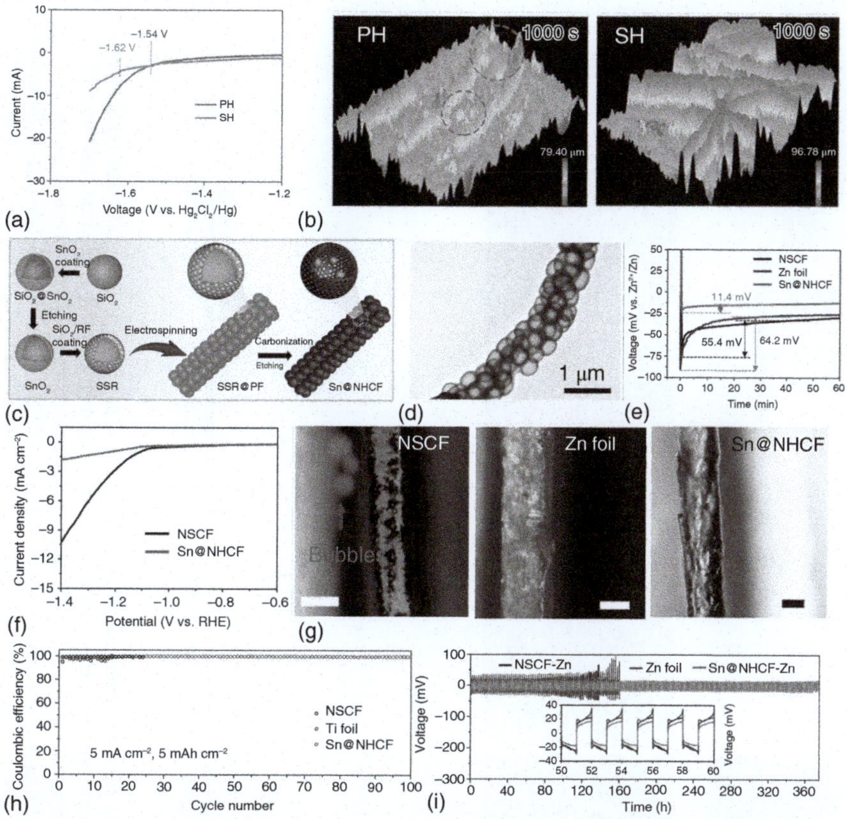

Figure 4.11 (a) Linear polarization curves of PH and SH in 3 M KCl electrolyte. (b) Super-depth surface images of zinc deposition on PH and SH for 1000 seconds. (c) Schematic illustration of the synthesis of Sn@NHCF. (d) TEM image of Sn@NHCF. (e) Voltage profiles of zinc plating on N-doped solid carbon fibers (NSCF), zinc foil, and Sn@NHCF. (f) Linear polarization curves of NSCF and Sn@NHCF. (g) Optical photographs of zinc deposition on NSCF, zinc foil, and Sn@NHCF for 5 minutes at a current density of 10 mA cm^{-2}. (h) CE and (i) voltage curves of NSCF, zinc foil, and Sn@NHCF anodes. Source: (a, b) Reproduced from Yin et al. [43] with permission. Copyright 2020 Wiley-Blackwell; (c–i) Reproduced from Yu et al. [135] with permission. Copyright 2022 American Association for the Advancement of Science.

layer on the substrate surface is effective to enhance the cycling life of zinc metal anodes (Figure 4.11h,i).

4.2.2.4 Regulating Interface Orientation

During the initial stage of electrodeposition, zinc nucleates randomly along different crystallographic orientations on the substrate. Zinc crystals are formed through the (001) planes with the plating process due to their hexagonal, close-packed lattice structure, resulting in randomly hexagonal zinc sheet growth with different orientations [136]. Among them, those grown perpendicular to the substrate form large dendrites that eventually pierce the separator, resulting in rapid battery failure. Conversely, if the plated zinc plate is formed parallel to the substrate, a uniform, and

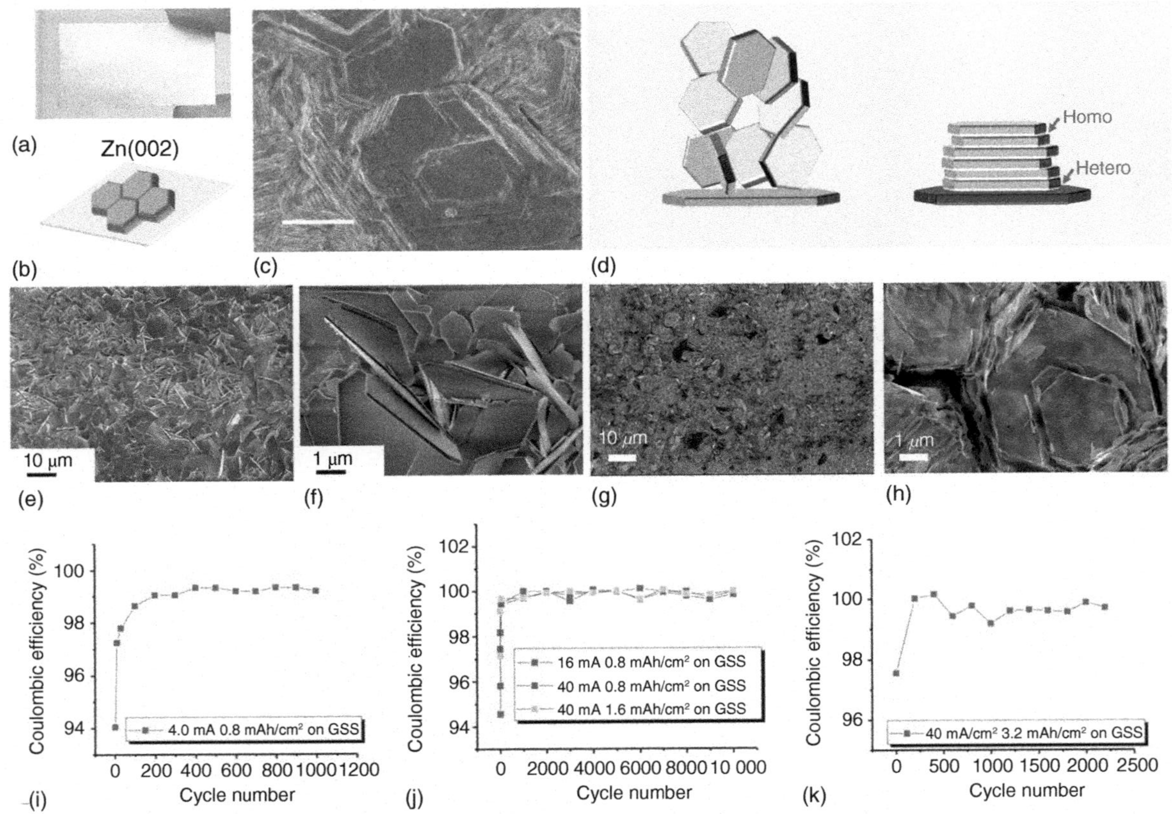

Figure 4.12 (a) Optical photograph of (002) preferred zinc after deposition. (b) Schematic illustration of (002) zinc structure on substrate. (c) SEM image of the hexagonal-like zinc plates after zinc deposition. (d) Scheme illustrating the design principle of epitaxial metal electrodeposition. SEM images of zinc deposition on (e, f) bare stainless steel and (g, h) graphene-coated stainless steel. (i–k) CE of the graphene-coated stainless steel substrate at different current densities and capacities. Source: (a–c) Reproduced with permission from Yuan et al. [138], © 2020/John Wiley & Sons. (d–k) Reproduced from Zheng et al. [137] with permission. Copyright 2019 John Wiley and Sons Ltd. Source (E–H): Reproduced with permission from ref. [137], © 2019/ American Association for the Advancement of Science – AAAS.

flat zinc deposition will be achieved [137]. Therefore, zinc crystals are expected to grow along the (002) direction to form zinc plate deposits parallel to the substrate surface (Figure 4.12a–c) [138, 139].

The regulated crystallographic orientation of the substrate makes it feasible to induce uniform epitaxial growth of zinc. Typically, substrates for epitaxial growth have a crystal structure with a well-defined surface lattice and similar crystal orientation to the deposited layer (Figure 4.12d) [137]. During epitaxial electrodeposition, a coherent or semi-coherent interface (heteroepitaxy) initially forms between the substrate and the deposited layer, which significantly affects subsequent deposition. In other words, when the first atomic layer is formed by heteroepitaxy, the subsequent depositions are conducted on the bulk lattice in a homoepitaxial growth direction [140]. Archer et al. found a low lattice mismatch of ~7% between the (002) planes of graphene and the (002) planes of zinc, suggesting that graphene is a good substrate for epitaxial deposition of zinc [137]. It can be seen from the SEM images that zinc deposition on graphene-coated stainless steel produces more uniform and compact flakes parallel to the substrate, in contrast to the random deposition on bare stainless steel (Figure 4.12e–h). During the complete stripping process, the zinc on the substrate is oxidized and dissolved into the electrolyte, and a stable structure for substrates is maintained to guide the next uniform zinc plating. As a result, regulating the crystal orientation of the substrate to match the deposition layer can fundamentally solve the dendrite problem, which allows reversible zinc electroplating/stripping behavior for thousands of cycles at moderate and high current densities (Figure 4.12i–k).

Most researchers focus on constructing substances with zinc epitaxial growth to promote uniform zinc growth during cycling but overlook zinc metal affinity, which increases nucleation barriers. Besides, the fabrication process is complicated for epitaxially grown substrate surfaces whose crystal orientation is uniform and parallel to the zinc (002) facet. In our previous work, the facet-selective zinc deposition behavior on polycrystalline copper foil with random crystal orientation was revealed by theoretical simulations and experimental observations [141]. The metal affinity of the substrate plays a crucial role in the dense and thin-plated anode. Accordingly, a substrate with both high zinc affinity and uniform crystal orientation was constructed by chemical tin plating technology assisted by low-temperature annealing to improve the cycling stability of electroplated zinc anodes in practical batteries. The obtained uniform-oriented substrates with high zinc affinity show highly reversible zinc plating/stripping behavior, compact deposition appearance, and enhanced hydrogen evolution resistance. The strategy of orientation regulation and metal affinity design of interfaces can also be applied to lithium metal anodes, indicating the tremendous potential of as-fabricated substrates in high-performance metal anodes for practical applications.

References

1 Fu, J., Cano, Z.P., Park, M.G. et al. (2017). *Advanced Materials* 29: 1604685.

2 Li, P., Jin, Z., and Xiao, D. (2018). *Energy Storage Materials* 12: 232–240.

3 Geng, M. and Northwood, D.O. (2003). *International Journal of Hydrogen Energy* 28: 633–636.

4 Wang, F., Borodin, O., Gao, T. et al. (2018). *Nature Materials* 17: 543–549.

5 Huang, J., Guo, Z., Ma, Y. et al. (2019). *Small Methods* 3: 1800272.

6 Mainar, R.A., Leonet, O., Bengoechea, M. et al. (2016). *International Journal of Energy Research* 40: 1032–1049.

7 Goff, A.H.-L., Joiret, S., Saïdani, B., and Wiart, R. (1989). *Journal of Electroanalytical Chemistry and Interfacial Electrochemistry* 263: 127–135.

8 Powers, R.W. and Breiter, M.W. (1969). *Journal of the Electrochemical Society* 116: 719.

9 Jia, X., Liu, C., Neale, Z.G. et al. (2020). *Chemical Reviews* 120: 7795–7866.

10 Krężel, A. and Maret, W. (2016). *Archives of Biochemistry and Biophysics* 611: 3–19.

11 Wang, W., Huang, G., Wang, Y. et al. (2022). *Advanced Energy Materials* 12: 2102797.

12 Cai, Z., Ou, Y., Wang, J. et al. (2020). *Energy Storage Materials* 27: 205–211.

13 Zhang, Q., Luan, J., Fu, L. et al. (2019). Angewandte Chemie. *International Edition* 58: 15841–15847.

14 Parker, J.F., Nelson, E.S., Wattendorf, M.D. et al. (2014). *ACS Applied Materials & Interfaces* 6: 19471–19476.

15 Liu, J., Guan, C., Zhou, C. et al. (2016). *Advanced Materials* 28: 8732–8739.

16 Guo, W., Cong, Z., Guo, Z. et al. (2020). *Energy Storage Materials* 30: 104–112.

17 Parker Joseph, F., Chervin Christopher, N., Pala Irina, R. et al. (2017). *Science* 356: 415–418.

18 Wang, S.-B., Ran, Q., Yao, R.-Q. et al. (2020). *Nature Communications* 11: 1634.

19 Parker, J.F., Chervin, C.N., Nelson, E.S. et al. (2014). *Energy & Environmental Science* 7: 1117–1124.

20 Zhang, Q., Yang, Z., Ji, H. et al. (2021). *SusMat* 1: 432–447.

21 Ko, J.S., Geltmacher, A.B., Hopkins, B.J. et al. (2019). *ACS Applied Materials & Interfaces* 2: 212–216.

22 Pei, A., Zheng, G., Shi, F. et al. (2017). *Nano Letters* 17: 1132–1139.

23 Wang, S.-H., Yin, Y.-X., Zuo, T.-T. et al. (2017). *Advanced Materials* 29: 1703729.

24 Liu, B., Wang, S., Wang, Z. et al. (2020). *Small* 16: 2001323.

25 Wu, L. and Dong, Y. (2021). *Energy Storage Materials* 41: 715–737.

26 Wu, Y., Wang, M., Tao, Y. et al. (2020). *Advanced Functional Materials* 30: 1907120.

27 Zeng, Y., Zhang, X., Qin, R. et al. (2019). *Advanced Materials* 31: 1903675.

28 Zhou, J., Xie, M., Wu, F. et al. (2022). *Advanced Materials* 34: 2106897.

29 Zhou, Y., Wang, X., Shen, X. et al. (2020). *Journal of Materials Chemistry A* 8: 11719–11727.

30 Dong, W., Shi, J.-L., Wang, T.-S. et al. (2018). *RSC Advances* 8: 19157–19163.

31 Li, C., Shi, X., Liang, S. et al. (2020). *Chemical Engineering Journal* 379: 122248.

32 Baumann, A.E., Burns, D.A., Liu, B., and Thoi, V.S. (2019). *Communications Chemistry* 2: 86.

33 Kang, Z., Wu, C., Dong, L. et al. (2019). *ACS Sustainable Chemistry & Engineering* 7: 3364–3371.
34 Zhang, Y., Xia, X., Liu, B. et al. (2019). *Advanced Energy Materials* 9: 1803342.
35 Qiu, B., Xing, M., and Zhang, J. (2018). *Chemical Society Reviews* 47: 2165–2216.
36 Pu, J., Li, J., Zhang, K. et al. (1896). *Nature Communications* 2019: 10.
37 Zhai, P., Wei, Y., Xiao, J. et al. (2020). *Advanced Energy Materials* 10: 1903339.
38 Guo, S., Pan, X., Gao, H. et al. (2010). *Chemistry – A European Journal* 16: 5379–5384.
39 Wang, Z., Huang, J., Guo, Z. et al. (2019). *Joule* 3: 1289–1300.
40 Li, G., Liu, Z., Huang, Q. et al. (2018). *Nature Energy* 3: 1076–1083.
41 Zhang, Q., Luan, J., Tang, Y. et al. (2020). *Angewandte Chemie International Edition* 59: 13180.
42 Yang, Q., Liang, G., Guo, Y. et al. (2019). *Advanced Materials* 31: 1903778.
43 Yin, Y., Wang, S., Zhang, Q. et al. (2020). *Advanced Materials* 32: 1906803.
44 Zhang, Q., Luan, J., Tang, Y. et al. (2018). *Journal of Materials Chemistry A* 6: 18444–18448.
45 Sagane, F., Ikeda, K.-I., Okita, K. et al. (2013). *Journal of Power Sources* 233: 34–42.
46 Zhang, R., Chen, X.-R., Chen, X. et al. (2017). *Angewandte Chemie International Edition* 56: 7764–7768.
47 Zhang, R., Cheng, X.-B., Zhao, C.-Z. et al. (2016). *Advanced Materials* 28: 2155–2162.
48 Lu, L.-L., Ge, J., Yang, J.-N. et al. (2016). *Nano Letters* 16: 4431–4437.
49 Zhao, Z., Zhao, J., Hu, Z. et al. (2019). *Energy & Environmental Science* 12: 1938–1949.
50 Desai, D., Wei, X., Steingart, D.A., and Banerjee, S. (2014). *Journal of Power Sources* 256: 145–152.
51 Zheng, J. and Archer Lynden, A. (2021). *Science Advances* 7: eabe0219.
52 Qiu, H., Du, X., Zhao, J. et al. (2019). *Nature Communications* 10: 5374.
53 Suo, L., Oh, D., Lin, Y. et al. (2017). *Journal of the American Chemical Society* 139: 18670–18680.
54 Zhang, Z., Said, S., Smith, K. et al. (2021). *Journal of Materials Chemistry A* 9: 15355–15362.
55 Zhou, M., Guo, S., Li, J. et al. (2021). *Advanced Materials* 33: 2100187.
56 He, P. and Huang, J. (2021). *ACS Energy Letters* 6: 1990–1995.
57 Rehnlund, D., Ihrfors, C., Maibach, J., and Nyholm, L. (2018). *Materials Today* 21: 1010–1018.
58 Garcia, G., Ventosa, E., and Schuhmann, W. (2017). *ACS Applied Materials & Interfaces* 9: 18691–18698.
59 Aslam, M.K., Niu, Y., Hussain, T. et al. (2021). *Nano Energy* 86: 106142.
60 Chazalviel, J.N. (1990). *Physical Review A* 42: 7355–7367.
61 Biswal, P., Stalin, S., Kludze, A. et al. (2019). *Nano Letters* 19: 8191–8200.
62 Hou, Z., Gao, Y., Zhou, R., and Zhang, B. (2022). *Advanced Functional Materials* 32: 2107584.
63 Yang, Z., Zhang, Q., Xie, C. et al. (2022). *Energy Storage Materials* 47: 319–326.

64 Jia, H., Qiu, M., Lan, C. et al. (2022). *Advanced Science* 9: 2103952.

65 Friel, J.J. (1986). *Corrosion* 42: 422–426.

66 Wang, J., Cai, Z., Xiao, R. et al. (2020). *ACS Applied Materials & Interfaces* 12: 23028–23034.

67 Sears, G.W. (1955). *Acta Metallurgica* 3: 367–369.

68 Asgari, H., Toroghinejad, M.R., and Golozar, M.A. (2008). *Journal of Materials Processing Technology* 198: 54–59.

69 Ramanauskas, R., Quintana, P., Maldonado, L. et al. (1997). *Surface and Coating Technology* 92: 16–21.

70 Wang, H., Zhou, S., Chen, C., and Wang, Q. (2006). *Chinese Journal of Chemical Engineering* 14: 551–554.

71 Wang, J., Zhang, B., Cai, Z. et al. (2022). *Scientific Bulletin* 67: 716–724.

72 Neyts, E.C., Ostrikov, K., Sunkara, M.K., and Bogaerts, A. (2015). *Chemical Reviews* 115: 13408–13446.

73 Xu, W., Zhao, K., Huo, W. et al. (2019). *Nano Energy* 62: 275–281.

74 Guo, Z., Fan, L., Zhao, C. et al. (2022). *Advanced Materials* 34: e2105133.

75 Kang, L., Cui, M., Jiang, F. et al. (2018). *Advanced Energy Materials* 8: 1801090.

76 Zheng, G., Wang, C., Pei, A. et al. (2016). *ACS Energy Letters* 1: 1247–1255.

77 Zhao, K., Wang, C., Yu, Y. et al. (2018). *Advanced Materials Interfaces* 5: 1800848.

78 He, H., Tong, H., Song, X. et al. (2020). *Journal of Materials Chemistry A* 8: 7836–7846.

79 Zhang, Q., Luan, J., Huang, X. et al. (2020). *Nature Communications* 11: 3961.

80 Xie, X., Liang, S., Gao, J. et al. (2020). *Energy & Environmental Science* 13: 503–510.

81 Zou, P., Zhang, R., Yao, L. et al. (2021). *Advanced Energy Materials* 11: 2100982.

82 Yan, H., Li, S., Nan, Y. et al. (2021). *Advanced Energy Materials* 11: 2100186.

83 Yang, H., Chang, Z., Qiao, Y. et al. (2020). *Angewandte Chemie International Edition* 59: 9377–9381.

84 Gallaway, J.W., Desai, D., Gaikwad, A. et al. (2010). *Journal of the Electrochemical Society* 157: A1279.

85 Koketsu, T., Ma, J., Morgan, B.J. et al. (2017). *Nature Materials* 16: 1142–1148.

86 Yan, C., Li, H.-R., Chen, X. et al. (2019). *Journal of the American Chemical Society* 141: 9422–9429.

87 Wang, F., Borodin, O., Ding, M.S. et al. (2018). *Joule* 2: 927–937.

88 Malik, R. (2017). *Joule* 1: 17–19.

89 Kim, J.Y., Kim, A.Y., Liu, G. et al. (2018). *ACS Applied Materials & Interfaces* 10: 8692–8701.

90 Park, J., Jeong, J., Lee, Y. et al. (2016). *Advanced Materials Interfaces* 3: 1600140.

91 Kim, J.Y., Liu, G., Shim, G.Y. et al. (2020). *Advanced Functional Materials* 30: 2004210.

92 Yu, Z. and Ang, C. (2001). *Journal of Applied Physics* 91: 794–797.

93 Prodromakis, T. and Papavassiliou, C. (2009). *Applied Surface Science* 255: 6989–6994.

94 Wu, K., Yi, J., Liu, X. et al. (2021). *Nano-Micro Letters* 13: 79.

95 Hao, J., Li, B., Li, X. et al. (2020). *Advanced Materials* 32: 2003021.
96 Cao, P., Zhou, X., Wei, A. et al. (2021). *Advanced Functional Materials* 31: 2100398.
97 Zhang, L., Zhang, B., Zhang, T. et al. (2021). *Advanced Functional Materials* 31: 2100186.
98 Deng, C., Xie, X., Han, J. et al. (2020). *Advanced Functional Materials* 30: 2000599.
99 Baugh, L.M. (1979). *Electrochimica Acta* 24: 657–667.
100 Cao, L., Li, D., Deng, T. et al. (2020). *Angewandte Chemie International Edition* 59: 19292–19296.
101 Gao, X., Zhou, Y.-N., Han, D. et al. (2020). *Joule* 4: 1864–1879.
102 Shi, X., Xu, G., Liang, S. et al. (2019). *ACS Sustainable Chemistry & Engineering* 7: 17737–17746.
103 Cui, M., Xiao, Y., Kang, L. et al. (2019). *ACS Applied Energy Materials* 2: 6490–6496.
104 Zhang, Y., Wang, G., Yu, F. et al. (2021). *Chemical Engineering Journal* 416: 128062.
105 Guo, W., Zhang, Y., Tong, X. et al. (2021). *Materials Today Energy* 20: 100675.
106 Han, D., Wu, S., Zhang, S. et al. (2020). *Small* 16: 2001736.
107 Li, S., Fu, J., Miao, G. et al. (2021). *Advanced Materials* 33: 2008424.
108 Yang, Z., Lv, C., Li, W. et al. (2021). *Small* 18: 2104148.
109 Grujicic, D. and Pesic, B. (2002). *Electrochimica Acta* 47: 2901–2912.
110 Hao, J., Li, X., Zhang, S. et al. (2020). *Advanced Functional Materials* 30: 2001263.
111 Zhu, M., Hu, J., Lu, Q. et al. (2021). *Advanced Materials* 33: 2007497.
112 Li, C., Liu, S., Shi, C. et al. (2019). *Nature Communications* 10: 1363.
113 Jiao, Y., Li, F., Jin, X. et al. (2021). *Advanced Functional Materials* 31: 2107652.
114 Du, H., Zhao, R., Yang, Y. et al. (2022). *Angewandte Chemie International Edition* 61: e202114789.
115 Mitha, A., Yazdi, A.Z., Ahmed, M., and Chen, P. (2018). *ChemElectroChem* 5: 2409–2418.
116 Yang, W., Du, X., Zhao, J. et al. (2020). *Joule* 4: 1557–1574.
117 Zhang, Q., Ma, Y., Lu, Y. et al. (2020). *Nature Communications* 11: 4463.
118 Ghosh, M., Vijayakumar, V., and Kurungot, S. (2019). *Energy Technology* 7: 1900442.
119 Ghosh, M., Vijayakumar, V., Anothumakkool, B., and Kurungot, S. (2020). *ACS Sustainable Chemistry & Engineering* 8: 5040–5049.
120 Cui, Y., Zhao, Q., Wu, X. et al. (2020). *Angewandte Chemie International Edition* 59: 16594–16601.
121 Zhou, S., Wang, Y., Lu, H. et al. (2021). *Advanced Functional Materials* 31: 2104361.
122 Zhao, R., Yang, Y., Liu, G. et al. (2021). *Advanced Functional Materials* 31: 2001867.
123 Gu, Y., Xu, H.-Y., Zhang, X.-G. et al. (2019). *Angewandte Chemie International Edition* 58: 3092–3096.

124 Wang, L.-P., Li, N.-W., Wang, T.-S. et al. (2017). *Electrochimica Acta* 244: 172–177.
125 Luan, J., Zhang, Q., Yuan, H. et al. (2019). *Advanced Science* 6: 1901433.
126 Yang, C.-P., Yin, Y.-X., Zhang, S.-F. et al. (2015). *Nature Communications* 6: 8058.
127 Hong, C., Yang, G., and Wang, C. (2021). *ACS Applied Materials & Interfaces* 13: 54088–54095.
128 Jian, Q., Wan, Y., Sun, J. et al. (2020). *Journal of Materials Chemistry A* 8: 20175–20184.
129 Xie, F., Li, H., Wang, X. et al. (2021). *Advanced Energy Materials* 11: 2003419.
130 Lee, J.-H., Kim, R., Kim, S. et al. (2020). *Energy & Environmental Science* 13: 2839–2848.
131 Chen, X., Chen, X.-R., Hou, T.-Z. et al. (2019). *Science Advances* 5: eaau7728.
132 Li, Q., Wang, Y., Mo, F. et al. (2021). *Advanced Energy Materials* 11: 2003931.
133 Ma, L., Li, Q., Ying, Y. et al. (2021). *Advanced Materials* 33: e2007406.
134 Yang, Q., Li, Q., Liu, Z. et al. (2020). *Advanced Materials* 32: 2001854.
135 Yu, H., Zeng, Y., Li Nian, W. et al. *Science Advances* 8: eabm5766.
136 Han, C., Li, W., Liu, H.K. et al. (2020). *Nano Energy* 74: 104880.
137 Zheng, J., Zhao, Q., Tang, T. et al. (2019). *Science* 366: 645.
138 Yuan, D., Zhao, J., Ren, H. et al. (2021). *Angewandte Chemie International Edition* 60: 7213–7219.
139 Cao, J., Zhang, D., Zhang, X. et al. (2020). *Journal of Materials Chemistry A* 8: 9331–9344.
140 Zhang, K., Yan, Z., and Chen, J. (2020). Joule. 4: 10–11.
141 Xie, C., Yang, Z., Zhang, Q. et al. (2022). *Research* 2022: 1.

5

Electrolytes for Aqueous Zinc-Ion Batteries

5.1 Development of Electrolytes for Aqueous Zinc-Ion Batteries

The electrolyte is the key to connecting the anode and cathode, and its design and optimization strategy are related to battery performance. Currently, lithium-ion batteries are dominant in the energy-storage market. However, their easy flammability resulting from organic electrolytes has jeopardized their application in different scenarios. In this context, the Zn-based battery, combined with the high theoretical capacity (820 $mAh\,g^{-1}$) and low electrochemical potential (−0.76 V vs. standard hydrogen electrode) of zinc metal [1], gradually stands out since non-flammable aqueous electrolytes serve inside this battery. Despite the side reactions related to water that pure organic solvents can eliminate [2], the much lower ionic conductivity of organic electrolytes than that of aqueous electrolytes will inevitably challenge the commercial application of zinc batteries, such as the inferior high-rate performance and the difficulty of quick response to a sudden change in electricity generation from unstable new energy resources [3]. Besides, most organic solvents are expensive, toxic, flammable, and neither safe nor environmentally friendly, which is inconsistent with the original intention of developing zinc batteries [4]. Therefore, unless the above problems with organic electrolytes can be well solved, aqueous electrolytes will still be the mainstream choice for zinc batteries.

The original Zn batteries (e.g. Zn–Mn and Zn–Ni batteries) are based on Zn dissolution/deposition at the anode side ($Zn + 4OH^- \leftrightarrow Zn(OH)_4^{2-} + 2e^- \leftrightarrow ZnO + 2OH^- + H_2O + 2e^-$) and proton intercalation/deintercalation at the cathode side ($MnO_2 + H_2O + e^- \leftrightarrow MnOOH + OH^-$ or $NiOOH + H_2O + e^- \leftrightarrow Ni(OH)_2 + OH^-$). They are generally assembled as a dry cell using a paste mixture of zinc chloride and ammonium chloride as an electrolyte [5]. The alkaline electrolyte has the advantages of high operating voltage, rapid reaction kinetics, and high ionic conductivity, so the alkaline aqueous electrolyte has gradually become the common electrolyte of zinc-based batteries, mainly including potassium hydroxide, sodium hydroxide, and lithium hydroxide [6]. Under the same conditions, the conductivity of cations in these electrolytes is $K^+ > Na^+ > Li^+$, which enables potassium hydroxide to prevail in an aqueous zinc battery [7]. In general, the potassium hydroxide electrolyte has

Aqueous Zinc Ion Batteries: Fundamentals, Materials, and Design, First Edition.
Haiyan Wang, Qi Zhang, Yixin Li, and Yougen Tang.

high ionic conductivity at room temperature since its concentration is between 26% and 40% (4.6–7.0 M) [8]. However, zincates, the discharge products at the Zn anode, dissociate seriously in the strong alkaline electrolyte. Simultaneously, self-corrosion and dendrite formation on the Zn anode lead to self-discharge, considerably crippling battery capacity in the end [9].

In addition to the alkaline electrolyte, aqueous zinc-ion batteries (AZIBs) have been developed in neutral or mildly acidic electrolytes in recent years [10]. Normally, the formula of these electrolytes consists of water as the solvent and Zn-containing salts as the solute. Compared with alkaline Zn batteries, AZIBs exhibit smaller hydrogen evolution and less passivation effect, but the low decomposition voltage of water solvent results in a narrow electrochemical window [11], which greatly hinders the application of high-voltage cathodes and significantly confines the energy density of batteries [12].

In an aqueous medium, composition and salt concentration have an important influence on the intrinsic properties and electrochemical performance of electrolytes. Besides, the choice of salt anions is also a key factor in battery performance since they can participate in the solvation configuration of zinc ions in aqueous zinc ion electrolytes, change the structure of the solvent shell, affect the composition of the electro-double layer, and even affect the ion flow on the electrode surface. Therefore, various salt anions have been evaluated and screened in recent years, such as $ZnSO_4$, $Zn(CF_3SO_3)_2$, $Zn(NO_3)_2$, $Zn(CH_3COO)_2$, $Zn(ClO_4)_2$, $ZnCl_2$, and ZnF_2. However, many of these solutes do not fully meet the requirements of electrolytes in practical ZIB applications [13].

$ZnSO_4$ salt has been the most widely used because of its low price, good economic benefit, and stable sulfate structure. In 1986, Yamamoto and Shoji jointly investigated the performance of the Zn–MnO_2 battery with various mild aqueous solutions and confirmed that the battery has good cycle stability and coulombic efficiency in $ZnSO_4$ electrolyte [14], due to the high electrochemical stability of SO_4^{2-} (reversible generation of discharge product $ZnSO_4[Zn(OH)_2]_3\cdot xH_2O$ and reversible reaction between Mn^{2+} and Mn^{4+}). However, during the charging and discharging process, the pH of the anode and cathode surfaces will change dynamically, and the increase in pH will form zinc hydroxyl sulfate (LDH–$Zn_4(OH)_6SO_4\cdot nH_2O$) on the negative electrode surface and then affect the properties of the battery. Also, it should be noted that although a by-product layer is generated at the zinc–electrolyte interface on the negative side, it will not block electrolytes due to its loose structure, which may provide researchers with ideas for optimal utilization [15].

Since cathode dissolution is serious in AZIBs, to a certain extent, $Zn(CF_3SO_3)_2$ can play a role in suppressing this problem. This is because the large volume of anion $[CF_3SO_3]^-$ can reduce the number of water molecules in the solvated shell structure of Zn^{2+}, avoid direct contact between the active substance and water, and accelerate carrier transmission and charge transfer rate of Zn^{2+}. Cheng and coworkers [16] found that increasing the electrolyte concentration of zinc trifluoromethyl sulfonate changed the ionic solvation and transport behavior, reduced ionic conductivity, improved solution viscosity, inhibited water activity and side reactions, and improved the cyclic stability of the electrode in aqueous solution. At the same time, using cationic defective spinel $ZnMn_2O_4$ as the positive electrode can allow rapid and repeated zinc intercalation/deintercalation, and 3 M $Zn(CF_3SO_3)_2$ can effectively alleviate Mn loss in spinel during the charge–discharge cycle to maintain a high battery capacity and cycle life. However, $Zn(CF_3SO_3)_2$ is about 15 times more expensive than $ZnSO_4$, which is a fatal disadvantage of its application in large-scale ESS.

In contrast to the above mainstream Zn salts, others seem to be less competitive. The nitrate ion in $Zn(NO_3)_2$ has a strong oxidation ability and is easy to cause negative electrode corrosion and positive electrode dissolution [17]. $Zn(CH_3COO)_2$ exhibits low ionic conductivity due to its incomplete ionization. Even though it has a large group, it is still structurally unstable, which can also interact with the positive electrode and cause rapid dissolution [18]. The stereoscopic structure of perchlorate in $Zn(ClO_4)_2$ reduces the oxidizability of chlorine ions, but ZnO by-products are formed on the surface of $Zn(ClO_4)_2$ when perchlorate contacts the negative electrode [19], which affects the reaction kinetics and leads to a large deposition/dissolution overpotential of zinc ions. Zinc halide salts can be alternative electrolytes due to their low price, low oxidizability of chloride and fluoride ions, and few negative side reactions [20]. However, ZnF_2 has low solubility and poor cycling performance in aqueous systems, and the excessively narrow anodic potential window of $ZnCl_2$ leads to continuous decomposition of electrolytes at high voltage, which limits their further practical applications.

5.1.1 Functional Electrolyte Additives

The intrinsic effect of electrolytes is finalized after the selection of the electrolyte type. To further the performance of electrolytes, it is natural to introduce functional additives appropriately, including organic additives and inorganic additives.

Organic additives can participate in the solvation of zinc ions to varying degrees due to differences in intrinsic structure and functional group polarity, including $-NH_2$, —OH, —COOH, —SH, and other groups with strong polarity. These groups have a higher affinity for zinc ions than the coordination between water and zinc ions, which can deprive water of the coordination position needed to effectively regulate the solvation structure of zinc ions [4]. Qiao and coworkers [21] found that polyols had a large molecular size and required a high energy barrier when inserted into a Zn^{2+} shell to break the solvation equilibrium. Comparatively, methanol with the smallest molecular size was more suitable as an electrolyte additive to adjust the solvation structure (Figure 5.1a) [24]. The introduction of methanol as an additive destroyed the coordination balance in the solvent shell of zinc ions, and methanol molecules replaced some water molecules for coordination, reducing the number of molecules on the electrode surface, inhibiting hydrogen evolution and corrosion side reactions, and promoting the improvement of the uniformity of zinc deposition.

Regardless of molecular size, the abundance of hydroxyl groups in polyols makes them similar in solubility to monohydric alcohol, and ethylene glycol can be miscible with H_2O in any proportion and thus can also be used as an electrolyte additive [25]. For example, with the help of experiments and theoretical

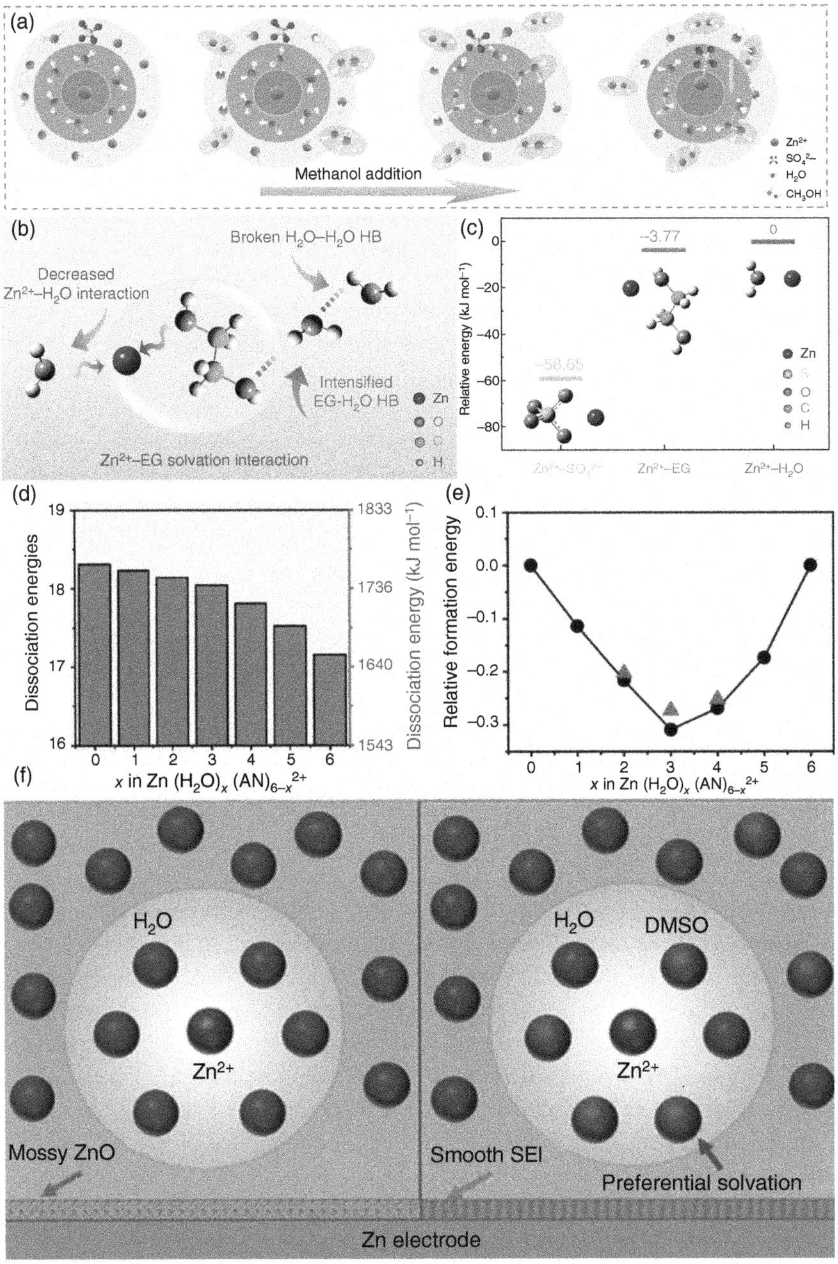

Figure 5.1 (a) Schematic of changes in the Zn^{2+} solvent sheath, together with methanol addition. (b) Schematic illustration of a possible mechanism of how the Zn^{2+}–EG solvation interaction impacts the chemistry of the hybrid electrolyte. (c) Relative binding energy for Zn^{2+} with different species obtained from DFT calculations. (d) Dissociation energies of the $Zn(H_2O)_x(AN)_{6-x}{}^{2+}$-complexes. (e) Pseudo-phase diagram of $Zn(H_2O)_x(AN)_{6-x}{}^{2+}$. (f) Scheme of Zn^{2+}-solvation structure and zinc surface passivation in H_2O (left) and H_2O-DMSO (right) solvents. Source: (a) Hao et al. [21]/John Wiley & Sons. (b, c) Chang et al. [22]/Royal Society of Chemistry. (f) Cao et al. [23]/American Chemical Society.

calculations, Li and coworkers [22] found that zinc ions were coordinated with EG first and then with H_2O. After the coordination of zinc and EG occurs, the electrostatic potential of zinc ions decreases significantly, and the solvation of water is also obviously weakened (Figure 5.1b,c) [26]. Obviously, alcohol additives modify the solvent shell of zinc ions, change the diffusion kinetics of zinc ions, and thus affect the nucleation overpotential and deposition morphology. Among many other organic solvent additives, including acetonitrile (AN) [27], dimethyl carbonate (DMC) [28], diethyl ether [29], 1,4-dioxane (DX) [30], 1,2-dimethoxy ethane (DME) [31], trimethyl phosphate (TMP) [2], and dimethyl sulfoxide (DMSO) [32], AN is unique in regulating zinc coordination. According to molecular dynamics (MD) simulation, the affinity between the AN molecule and Zn^{2+} is stronger than that of other molecules. Therefore, all H_2O molecules can be replaced to form the $[Zn(AN)_6]^{2+}$ complex, thus changing the dissociation kinetics and affecting the electrodeposition behavior (Figure 5.1d,e) [33]. In addition, DMSO has also been identified as a superior electrolyte additive because of its high dielectric constant, high donor number, and low cost. In the configuration of Zn^{2+} coordination ions, DMSO is preferentially solvated with Zn^{2+}, limiting the formation of the Zn-H_2O solvent shell, and it can also interact with H_2O to reduce activity and inhibit the decomposition of H_2O (Figure 5.1f). Beyond the solvation effect, DMC introduced into the $Zn(CF_3SO_3)_2$ electrolyte can form $ZnCO_3$ on the zinc anode, achieving reversible Zn^{2+} plating/stripping and extending the service life [34].

In general, the above organic solvents have good solubility in water and can preferentially solvate with Zn^{2+}, reducing the activity of the free H_2O molecules, but increasing the Zn^{2+} solvent energy barrier, which requires a higher nucleation overpotential to provide energy and change zinc deposition morphology. Similarly, the increase in the radius of the complex also requires more energy to break up the solvent shell.

Organic additives cannot fully apply to the optimization of electrolytes, so organic/inorganic compound electrolytes with more function can be used as a direction of exploration. Nonionic low-foam surfactant, fatty methyl ethoxylate (FMEE), can homogenize electrolytes, reduce activation energy, and inhibit the formation of zinc dendrites. Gaseous silicon dioxide (SiO_2) exhibits a shear thickening effect, which significantly increases the mechanical strength of the electrolyte, thus inhibiting dendrite growth [35]. Huang et al. [36] mixed these two additives with 1 M $ZnSO_4$ and 0.1 M $MnSO_4$ electrolytes to obtain the composite electrolyte. The cyclic stability of the Zn/Zn symmetric battery reached 1500 h, and the Zn corrosion rate of the composite electrolyte was reduced by 10 times compared with the original electrolyte. The initial coulombic efficiency increases from 92.3% to 99.5%, and the corresponding Zn–MnO_2 battery also has good cyclic stability (the single-cycle capacity loss is less than 0.002%).

Due to their small size and simple structure, inorganic additives can explore the specific mechanism of interface reaction optimization from a microscopic perspective. Based on the principle of electrostatic shielding, cations with lower reduction potential can be added to the electrolyte to inhibit the formation of metal dendrites (e.g. Ce^{3+}, La^{3+}, and Na^+ in $ZnSO_4$) [37]. Under the action of the electric field, high-value cations (Ce^{3+} and La^{3+}) migrate to the electrode surface and preferentially occupy nucleation sites with high activity. Zinc ions were reduced and deposited at nucleation sites with low initial activity. With continuous zinc deposition, cations leave the initial position and migrate to sites with stronger electric fields. The original high-active site was activated to accept zinc deposition, and the overall pattern of progressive nucleation was displayed through an electrostatic shielding mechanism [38].

Other than the above strategies, zincophilic additives that can generate a protective layer via pure adsorption or in situ reaction [39] are popular choices as well.

Saccharin is a good interface stabilizer. After being introduced into the $ZnSO_4$ electrolyte, it can produce adsorption after taking off protons, thus making the electrode/electrolyte interface short/free of water, which restrains zinc corrosion and hydrogen evolution reactions (HERs). In the meantime, this could inhibit the local proliferation and nucleation of zinc ions at places with low surface energy that may induce zinc dendrites [39].

On account of the electrostatic attraction, some cationic particles are prone to occupy the electrolyte/anode interface. The positively charged amino acids (arginine and glutamic acid) have high adsorption energy, leading to an adaptive zinc electrolyte interface with strong steric hinders, thus effectively hindering the adsorption of H_2O (Figure 5.2a–d). Polar groups of amino acids transport zinc ions to the electrode surface through coordination and achieve uniform zinc deposition through electrostatic repulsion and homogeneous ion flow [40]. In addition, the cations in some chlorine salts are large in volume, and they snatch water in the solvation structure via strong interaction and form $[Zn(H_2O)_6]^{2+}$, which significantly inhibits the growth of dendrites and side reactions. Also, tetrabutylammonium sulfate as a cationic surfactant [41] can form a cationic layer on the anode surface due to its butyl cation, which can shield the diffusion of Zn^{2+} and make zinc deposition more uniform.

Water molecules preferentially contact the electrode when zinc ions with solvent shell migrate to the electrode surface, which is the root of hydrogen evolution and corrosion side reactions. The introduction of an additive with strong electron-donating ability reconstructed the solvent shell of zinc ions and tended to transfer electrons to the original solvated structure, thus improving the LUMO level of the solvated structure and inhibiting the reaction of bound water. At the same time, the additive layer at the electrode–electrolyte interface can block the contact between water and the electrode or generate a stable solid electrolyte interface (SEI) via in situ reaction during the electrochemical process, thus affecting the subsequent zinc ion migration/nucleation. However, the use of additives also has some drawbacks. Higher intermolecular adsorption greatly increases the solvation energy barrier of zinc ions, which may lead to ionic diffusion kinetic hysteresis. In addition, the stability and cost of additives also need to be considered [42].

5.1.2 High-Concentration Electrolyte (Water in Salt)

For AZIBs, their lower energy density has put them at a disadvantage because of the narrow electrochemical stability window (ESW) of aqueous electrolytes. In this

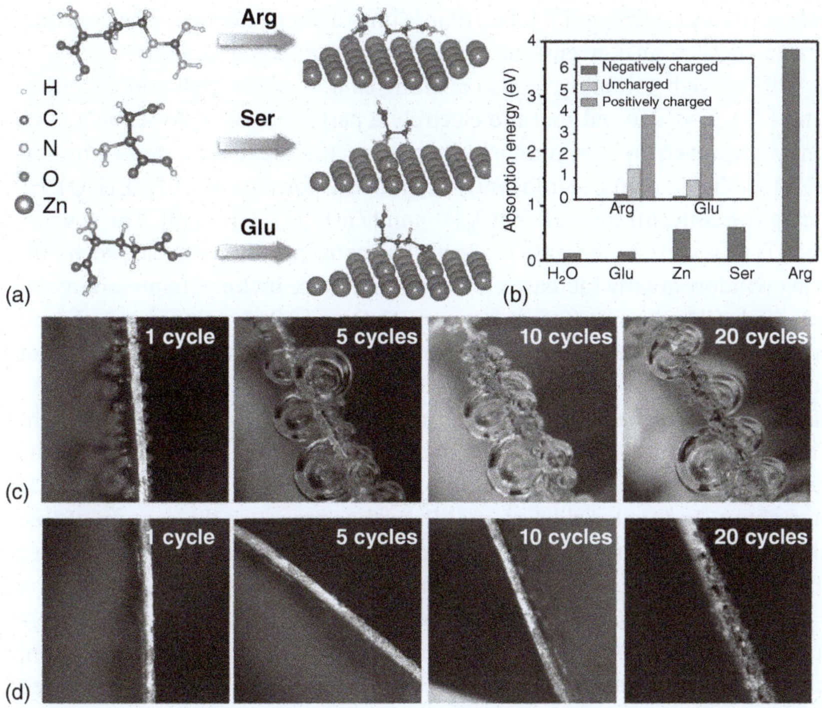

Figure 5.2 (a) DFT calculation models of Arg, Ser, and Glu. (b) Absorption energy of H_2O, Glu, Zn, Ser, and Arg on Zn surface in mildly acidic electrolyte. Inset shows absorption energy of Arg and Glu with positively charged, uncharged, and negatively charged on Zn surface. In situ optical microscope image of Zn electrode at 1st, 5th, 10th, and 20th cycles (c) in bare $ZnSO_4$ electrolyte and (d) in $ZnSO_4$ + Arg electrolyte. Source: Lu et al. [40]/John Wiley & Sons.

context, researchers tend to reduce the amount of water in an electrolyte. By this logic, water-in-salt electrolytes have been emerging since they can significantly extend the ESW, thereby enhancing the energy density of AZIBs. The principle is that the equilibrium potential of H_2O/H_2 will decrease along with increasing salt concentration in the electrolyte, which directly inhibits the HER [43]. And the electrolyte pH becomes more neutral. At the same time, the densely populated anions can forcefully intrude on the solvent shell structure of Zn^{2+}, which will reduce the coordination number of H_2O around Zn^{2+}. All of these inevitably contribute to reducing the possibility of anodic corrosion [44]. Beyond that, thanks to the special solvation structure and dense distribution of Zn^{2+}, dendrite formation can be effectively suppressed, and the coulomb efficiency of electroplating/stripping can be critically improved [45].

Zinc sulfate is the most commonly used aqueous electrolyte, which can effectively improve the reversibility of the zinc anode by increasing the concentration of the electrolyte. For example, in 3 M $ZnSO_4$ [46] or double-salt electrolytes (0.1 M $MnSO_4$ and 4.2 M $ZnSO_4$) [47], increasing the electrolyte concentration will push out H_2O in the solvated Zn^{2+} structure, change the hydration ion configuration, and achieve higher cycling performance and longer service life (>5000 cycles and >8000 cycles at currents of 10 and 20 mA cm^{-2}). But that is only a slight increase in electrolyte concentration, not a high concentration of salt, let alone "water-in-salt."

$ZnCl_2$, with its high solubility, is also a strong candidate for high-concentration electrolytes [48]. It was found that the electrolyte performance of 20 M $ZnCl_2$ was significantly improved by continuously increasing the electrolyte concentration and that 30 M $ZnCl_2$ could significantly inhibit the formation of $[Zn(H_2O)_6]^{2+}$ and increase the concentration of $[ZnCl_4]^{2-}$ and $[Zn(H_2O)Cl_4]^{2-}$ [49]. The specific mechanism is that with an increase in electrolyte concentration, the electrochemical working window greatly increases. Zinc ions are more inclined to preferentially coordinate with Cl^-; the interaction between H_2O and Zn^{2+} was weakened, and the activity of H_2O was decreased, which inhibits the HER and the generation of by-products.

In addition to the basic high-concentration zinc electrolyte, other high-concentration salts can also be added to obtain complex high-concentration salts to regulate the ionic structure. For example, add 5 M LiCl to the 30 M $ZnCl_2$ mentioned above. In this hybrid system, Li^+ further promotes the capture of free water molecules and limits the solvation process of Zn^{2+}, thus providing higher coulombic efficiency [50]. If $Zn(TFSI)_2$ is used as the electrolyte, it can show a lifetime of more than 800 h at 3 M at a low current density. Some researchers introduced LiTFSI into the electrolyte solution. With the increase in concentration, they found that the number of coordinated H_2O molecules around Zn^{2+} decreased sharply, but Zn-TFSI was completely coordinated. When adding 20 M LiTFSI, the solvated structure could be significantly adjusted, the HER and water consumption could be inhibited, and the overall stability of the battery could be greatly improved (Figure 5.3a–d) [16].

On the other hand, other metal ions with a high affinity for water molecules can be introduced into the electrolyte to capture part of the water molecules and form a shared solvation structure to induce uniform zinc deposition. However, high-concentration electrolytes also have some disadvantages, such as high-voltage polarization under low current density, high-concentration electrolytes often mean high viscosity and low ionic conductivity, resulting in unsatisfactory electrochemical performance. These factors will greatly restrict the application of high-concentration electrolytes [51].

5.1.3 Hydrogel Electrolyte

To deal with various problems with aqueous electrolytes at the same time, some researchers have gradually focused on polymer electrolytes with good effects in zinc battery systems. As the name implies, polymer electrolyte is composed of polymer, salt, additives, etc., based on the polymer skeleton to form zinc salt electrolyte [52], which has demonstrated different effects of self-healing [53], anti-freezing [54] and so on. Compared with the traditional aqueous electrolyte liquid phase, gel electrolytes have unique advantages: (i) gel electrolytes can avoid electrolyte leakage and increase the use convenience [55]; (ii) the structural integrity and electrochemical performance of the device can be maintained under dynamic

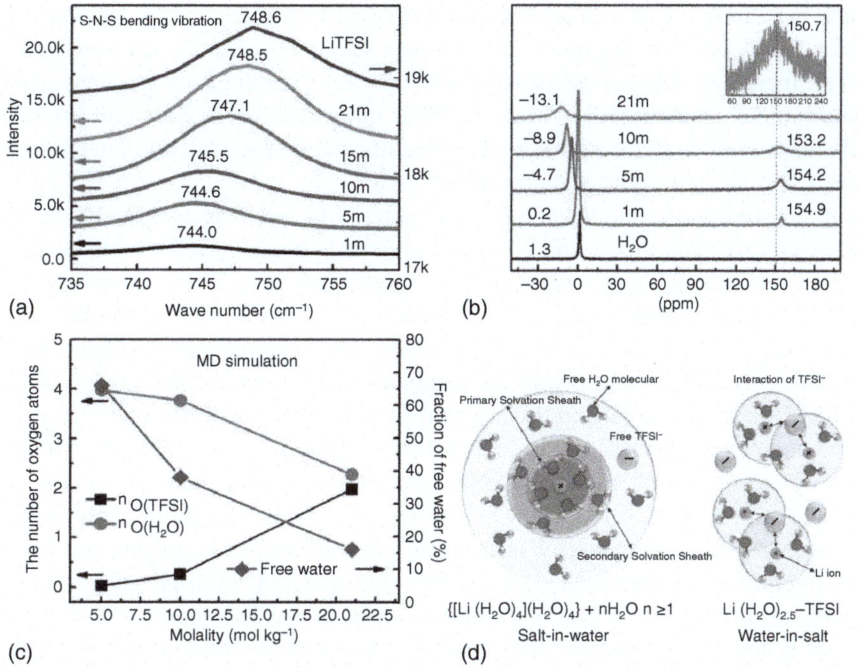

Figure 5.3 The effect of LiTFSI concentration on ion–solvent and ion–ion interactions. (a) Progression of Raman vibration at ~744 cm^{-1} with salt concentration. (b) Change of chemical shifts for ^{17}O nuclei in solvent (water) and anion (THSI). The inset shows the $TFSI^-$ peak at 21 M. (c) Numbers of water and $TFSI^-$ oxygen atoms within the Li^+ primary solvation sheath of 0.27 nm and the fraction of "free" water not bound to any Li^+, all obtained from molecular dynamics simulations. (d) Evolution illustration of the Li^+ primary solvation sheath in diluted and water-in-salt solutions. Source: Suo et al. [45]/American Association for the Advancement of Science.

external forces [55]; (iii) gel electrolytes can play the role of separator, greatly simplifying the device composition of flexible ZIBs [56]; (iv) gel electrolytes can alleviate the dissolution of cathode-active materials and inhibit the dendrite growth and corrosion reaction of the anode [57]. At present, various polymers, such as gelatin, sodium alginate, xanthan gum, polyvinyl alcohol, Polyacrylamide (PAM), and polyethylene glycol diacrylate, can be used as electrolytes for ZIBs.

Gelatin is a biocompatible, hydrophilic polymer widely used in the food and pharmaceutical industries. It can be used as an important electrolyte gel additive because of its good environmental compatibility and salt resistance. Chi and Liu et al. [58] developed a simple self-supporting gelatin-based hydrogel electrolyte with relatively low mechanical strength but high ionic conductivity because of its high porosity and high electrolyte retention ability. In flexible Zn–$LiMn_2O_4$ cells, the electrolyte can inhibit dendrite growth and zinc anode corrosion reactions, and achieve a lower self-discharge rate and higher cyclic stability. After treatment in the high-concentration inorganic salt solution, the electrolyte is further strengthened through the dehydration effect and polymer chain bundle strengthening [58].

Algin is a carbohydrate extracted from brown seaweed. It can induce gelation of the solution by crosslinking multivalent cations such as Ca, Cu, Pb, Cd, and Ba at room temperature. The gel electrolyte with high ionic conductivity was prepared by Zhou and coworkers [59], which can limit the migration of Zn^{2+} and inhibit the dendrite growth and side reactions of the zinc anode based on the interaction between alginate carboxylic acid groups and Zn^{2+}.

Xanthan gum is a complex extracellular polysaccharide that consists of α,β-1, 4-linked glucan main chain and a trisaccharide side chain linked to alternating D-glucose residues, which promotes high salt tolerance [55]. The hydroxyl group on the side chain of xanthan gum can attract a certain number of water molecules, and the long carbon chain can enhance the interaction between xanthan gum and water molecules. These advantages are conducive to high ionic conductivity, high solubility, and high viscosity, indicating that xanthan gum has great potential as a gel polymer electrolyte [60]. Di and coworkers [61] prepared a flexible Zn–MnO_2 battery using MnO_2/CNT composite film as the cathode and the mixture of xanthan gum, 2 M $ZnSO_4$, and 0.1 M $MnSO_4$ as the electrolyte. Compared with other commonly used gel polymer electrolytes, xanthan gum-based electrolytes are more stable and uniform (Figure 5.4a–d). Kaltenbrunner et al. [62] reported a dry gel battery with a xanthan gel electrolyte. In this cell, a highly elastic carbon black silicone oil paste is added to the acrylic elastomer as a flexible collector. The results show that xanthan gum has high salt tolerance, hydrophilicity, and ionic conductivity, but its mechanical durability is low. The elongation of the gel cell will lead to a disconnection between the electrode and the collector, thus reducing the discharge capacity [63]. To solve this problem, stretchable fluid collection is a promising strategy to improve the mechanical strength of xanthan gum electrolyte-based ZIBs.

Polyacrylamide (PAM, PAAm)-based polymer electrolyte has good compatibility and excellent strength. PAM has a large number of hydrophilic groups and amide groups (—$CONH_2$), which is conducive to regulating the ion migration of water molecules and network structures [64]. However, the mechanical properties of pure PAM are generally poor. Chemical crosslinking is an effective way to alleviate this problem. Crosslinking PAM electrolytes with excellent mechanical properties can be synthesized by acrylamide monomers and crosslinking agents. Strong covalent bonds can be found between the groups of the PAM polymer chain and the cross-linking agent, so it is believed that the cross-linking agent can unify the network structure of PAM. Li et al. [65]. prepared a rechargeable, waterproof, customizable, and scalable Zn–MnO_2 battery. Zn and MnO_2 coatings on double-helix carbon nanotubes provide high strength and roughness. Due to its good compatibility with zinc sulfate and manganese sulfate, the electrolyte has high ionic conductivity.

Gel–polymer electrolytes based on polyvinylidene fluoride (PVDF) are favorable for ZIB applications as well due to their porous and semicrystalline structure. In general, PVDF-based polymer electrolytes have high negative stability as a result of the strong electron interaction of the —CF_2 group and the high dielectric constant, which can expand the isolation of metal salts and increase the number of charge carriers, thus achieving high ionic conductivity [66]. However, similar

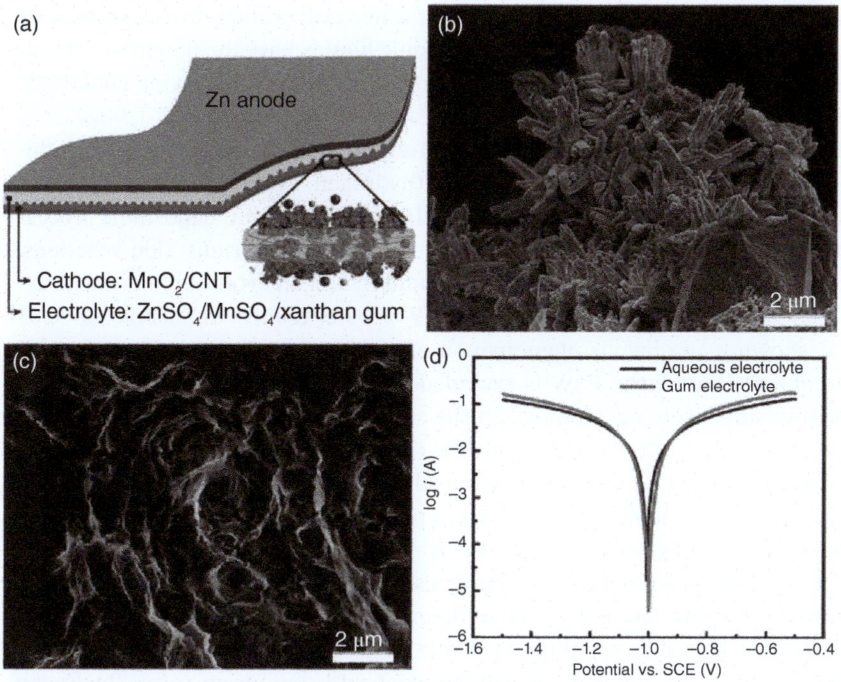

Figure 5.4 (a) Schematic showing the structure of gum Zn–MnO_2 battery, which consists of a zinc foil as the anode, a $ZnSO_4/MnSO_4$/xanthan gum film as the separator and the electrolyte, and a MnO_2/CNT film as the cathode, respectively. The pop-up shows the structure of the MnO_2/CNT electrode, where a porous MnO_2 layer is coated on a CNT thin film. (b) SEM of the Zn foil after 50 cycles at 5 C in 2 M $ZnSO_4$/0.1 M $MnSO_4$ aqueous electrolyte. (c) SEM of the Zn foil after 1000 cycles at 5 C in 2 M $ZnSO_4$/0.1 M $MnSO_4$/20 wt% xanthan gum electrolyte. (d) Tafel plots of the Zn anode in 2 M $ZnSO_4$/0.1 M $MnSO_4$ aqueous and gum electrolytes. Source: Zhang et al. [61]/Royal Society of Chemistry.

to polyethylene oxide polymer (PEO)-based gel polymer electrolytes, PVDF-based gel electrolytes also have weak mechanical strength. Kumar et al. [67] prepared a Zn-γ-MnO_2 battery using a gel electrolyte composed of PVDF, propylene carbonate (PC), ethylene carbonate (EC), and zinc trifluoromethanesulfonic acid. Adding polyethylene glycol dimethyl ether to PVDF–hexafluoropropylene (HFP) can further provide ionic conductivity, and PVDF–HFP-based electrolyte can dissolve zinc salts and stabilize PVDF–HFP.

5.1.4 Ionic Liquids

Ionic liquids (salts with a low melting point at room temperature or below, composed of large asymmetric cations or weak coordination anions) are a new kind of solvent [68], which have the characteristics of nonflammability [69], high-temperature stability [70], good chemical stability, adjustable viscosity/conductivity [71], low steam pressure, wide chemical stability window [72] and so on. Ionic liquids have been regarded as promising liquid solvents in battery systems. On the one hand, the use of ionic liquids can avoid the problem of water evaporation, thus significantly extending the battery life [73]. Additionally, ionic liquids have the potential to eliminate hydrogen evolution and induce metal deposition, thus improving coulombic efficiency and battery life [74].

The ionic liquid-based zinc electrolyte (1-ethyl-3-methylimidazolium tetrafluoroborate) can inhibit the side reactions (hydrogen evolution, corrosion, and by-products) and zinc dendrite growth in ZIBs [75]. The zinc deposition process in mainstream electrolytes is continuously disturbed by serious side reactions, including dendrite formation, Zn corrosion, and hydrogen evolution. For instance, in the $ZnSO_4$ or $Zn(BF_4)_2$ electrolyte, the ESW of water relative to Zn/Zn^{2+} is −0.05 to 1.7 V, showing an obvious phenomenon of hydrogen evolution. However, with the use of ionic liquid, the ESW is extended to −0.75 to 2.8 V, indicating that the HER can be effectively constrained. In the solvation configuration of Zn^{2+}, BF_4^- ions occupy the solvent shell to form $[Zn\ (BF_4)]^{3-}$, which is different from the ions formed by $Zn-H_2O$ and can limit the adverse reactions associated with water molecules. At the same time, the abundant polar groups in ionic liquids can have coordination with zinc ions, promote the transport kinetics of zinc ions, alleviate the concentration polarization of electrolytes, prevent zinc ions from gathering at favorable nucleation sites, and promote the uniform deposition of zinc ions.

Ionic liquids also play an important role in electrodeposition. The positively charged C—H in the ionic liquid will interact with anions to form "hydrogen bonds," and the strength of such interaction is $OAc^- > OMs^- > TfO^- > TFSI^-$. This indicates that the interface environment between the electrode and ionic liquid can be modified and regulated by anions (Figure 5.5a,b) [76]. Cations can also affect the electrodeposition process, and the grain can be further refined when 1-butyl-1-methylpyrrolidinium trifluoromethyl sulfonate is used as an electrolyte [77]. The application of additives can also change the interface structure and affect the nucleation and deposition of zinc. In pure ionic liquid electrolytes, the innermost layer of the anode surface is mainly ionic liquid pairs. After adding zinc salt additives (e.g. $Zn(TfO)_2$, ZnO), zinc ions and ionic liquid form a new ion pair, which promotes the electric field distribution at the interface to be more uniform. In short, ionic liquids can effectively regulate zinc deposition and prevent hydrogen decomposition and corrosion. Since some ionic liquids are relatively easy to prepare and their cost is controllable, they have considerable potential for further research and application in AZIBs [78].

5.1.5 Deep Eutectic Solvents

The chemical reaction process lies in the reactions of atoms and groups. The donor can provide a chemical group, while the corresponding binding group is called the acceptor. In terms of the concept of hydrogen bonds, there are hydrogen bond donors and hydrogen bond acceptors. In 2001, Abbott et al. [79] first proposed the concept of deep eutectic solvents (DESs). This kind of solvent is

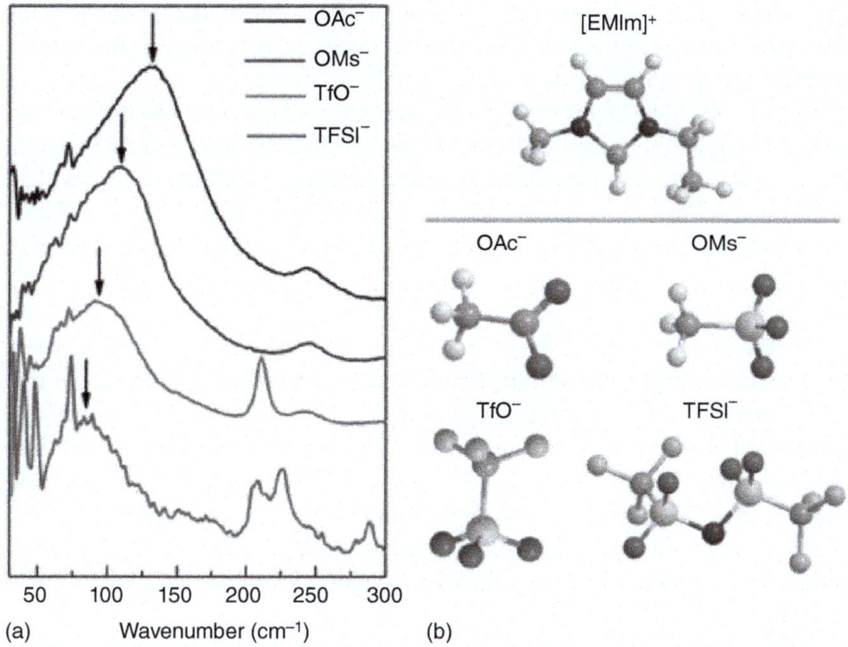

Figure 5.5 (a) Far-IR spectra of [EMIm]TFSI, [EMIm]TfO, [EMIm]OMs, and [EMIm]OAc between 30 and 300 cm^{-1}. (b) Chemical structures of cations and anions Source: Liu et al. [76]/American Chemical Society.

mainly composed of hydrogen bond suppliers (polyols, urea, and carboxylic acid) and hydrogen bond acceptors (quaternary ammonium salt) to form a binary and ternary eutectic solvent [80], generally using Lewis acid and base in an eutectic molar ratio mixture [81]. Based on this high degree of freedom design, various eutectic solvents of metal salts have been developed and applied to electrolytic polishing and electroplating [82]. In recent years, it has been found that the system of DESs can be introduced into ZIBs to inhibit zinc dendrites and side reactions. DESs have an adjustable composition and abundant intermolecular forces, so they can adapt to the presence of high concentrations of ions and thus be used as a substitute for high concentrations of electrolytes [83]. At the same time, the coordination effect of the DESs and the type of hydrogen bonds (ions or molecules) make the system highly hygroscopic [84], which can make it miscible with water and separate water molecules from salt. These favorable characteristics can be applied to regulate the solvation structure of Zn^{2+}, which is considered a new solvent system similar to ionic liquid and is conducive to the further development of ZIBs [85].

Under the conditions of air and moderate stability, the DESs formed by quaternary ammonium salts and hydrogen bond donors can also remain stable, which can be used for metal plating and stripping [86]. The combination of choline chloride/urea with anhydrous $ZnCl_2$ can produce electrodeposited zinc with a nano-porous structure [87]. In this system, zinc ions tend to nucleate with small grains, which can inhibit dendrite growth in the long-term cycle.

Referring to the idea of increasing the concentration of zinc electrolyte, the proportion of DES is increased to form water-in-DES. The solvent of this system retains most of the eutectic characteristics and introduces H_2O molecules, which can make the electrolyte have good ionic conductivity, while H_2O will be confined in the interface composed of zinc salt, DESs, and water, avoiding the side reaction of water [88].

By changing the functional groups of the DES, a new solvation structure was formed by the coordination of acetamide and $TFSI^-$ with Zn^{2+}, which induced the preferential reduction decomposition of $TFSI^-$ and formed a SEI film on the surface of the zinc anode after the initial cycle [89]. The stable SEI can eliminate direct contact between the active anode and electrolyte, thus effectively inhibiting interface side reactions (such as hydrogen evolution and passivation). More importantly, once SEI is formed, stable and uniform Zn^{2+} transport can be achieved on the Zn anode with a low diffusion barrier, which is conducive to reversible stripping/plating of zinc. However, the ionic conductivity of amide DESs is poor, which hinders its further development.

Until now, researchers have accumulated a lot of achievements in the positive electrode, negative electrode, and electrolyte of water ZIB. However, because these studies are independent or even contradictory, the internal mechanism of electrolyte and electrode interaction is not clear, and there is no unified guideline to include all of them.

5.2 Issues and Solutions of Electrolytes for Aqueous Zinc-Ion Batteries

Currently, AZIBs are meeting with the following challenges: (i) Cathode dissolution. Some cathode materials will continuously dissolve when they contact an aqueous electrolyte due to their weak water solubility, leading to the rapid decay of battery capacity [90]. (ii) Water decomposition, corrosion, and passivation. The HER easily occurs in weakly acidic Zn aqueous solutions, and the increase in pH value will lead to side reaction corrosion of the electrode in the discharge process, and the produced by-products will passivate the electrode [91]. (iii) Dendrite growth. The uneven distribution of electrons and ions results in dendrite growth on the anode surface [92]. Notably, these problems can be solved by adjusting the electrochemical behavior of water or zinc ions at the solid–liquid interface to stabilize the electrode and avoid side reactions [75]. Specific countermeasures involve pH regulation, concentration adjustment, solvent change, and the introduction of additives. The following sections will start with the above problems and phenomena, followed by deeply exploring their internal mechanisms and putting forward some effective solutions (Figure 5.6).

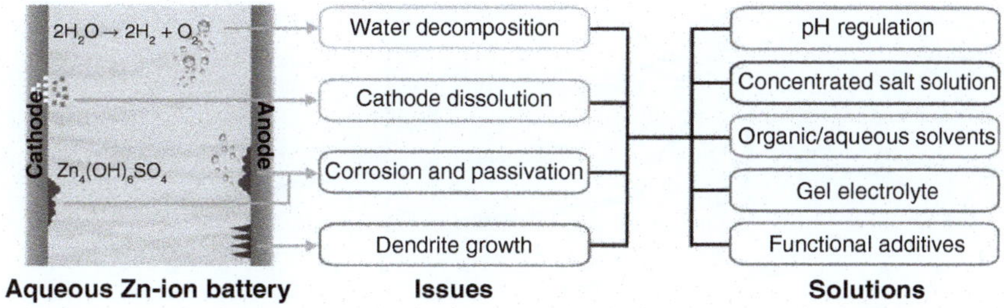

Figure 5.6 The current issues and the corresponding solutions using electrolyte engineering for AZIBs. Source: Zhang et al. [13]/John Wiley & Sons/Licensed under CC BY 4.0.

5.2.1 Cathode Dissolution

Most of the cathodes or intermediates produced in the charge–discharge cycling are soluble in weakly acidic electrolytes of AZIBs, which may just make the electrode structure disordered and lead to the rapid deterioration of electrochemical performance [93]. In some existing studies, a common strategy to inhibit the dissolution of cathodes is presalting (introducing some inorganic salts identical to the cathode elements into electrolytes in advance) to adjust the dissolution equilibrium [94]. For the MnO_2 cathode, soluble Mn^{2+} is generated by the equilibrium reaction (5.1).

$$MnO_2 + 2H_2O + 2e^- \leftrightarrow Mn^{2+} + 4OH^- \tag{5.1}$$

For the above reversible reaction, its equilibrium constant at a certain temperature is expressed as Eq. (5.2).

$$K = \frac{a_{Mn^{2+}} \cdot a_{OH^-}^4}{a_{MnO_2} \cdot a_{H_2O}^2} \tag{5.2}$$

where K, $a_{Mn^{2+}}$, a_{OH^-}, a_{MnO_2} and a_{H_2O} represent chemical equilibrium constants, Mn^{2+} activity, OH^- activity, MnO_2 activity, and H_2O activity, respectively. Accordingly, the addition of $MnSO_4$ in the $ZnSO_4$ electrolyte can not only make the electrolyte reach the Mn^{2+} dissolution equilibrium before the electrochemical reaction to suppress MnO_2 dissolution but also prevent the formation of Mn vacancies for Zn^{2+} insertion since excessive Zn^{2+} insertion may result in an unstable crystal structure [95].

In comparison with the bare $ZnSO_4$ electrolyte, the electrochemical behavior of the MnO_2 cathode in the $ZnSO_4/MnSO_4$ electrolyte remains unchanged, indicating that the addition of $MnSO_4$ did not introduce additional side reactions. Therefore, the inhibition of battery capacity decline can be achieved by adding $MnSO_4$ at an appropriate concentration (above the theoretical equilibrium concentration of Mn^{2+}, 0.089 M) to prevent the dissolution of MnO_2 (Figure 5.7a–h) [90]. For vanadium cathode solutions, a similar strategy can be used to modify the electrolyte. Niu and coworkers [37] reported that the introduction of Na_2SO_4 into the $ZnSO_4$ electrolyte can maintain the long-term stability of NaV_3O_8. By using this facile strategy, NVO nanobelts as the positive electrode can provide a

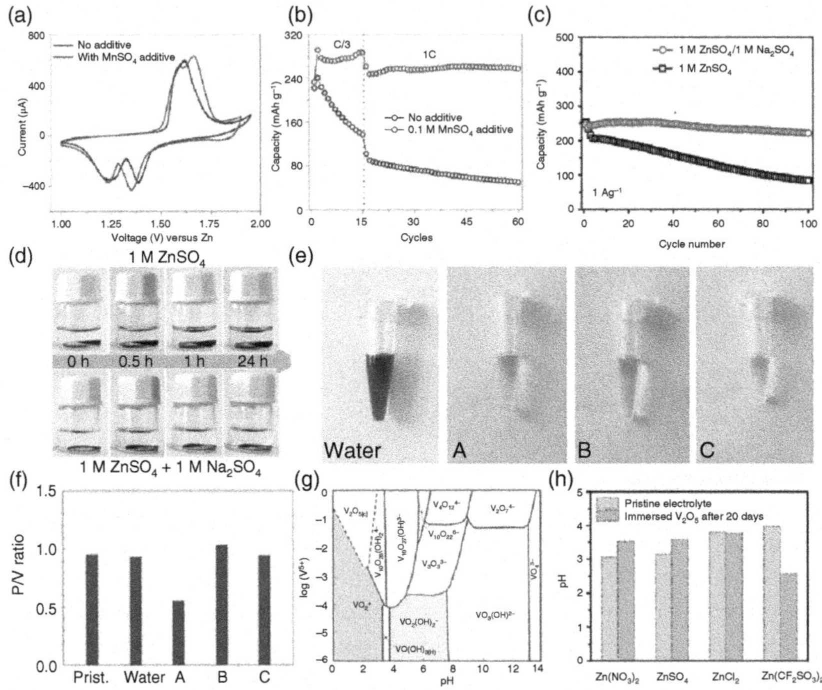

Figure 5.7 The cathode dissolution tests. Comparison of CV (a) curves and (b) cycling performances of Zn–MnO_2 cells using normal $ZnSO_4$ electrolyte and $ZnSO_4$ electrolyte with $MnSO_4$ additive. (c) The electrochemical performance of Zn–NaV_3O_8 cells cycled in $ZnSO_4$ and $ZnSO_4/Na_2SO_4$ electrolytes. (d) Change in color of $ZnSO_4$ and $ZnSO_4/Na_2SO_4$ electrolytes with NaV_3O_8 cathodes immersed. (e) Digital photograph of supernate of the liquid mixed with $VOPO_4$ powder and different electrolytes after three days. (f) P/V ratios of the remaining $VOPO_4$ powder (A: 3 M $Zn(OTf)_2$, B: 3 M $Zn(OTf)_2$/0.8 M H_3PO_4, C: 13 M $ZnCl_2$/0.8 M H_3PO_4). (g) The pH-log[V^{5+}] plot in the V_2O_5–H_2O system. (h) Variation of pH value before and after V_2O_5 powder soaking in $Zn(CF_3SO_3)_2$, $ZnSO_4$, $ZnCl_2$, and $Zn(CF_3SO_3)_2$ electrolytes for 20 days. Source: (a, b) Pan et al. [90]/Springer Nature. (c, d) [37]/Springer Nature/Licensed under CC BY 4.0. (e, f) Shi et al. [96]/John Wiley & Sons. (g) Lu et al. [97]/John Wiley & Sons. (h) Adapted from Lu et al. [97].

reversible capacity of 380 mAh g^{-1} (corresponding energy density of 300 Wh kg^{-1}) and a high-capacity retention of 82% after 1000 cycles at 4 A g^{-1}. This is because the Na_2SO_4 additive not only limited the continuous dissolution of NVO by changing the dissolution equilibrium of sodium ions in NVO but also inhibited the growth of zinc dendrites by a fine electrostatic shield mechanism (the reduction potential of Na^+ was lower than that of Zn^{2+}). Similarly, Mai and coworkers [98] reported a Zn/NVO aqueous battery with $Na_{0.33}V_2O_5$ nanowires as the positive electrode, 3 M $Zn(CF_3SO_3)_2$ electrolyte, and zinc as the negative electrode. The NVO electrode could demonstrate a high capacity of 367.1 mAh g^{-1} at 0.1 A g^{-1}, long-term cyclic stability with a capacity retention of over 93% for 1000 cycles, as well as excellent rate performance. The improvement of electrical conductivity owing to the intercalation of sodium ions between the $[V_4O_{12}]_n$ layers was investigated in a single nanowire device. As revealed, the improved performance of the electrode came from the stable-layered structure based on the support of intrinsic ions and high conductivity.

"Water-in-salt" electrolyte is another strategy to inhibit cathode dissolution. As we discussed in Section 5.1.2, all the water molecules participate in the formation of the solvation sheath and bind firmly with ions so that there are no free water molecules to dissolve soluble intermediates from the cathode in the high-concentration salt solution [99]. Based on this, cathode dissolution can be more effectively suppressed by combining the addition of salt and "water-in-salt" to simultaneously adjust the solubility equilibrium and reduce the activity of water molecules. Sun and coworkers [96] developed a high-concentration electrolyte (13 M $ZnCl_2$) with 0.8 M H_3PO_4 and found that the supernate remained colorless all the time in the mixture of $VOPO_4$ powder and modified electrolyte and remained unchanged after standing for three days, which fully indicated that the inhibition of cathode $VOPO_4$ dissolution was achieved. Wang et al. [100] studied the influence of mixed electrolytes of high-concentration lithium salt solution and zinc salt solution on battery performance. When LiTFSI concentration was high (≥20 M), Zn^{2+} was surrounded by $TFSI^-$ instead of water, which effectively prevented H_2 precipitation and achieved reversible dendrite-free plating/stripping. Zn/$LiMn_2O_4$ batteries show excellent cyclic performance (CE: 99.9%, with capacity retention over 85% after 4000 cycles), while the Zn/O_2 system keeps a high energy density of 300 Wh kg^{-1} for 200 cycles. Other than that, Liu and coworkers [101] prepared a high-concentration double-complexing electrolyte by introducing rich buffer pairs of NH_3 and NH^{4+}–NH_3, realizing a double-complexing effect between Zn^{2+} and Mn^{2+} and high pH stability. Benefiting from this strategy, Zn–Mn batteries can serve for 4500 cycles (CE ≈ 100%) and achieve 200 cycles at 10 mAh cm^{-2}.

The inhibition of cathode dissolution is not limited to the above two methods. Adjusting the pH value is also a good way to convert dissolved ions into compounds suitable for zinc ion intercalation. It is well known that the dissolution of crystalline V_2O_5 in water strongly depends on the pH value and the content of V^{5+} species. The dissolution reactions for different acidities and V^{5+} species are shown as follows:

$$V_2O_5 + 2H^+ \rightarrow 2VO_2^+ + H_2O\ (\text{pH} < 3.2 \text{ and } [V^{5+}] < 10^{-1}\ \text{M}) \tag{5.3}$$

$$V_2O_5 + 3H_2O \rightarrow 2VO_2(OH)_2^- + 2H^+\ (3.8 \leq \text{pH} \leq 7.8 \text{ and } [V^{5+}] < 10^{-4}\ \text{M}) \tag{5.4}$$

A further reaction can happen in an aqueous solution with Zn salt.

$$2[VO_2(OH)_2]^- + 3Zn^{2+} + 3H_2O \rightarrow Zn_3V_2O_7(OH)_2 \cdot 2H_2O + 4H^+ \tag{5.5}$$

Lu et al. [97] compared the change in pH value in several electrolytes, including $Zn(NO_3)_2$, $ZnSO_4$, $ZnCl_2$, and $Zn(CF_3SO_3)_2$ (all are 3 M), after V_2O_5 powder immersion. It is found that a 3 M $Zn(CF_3SO_3)_2$ electrolyte with an initial pH = 3.95 is favorable to Eq. (5.4). The large volume and low charge of $[CF_3SO_3]^-$ reduce the solvation effect and activate the reaction of the zinc ion with $[VO_2(OH)_2]^-$. Therefore, 3 M $Zn(CF_3SO_3)_2$ electrolyte can promote the progress of Eq. (5.5), which means that an in situ transformation of V_2O_5 into $Zn_3V_2O_7(OH)_2{\cdot}2H_2O$ (ZVO) nanosheet clusters was achieved. During the electrochemical discharge process, ZVO for co-Zn^{2+}/H^+ intercalation/deintercalation is completely converted from V_2O_5 owing to the continuous intercalation of H^+ into ZVO. ZVO becomes the only medium for Zn^{2+}/H^+ storage, which accelerates ion transport and promotes the optimal utilization of active materials. This method of in situ conversion using intermediate ions provides a new idea for inhibiting cathode dissolution.

Recently, the solvation sheath structure with Zn^{2+} ions and six water molecules was pointed out as being able to influence the cathode dissolution. It is difficult for the $[Zn(H_2O)_6]^{2+}$ solvation sheath to completely desolvate on the interface between the cathode and electrolyte due to the strong interaction between Zn^{2+} and water molecules. The partially desolvated $[Zn(H_2O)_4]^{2+}$ complex can be intercalated into the internal lattice of the cathode, which aggravates the structure decomposition of cathode materials owing to excessive co-insertion of water.

However, a poor amount of solvation water has a favorable tendency; for example, two water molecules can act as lubricants to promote the diffusion of zinc ions at the electrode interface and inside [102]. Hou et al. reported highly reversible $[Zn(H_2O)_2]^{2+}$ complex insertion/extraction into/from birnessite cathode through engineering a Zn^{2+} primary solvation sheath with strong-field ligand urea [103]. The results show that the $[Zn(H_2O)_2(NH_3)_3]^{2+}$ solvation sheath can de-solvate three urea's into the $[Zn(H_2O)_2]^{2+}$ complex in the electrolyte with the formation of a cathode–electrolyte interphase (CEI) to suppress Mn dissolution. Generally, cathode dissolution mainly occurs at the solid–liquid interface between the electrode and electrolyte. Hence, it is a good idea to construct the CEI at the solid–liquid interface, which can not only reduce the water activity of the cathode surface but also desolvate the solvation molecules on the cathode–electrolyte interface instead of on the cathode surface. Besides, CEI can be formed in situ to inhibit Mn dissolution and maintain the integrity of the cathode structure during the reversible insertion of the complex into layered birnessite. In other words, the presence of CEI greatly weakens the active water on the cathode surface and effectively inhibits the dissolution of the cathode. According to the above examples, it can be concluded that reducing the insertion of coordination water can reduce the disintegration of the cathode lattice structure caused by the co-insertion of zinc ions and water ligands. Therefore, CEI plays an important role in preventing cathode dissolution and improving the electrochemical performance of aqueous ZIB.

In short, the addition of salt to the electrolyte cannot prevent the slow erosion of the cathode structure during the long-time cycling because the cathode dissolution and deposition still occur dynamically at the same rate in a saturated salt solution. And "water-in-salt" keeping slow ion diffusion kinetics is not conducive to the high-rate performance of cathode dissolution, which may make it not a good choice to inhibit cathode dissolution. Although the dissolved ions can be transformed into a new species for Zn^{2+}/H^+ ion co-intercalation by pH modification, the conversion reaction with water may change the local concentration of the electrolyte to induce undesirably parasitic reactions, thus affecting the overall battery performance.

Therefore, it is better to further develop a cathode material that is completely insoluble in the aqueous electrolyte.

5.2.2 Water Decomposition

The thermodynamic decomposition voltage of water is only 1.23 V, corresponding to the hydrogen (HERs) and oxygen evolution reactions (OERs) ($2H_2O \rightarrow 2H_2 \uparrow + O_2 \uparrow$). When the AZIBs work at excessive voltage, the continuous gas generation increases the internal pressure and eventually bursts the sealed battery from the inside. Water decomposition occurs at about 2 V in practice due to the overpotential caused by the slow reaction kinetics of oxygen evolution involving the multi-electron transfer process [104]. Nevertheless, the operating voltage is still limited, and the energy density of AZIBs is insufficient, which is far lower than that of nonaqueous batteries. The specific water decomposition voltage, which can achieve enough gas generation rate to affect the battery's performance, depends on the pH value and the composition of electrolytes [39].

The theoretical decomposition voltage of water is independent of pH value, but the decrease in pH value of the electrolyte positively shifts the potentials of both HER (Eq. (5.6)) and OER (Eq. (5.7)) [40]. HER is insensitive to pH change because it can easily occur when the applied voltage is slightly higher than the theoretical decomposition voltage of water due to its simple reaction process and low thermodynamic cost. However, the positive shift in potential results in a higher OER overpotential, which thus increases the actual water decomposition voltage [105]. Qiao and coworkers investigated the impact of pH values on the electrochemical properties of Zn–MnO_2 cells [106]. The potentials of HER and OER are significantly suppressed to −1.06 and 1.35 V (vs. Ag/AgCl) when 0.1 M H_2SO_4 is added to the electrolyte, corresponding to a maximum stable operating voltage of 2.41 V (Figure 5.8a). A higher OER potential in the weakly acidic electrolyte can also ensure the full oxidation of the MnO_2 cathode [107]. It is found that the symmetric cell displays poor reversibility in the excessively acidic electrolyte (Figure 5.8b) since the nonnegligible HER is accompanied during the Zn electrodeposition. HER even occurs spontaneously in the acidic AZIBs, and the excessively acidic environment can accelerate the Zn dissolution, which is the limitation of expanding the electrochemical window by pH alteration [108].

$$2H^+ + 2e^- \rightarrow H_2 \uparrow \tag{5.6}$$

$$2H_2O - 4e^- \rightarrow O_2 \uparrow + 4H^+ \tag{5.7}$$

The other way to expand the electrochemical window of electrolytes is to reduce the activity of water molecules. In a low-concentration normal electrolyte, most of the water molecules are dissociative, and only a small portion integrates with ions to form the solvation sheath. When increasing the salt concentration, more water molecules can participate in the solvation process, which can decrease activity and increase the overpotential of water decomposition [109]. Taking the concentrated $NaClO_4/Zn(CF_3SO_3)_2$ electrolyte as an example (Figure 5.9a),

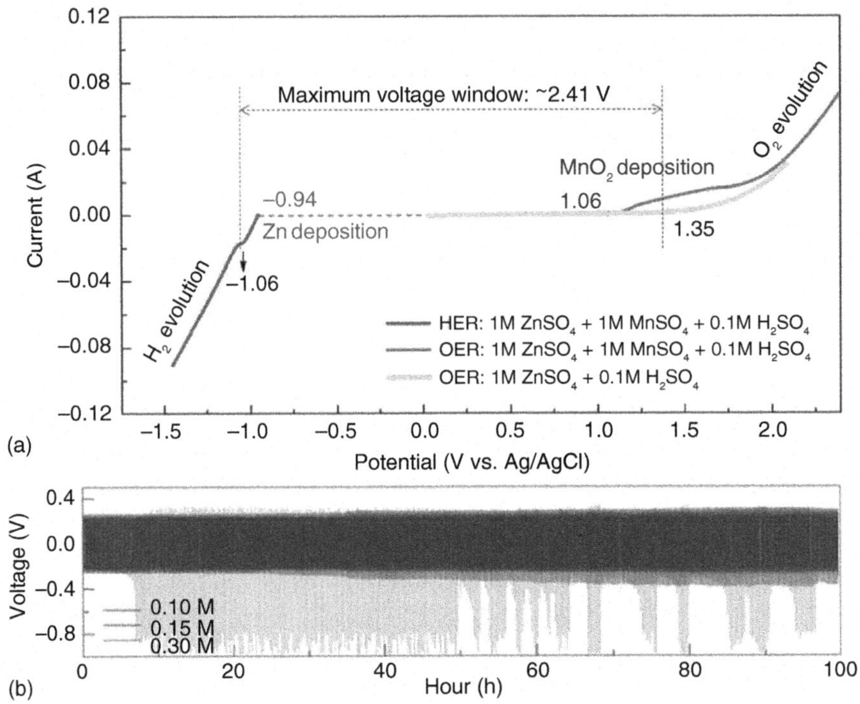

Figure 5.8 The expanded electrochemical window. (a) The expanded operating voltage of the Zn–MnO_2 battery with the addition of 0.1 M H_2SO_4. (b) Electrochemical performance of Zn–Zn symmetric cells cycled in the electrolytes with 0.10, 0.15, and 0.30 M H_2SO_4 at 20 mA cm^{-2}. Source: (b) Chao et al. [106]/John Wiley & Sons.

the electrochemical window can be significantly expanded in the $Zn(CF_3SO_3)_2$ electrolyte with a high $NaClO_4$ concentration owing to the higher kinetic barrier of water molecules in the solvation sheath compared with free water molecules [110, 112]. The wide electrochemical window in the above method is achieved by reducing the activity of water molecules in the electrolyte. In addition to this, a binary hydrate-melt $ZnCl_2/Zn(OAc)_2$ electrolyte is designed to suppress the HER by adjusting the Zn^{2+} solvation structure [113]. In the solvation structure of the hydrate-melt $ZnCl_2/Zn(OAc)_2$ electrolyte, the carboxylate group in OAc^- will coordinate with the Zn^{2+}, which will weaken the interaction between Zn^{2+} and H_2O molecules to achieve higher ionization energy of H_2O molecules. Simultaneously, these carboxylate groups of OAc^- can serve as H-bond acceptors to construct H-bonds with H_2O molecules in their neighboring solvation structures, forming a cross-linking H-bond network. Such a cross-linking H-bond network further suppresses the water activity in $ZnCl_2/Zn(OAc)_2$ electrolytes. However, the additional demand for large amounts of salt will increase the cost, and the resulting higher electrolyte density will also reduce the overall energy density of the battery. Similar to OAc^-, some oxygen-containing chemical bond organic compounds can be employed as electrolyte additives to inhibit water reactivity by constructing H-bonds with water molecules, which can form robust H-bonds

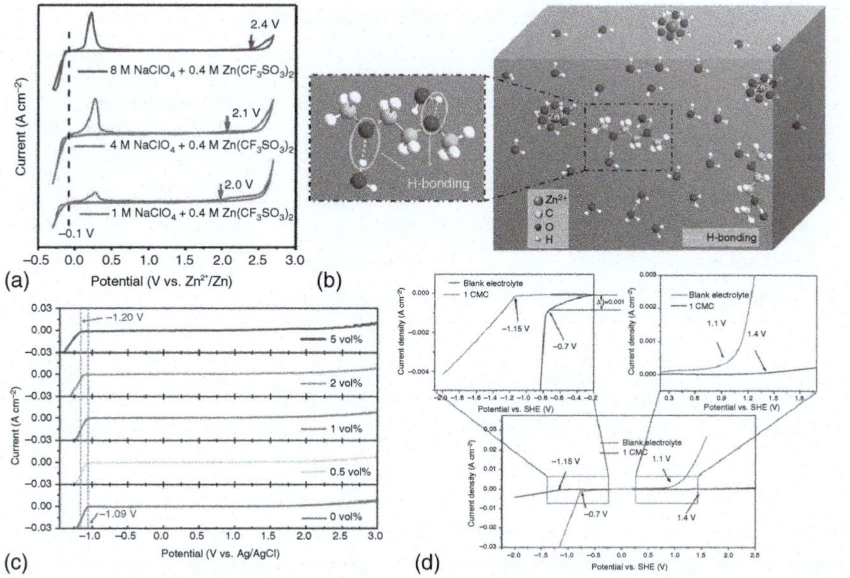

Figure 5.9 The water decomposition performances. (a) CV curves measured in the electrolytes containing $NaClO_4$ and $Zn(CF_3SO_3)_2$ with different concentrations. (b) Schematic illustration of suppressing water reactivity via constructing H-bonding with free water by adding low-concentration DME. (c) Electrochemical windows in 2 M $ZnSO_4$ solution with various volume fractions of DME. (d) The electrochemical window of blank electrolyte and the modified electrolyte with the addition of sodium dodecyl sulfate (SDS) at the critical micelle concentration (CMC). Source: (a) Li et al. [110]/American Chemical Society. (b, c) Cui et al. [94]/With permission of Elsevier. (d) Hou et al. [111]/Royal Society of Chemistry.

with water to reduce free water molecules, thus suppressing hydrogen evolution in the wide voltage window. DME additive with hydrophilic hydration ether oxygen was proposed by Cui et al. to restrict HER in the Zn–salt electrolyte caused by dissociative solvent water molecules (Figure 5.9b) [94]. As shown in Figure 5.9c, at the optimal concentration of DME in the aqueous electrolyte, the onset potential for HER in the weakly acidic electrolyte with 5 vol% DME additives was reduced to −1.20 V (vs. Ag/AgCl) in comparison with the blank electrolyte (−1.09 V). These results can be attributed to the intensive H-bond connection between ether oxygen in the DME additive and free water in the aqueous electrolyte. Plenty of water molecules can be bundled in the surroundings of DME to decrease active water participating in HER, which enables a wide electrochemical window in the weakly acidic electrolyte for AZIBs. Due to the colloidal property, water activity is massively blocked. Therefore, through an in situ gel-forming approach, Qian and coworkers designed a colloid–polymer electrolyte (hereafter called CPE), which could enable a manifesting shift for HER from ~1.2 to ~1.5 V (vs. Ag/AgCl) [114]. It is only necessary to limit the contact between water molecules and electrodes since water decomposition occurs at the solid–liquid interface. The stable voltage window of the aqueous electrolyte could be further extended to 2.5 V (Figure 5.9d) with the

Table 5.1 Summary of recently reported electrolyte engineering for expanding the electrochemical window of AZIBs.

Electrolyte	Strategies	Electrochemical window variation	Refs.
$ZnSO_4 + MnSO_4 + H_2SO_4$	pH regulation	2.0–2.41 V	[105]
$Na_2SO_4 + ZnSO_4$ + sodium dodecyl sulfate	Additive	1.8–2.5 V	[110]
$LiNO_3$-based aqueous electrolyte	High concentration	2.2–2.55 V	[112]
$NaClO_4 + Zn(CF_3SO_3)_2$	High concentration	2.1–2.5 V	[108]

addition of SDS when its concentration reached the CMC [111]. SDS is a common surfactant whose hydrophilic groups tend to adsorb at the electrode surface and whose alkyl chains are vertically distributed to form a hydrophobic layer. It becomes more difficult for water molecules to contact the electrolyte because extra energy is required to allow the water molecules to penetrate the hydrophobic layer, which contributes to the extra overpotential for the evolution of hydrogen or oxygen [115].

In this section, some promising methods to avoid water decomposition in the aqueous electrolyte have been discussed. The expanding electrochemical window of AZIBs using electrolyte engineering is summarized in Table 5.1. A wide voltage window to maintain the inherent stability of aqueous solvents during the electrochemical process is a prerequisite for high-performance AZIBs. Some other issues caused by the interaction between electrolyte and electrode are more fatal, which may decrease the coulombic efficiency and even induce battery failure. Therefore, suitable modification of electrolytes to match the cathode and anode is another key factor in achieving the long-term stability of AZIBs.

5.2.3 Corrosion and Passivation

Due to the reactivity of the interface between electrode and electrolyte, some unexpected reactions may occur during battery storage or operation [116]. Reactions (5.8) and (5.9) show the probable corrosion and hydrogen evolution at the Zn anode surface in the commonly used 1 M $ZnSO_4$ electrolyte with a pH value of about 5.

$$Zn + 2H^+ \rightarrow Zn^{2+} + H_2 \uparrow \text{ (Corrosion)} \tag{5.8}$$

$$2H^+ + 2e^- \rightarrow H_2 \uparrow \text{ (Hydrogen evolution)} \tag{5.9}$$

Electromotive force (E) calculated by the Nernst equation (5.10) can be used to judge whether this reaction is theoretically feasible.

$$E = E^\theta - \frac{RT}{2F} \ln \frac{a_{Zn^{2+}}}{a^2_{H^+}} \tag{5.10}$$

where E^θ, R, T, F, $a_{Zn^{2+}}$ and a_{H^+} represent standard electromotive force (−0.76 V), thermodynamic constant (8.31 J mol^{-1} K^{-1}), temperature (298 K), Faraday constant (95 600 C mol^{-1}), Zn^{2+} activity (1), and H^+ activity (10^{-5}). The calculated positive E (0.46 V) indicates that the corrosion reaction is spontaneous in

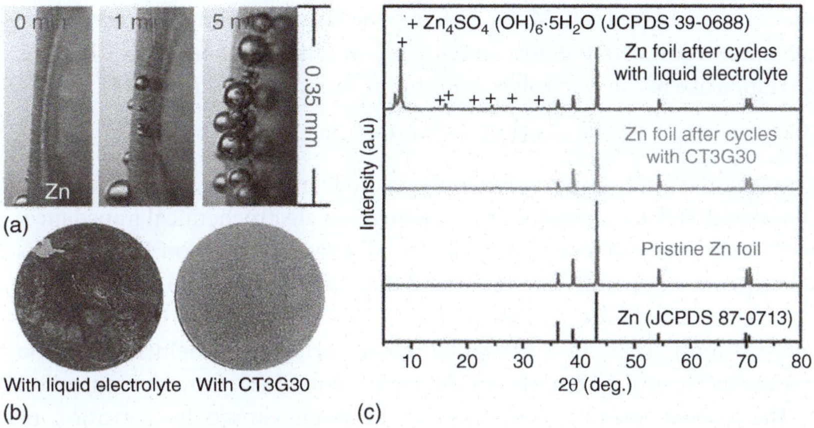

Figure 5.10 The characterization of corrosion and passivation. (a) In situ optical microscopic images of the hydrogen evolution on Zn anode surface. (b) Photograph observations and (c) XRD patterns of Zn foil with liquid and hydrogel electrolytes in Zn–Zn cell after 500 cycles. Source: (a) Shoji et al. [14]/Elsevier. (b) Ma et al. [63]/Royal Society of Chemistry. (c) Adapted from Ghorai et al. [60].

thermodynamics [117]. And the hydrogen production in this reaction is much less compared to the battery charge process under high applied voltage [118]. Figure 5.10a shows the slow gas generation process at the interface between the Zn anode and weakly acidic electrolyte [14]. However, the accumulation of hydrogen cannot be neglected because of the potential risk of battery failure from the continuously increased internal pressure during the long-time battery storage process [119].

The consumption of H^+ during corrosion increases the OH^- concentration and further leads to the formation of insoluble zinc hydroxide sulfate ($Zn_4(OH)_6SO_4 \cdot nH_2O$, ZHS), which is the main component of the passivation layer. Accordingly, it can be considered that corrosion and passivation occur simultaneously based on the reaction (5.11) [120].

$$3Zn + (6 + n)H_2O + ZnSO_4 \rightarrow Zn_4(OH)_6SO_4 \cdot nH_2O + 3H_2 \uparrow \tag{5.11}$$

Anode passivation can also be caused by electrostatic interaction during battery operation. During the stripping process, Zn^{2+} is transferred from Zn metal, and the anode surface is occupied by positive charges. Anions (${SO_4}^{2-}$ and OH^-) tend to aggregate at the interface and coordinate with Zn^{2+} to form ZHS, which is easily supersaturated in the electrolyte and precipitates on the anode surface [121]. In this case, the interfacial impedance is increasing as the insulating passivation layer emerges and continuous water consumption occurs, which is detrimental to the reversible Zn plating/stripping on the anode surface.

Meanwhile, cathode passivation could foster the appearance of ZHS in AZIBs as well. Taking the MnO_2 cathode as an example, the H^+/Zn^{2+} co-insertion reactions (5.12) and (5.13) are its energy-storage mechanisms in weakly acidic electrolytes.

$$MnO_2 + H^+ + e^- \leftrightarrow MnOOH \tag{5.12}$$

$$MnO_2 + xZn^{2+} + 2xe^- \leftrightarrow Zn_xMnO_2 \tag{5.13}$$

H^+ is preferentially inserted into MnO_2 owing to the faster H^+ diffusion and weaker electrostatic interaction with the host lattice [122]. Similar to anode passivation, the resulting pH increase of the electrolyte leads to the formation of ZHS (5.14).

$$3Zn^{2+} + 6OH^- + ZnSO_4 + nH_2O \leftrightarrow Zn_4(OH)_6SO_4 \cdot nH_2O \quad (5.14)$$

In this context, the ion transportation in the cathode may be blocked by the gradually accumulated ZHS layer, leading to an increase in electrochemical impedance. However, cathode passivation is not as severe as anode passivation because this conversion reaction is generally reversible and ZHS can dissolve into an electrolyte when H^+ is extracted from the cathode [123].

Based on the above analysis, detrimental anode corrosion is ineluctable in the common aqueous electrolyte because of the production of H^+ from the ionization of water. The shifted ionization equilibrium of water caused by corrosion or electrochemical reactions may further lead to the passivation of electrodes. Previous research proposed functional interfacial layers to modify the surface activity of the Zn anode, which could exhibit enhanced corrosion resistance during battery cycling [15]. However, the interfacial layer may lose contact with the Zn anode when a large amount of Zn is stripped, and the resulting void space between the anode and layer may deactivate its corrosion prevention. Therefore, it is more effective to modify the inherent characteristics of electrolytes to avoid anode corrosion and maintain continuous anode protection during the dynamical plating/stripping process [4]. There are two main methods of side reaction prevention by electrolyte modification: reducing the activity of water to inhibit its ionization and increasing the reaction overpotential to retard corrosion. The increase in salt concentration or the introduction of organic solvent in the aqueous electrolyte can decrease the activity of ionizable free water. Cui and coworkers proposed a novel water-in-DES electrolyte consisting of water in urea/LiTFSI/$Zn(TFSI)_2$ [88]. The water content is extremely low (6 wt%) in this highly concentrated aqueous electrolyte. Thus, the water molecules are all coordinated with the eutectic solvent network, and the reactivity of hydrogen evolution is restrained by a lower chance of contact with Zn (Figure 5.11a). There have been a great number of researches on gel electrolytes with high ion conductivity and adhesion to hamper contact between the aqueous solution and the Zn anode. Recently, Chen and coworkers fabricated a versatile hydrogel electrolyte for dendrite-free, noncorrosive, and long-life AZIBs [125]. The as-prepared CT3G30 hydrogel composed of cotton as the raw material of the polymetric framework and tetraethyl orthosilicate as the crosslinker, shows excellent adhesion with Zn electrodes and successfully limits free water molecules to induce side reactions. After 500 cycles, Zn foil against hydrogel remains smooth and unchanged in color, and no ZHS peak is identified in XRD analysis (Figure 5.10b,c). Gelatin microgels have been studied as potential corrosion inhibitors in self-healing coatings [126]. Liu et al. investigated the influence of the gelatin electrolyte additive on Zn corrosion in a mildly acidic aqueous solution. It is found that the gelatin additive at the selected concentration is capable of effectively suppressing Zn corrosion. Linear polarization curves show that the corrosion current of 2 M $ZnSO_4$

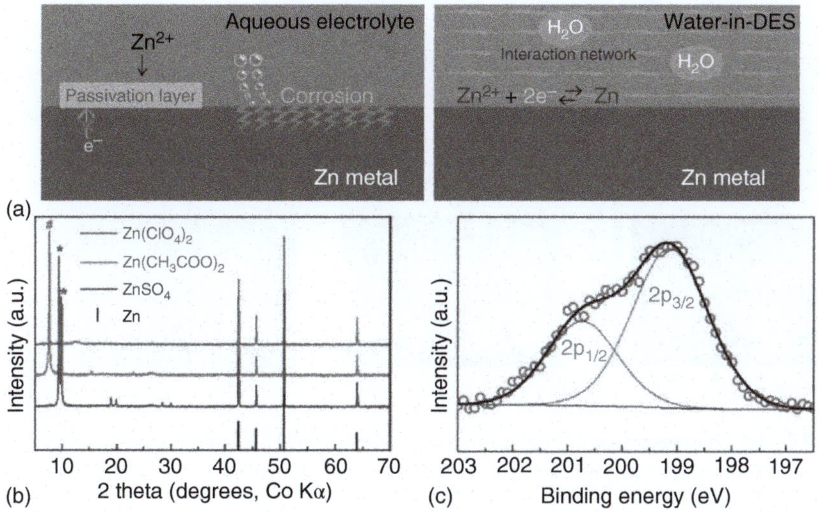

Figure 5.11 The interface reaction in the aqueous electrolyte. (a) The comparison of the different interfacial reactions on Zn anode in the traditional aqueous electrolyte and water-in-DES. (b) XRD patterns of Zn anodes cycled in $ZnSO_4$, $Zn(CH_3COO)_2$, and $Zn(ClO_4)_2$ electrolytes. (c) Cl 2p XPS spectrum of Zn anode after cycling in $Zn(ClO_4)_2$ electrolyte. Source: (a) Adapted from Borodin et al. [109]. (b, c) Wang et al. [124]/American Chemical Society.

electrolyte with a 5% gelatin additive is reduced by ~42 times compared with a pristine electrolyte [127].

An anticorrosive coating similar to the interfacial protective layer can be formed on the anode surface by chemical or electrochemical reactions. $Zn(ClO_4)_2$ aqueous solution is mildly acidic with a pH value of 4.9. When using $Zn(ClO_4)_2$ as the salt in an aqueous electrolyte, the Zn plating and stripping processes are also based on the typical Zn^{2+}/Zn redox reaction [124]. The lower intensity of the ZHS peak was detected in the Zn anode cycled in $Zn(ClO_4)_2$ electrolyte compared with $ZnSO_4$ and Zn $(CH_3COO)_2$ electrolytes (Figure 5.11b), which indicates the suppressed passivation reaction at the interface between the Zn anode and electrolyte by the controlled formation of a Cl^- containing protective layer on the anode surface (Figure 5.11c). Various other inorganic or organic additives have been proposed based on these methods, such as fumed silica, SDS, and thiourea, which also show great effect in inhibiting the corrosion and passivation of the stable Zn anodes [36, 124].

There is an easily overlooked issue that oxygen from the air dissolved in the aqueous electrolyte may be detrimental to the electrodes. When the oxygen is removed from the electrolyte, the corrosion potential is positively shifted and the anode surface remains smooth, indicating that the removal of dissolved O_2 can alleviate the serious corrosion [120]. This discovery brings a new viewpoint that researchers cannot just focus on improving the electrolyte's performance by additives but also observe the influence of substances inherently existing in the electrolyte, such as dissolved salt and gas.

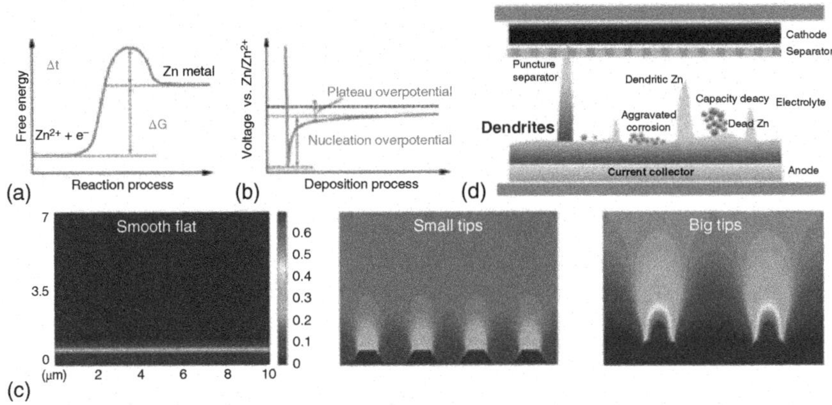

Figure 5.12 (a) Change in free energy during Zn nucleation. (b) Typical voltage profile during zinc deposition. (c) Electric field simulation of the plane with different roughness. (d) Schematic diagram of the influence of dendrites.

5.2.4 Dendrite Growth

The reaction mechanisms of the Zn anode for ZIBs with mild aqueous electrolyte involve the reversible plating/stripping of Zn^{2+} in the anode, which can be summarized as $Zn \leftrightarrow Zn^{2+} + 2e^-$. The quality of zinc deposition closely depends on both the Zn nucleation and growth processes [128]. In a nucleation process, Zn^{2+} needs to transcend a barrier (Figure 5.12a) that corresponds to the nucleation overpotential, as shown in Figure 5.12b [128, 129]. The Zn nucleation process can be driven by an electrical field and ion concentration. In terms of the electrical field, the intensity at the tips is much stronger than the rest area because the greater curvature there can trigger a higher surface charge density, which is known as the "tip effect". Zn near those tips will easily nucleate (Figure 5.12c). Ion concentration also plays a significant role in the nucleation process because the nucleation barrier will be reduced in areas with higher ion concentrations or faster ion transmission. Therefore, Zn dendrites are mainly caused by partial nucleation due to the uneven distribution of electrical fields and ions or a small number of nucleation sites due to high nucleation barriers on the interface. Overall, zinc dendrites are the result of uneven and loose Zn electrodeposition. Note that the partial accumulation of Zn deposition will increase electrode thickness and increase the risk of piercing the membrane, eventually leading to a short circuit in batteries. In addition, the detachment of loose Zn deposition will result in low coulombic efficiency, significantly decreasing the capacity of Zn anodes. Due to their high specific area, those falling Zn particles can react with electrolytes to produce byproducts that will increase electrode impedance, further deteriorating the battery's performance (Figure 5.12d). Therefore, dendrite-free Zn anodes are critical to the practical application of ZIBs.

There have been many anode modification strategies to uniformize electron and ion distribution, such as current collectors with three-dimensional porous structures and zincophilic decoration [130]. However, they can only guide the Zn nucleation process and a few Zn growth layers, which may be useless when a large amount of Zn is plated on the substrate because the Zn dendrites may still grow on the anode. The electrolyte modification not only influences the interfacial ion behavior but also participates in the Zn reduction reaction to continuously control the Zn deposition from the nucleation process to the subsequent growth process [131].

As discussed above, the electric field is relatively high at the tips of small Zn particles formed at the beginning of Zn deposition on the anode surface. The highly polarized diethyl ether shows a "self-healing electrostatic shield" effect, which can preferentially attach to Zn tips to avoid the two-dimensional Zn^{2+} diffusion to the high-electric-field region [29]. The Zn deposition is transferred to the flat regions with fewer diethyl ether molecules, which induces the smooth surface of Zn anodes. According to the electrostatic shield mechanism, cation additives with a low reduction potential (vs. Zn/Zn^{2+}) can preferentially be reduced on the Zn tips to avoid dendrite growth as well. Introducing a Na_2SO_4 additive with low reduction potential into the $ZnSO_4$ electrolyte not only suppresses Zn dendrite extension but also prevents cathode dissolution when the $NaV_3O_8{\cdot}1.5H_2O$ cathode is coupled with Zn to assemble full batteries [132]. Our group reported that uniform zinc deposition can be realized as well with high-valence cations (Ce^{3+} and La^{3+}) involved to homogenize electric fields (Figure 5.13a–h) [133]. It is demonstrated that such cations preferentially occupy the initial active sites and then compel zinc nucleation/growth at other relatively inert areas. Those occupied sites will be free to nucleate/grow because the cation additives are summoned away by the higher electric fields with the ongoing deposition process (Figure 5.13i), guiding zinc nucleation to progressive nucleation from instantaneous nucleation in a bare electrolyte (Figure 5.13j,k). Nucleation occurring at a high rate on a small number density of active sites is called "instantaneous" because the nucleation process arrests as nucleation sites are exhausted. Slow nucleation on a large number of active sites is "progressive" because nuclei continue appearing throughout the phase transformation process. Besides, some organics (such as sodium dodecylbenzene sulfonate and PAM) can be served as anchors to forcibly guide the Zn^{2+} transfer in the limited position of the anode surface, and the modified electrolytes with these organic additives also exhibit dendrite-free Zn deposition behavior [134].

Generally, the ion transmission from the electrolyte to the anode surface could be demonstrated by an electric double-layer structure (Figure 5.14a). Hydrated Zn^{2+} is first transferred near the anode surface by the electric field, then de-solvated in the compact layer, and finally deposited on the anode surface [135]. The energy consumption of Zn^{2+} de-solvation in the compact layer has an important influence on the quality of Zn deposition. Low energy consumption indicates fast transfer kinetics in the electric double layer, which can provide abundant nucleation sites and uniformize Zn deposition to reduce the risk of dendrite formation [12]. Besides, the Zn^{2+} adsorbed on the anode surface can diffuse laterally along the surface and be reduced to Zn at the low-energy barrier site, resulting in aggregated zinc deposition at the previous Zn particles to minimize the surface area exposed to the electrolyte [137]. This uneven Zn deposition may eventually lead to dendrite growth (Figure 5.14b) [136]. Therefore, the modification method to suppress

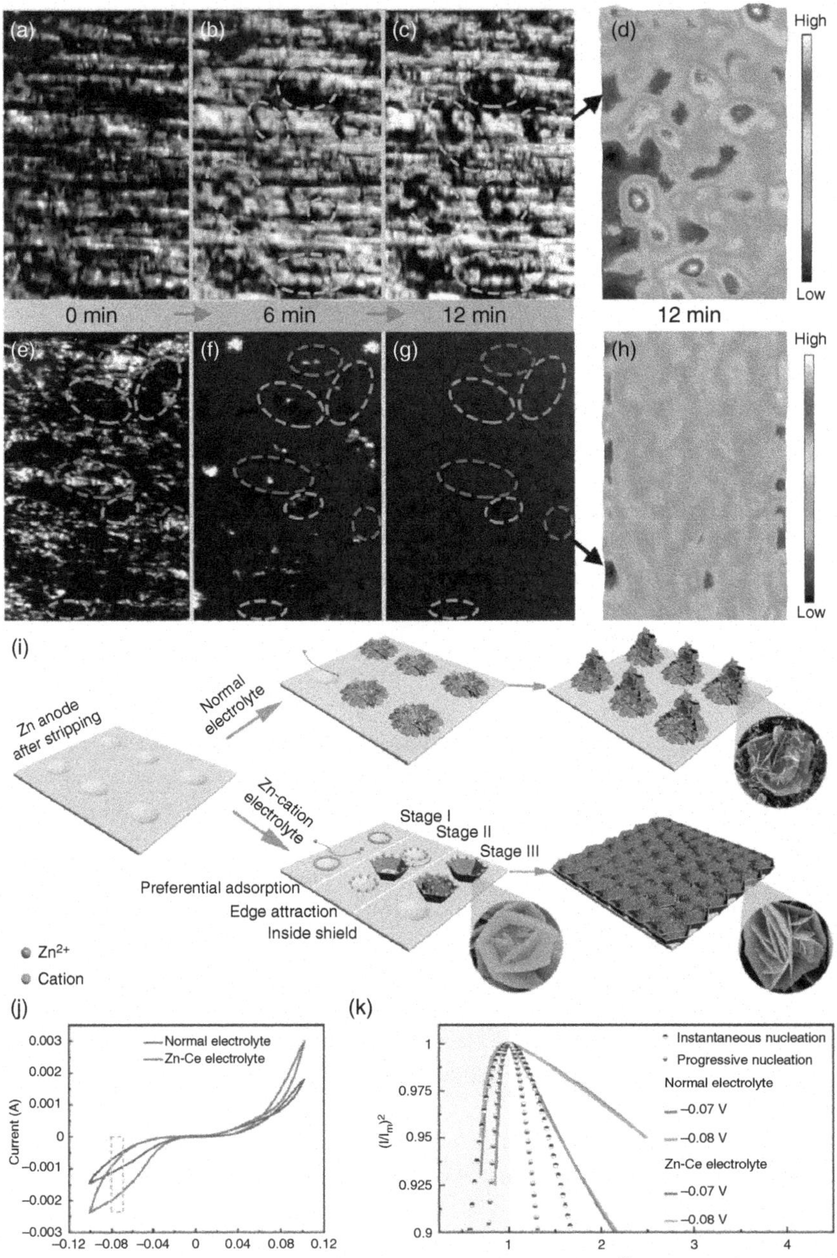

Figure 5.13 The electrolyte engineering for dendrite-free Zn anode. SEM images of the zinc anodes in the (a–c) normal and (e–g) Zn–Ce electrolytes after deposition at 5 mA cm^{-2} for 2, 6, and 12 min, respectively. Surface simulation of 12 min deposition in the (d) normal and (h) Zn–Ce electrolytes. (i) Schematic illustration of the zinc deposition on the zinc foil after stripping in the normal and Zn–Ce electrolytes. (j) Cyclic voltammetry (CV) curves of the Zn//Zn battery at 1 mV s^{-1}. (k) Dimensionless curves (I/I_m) vs. (t/t_m) for instantaneous and progressive nucleation, respectively (scatters), along with experimental data (thick line).

Zn dendrites should focus on optimizing the de-solvation process in an electric double-layer to reduce the charge transfer resistance and restraining the interfacial two-dimensional Zn^{2+} diffusion to prevent Zn aggregation. By applying this principle, our group employed an extremely low concentration of sulfolane to reshape the Helmholtz layer, achieving a dendrite-free and corrosion-free Zn anode. It was confirmed that sulfolane could occupy the inner Helmholtz layer before water and then anchor the Zn^{2+} from the outer Helmholtz layer to the direct deposit at which they contact (Figure 5.14c), eradicating the aggregation of zinc ions/atoms. In addition, the widened Helmholtz layer with reduced electric field intensity could generate more nucleation sites and narrow the nuclei radius. This facile strategy provides fresh and deep insights into functionalizing the double layer and brings new possibilities for realizing practically rechargeable Zn batteries. Besides, the TBA_2SO_4 (Figure 5.14d) additives can also be adsorbed on the Zn electrode surface to hinder Zn-ion migration and aggregation, enabling the formation of numerous tiny nucleation centers to induce uniform Zn deposition by limiting 2D Zn^{2+}-ion diffusion on the anode surface [41]. In addition, Zn^{2+} with fast reaction kinetics on the Zn anode surface may facilitate the inhomogeneous Zn deposition in the mild aqueous electrolyte. A long-chain PEO additive in $ZnSO_4$ electrolyte can effectively slow down Zn^{2+} transfer to regulate concentration distribution and electrolyte flux through interactions between Zn^{2+} and the ether group in PEO, which ensures homogeneous Zn^{2+} distribution on the interface between the Zn anode and electrolyte to achieve a uniform deposition/dissolution process [100]. Nevertheless, the organic additives introduced in the aqueous electrolyte remain on the anode surface and participate in the interfacial reactions, which may complicate the continuous Zn plating/stripping processes to some extent.

For the Zn^{2+} in the double layer, its solvation structure is critically important to the subsequent zinc deposition. A different Zn^{2+} solvation sheath can be obtained through electrolyte composition design, which can further affect the de-solvation process of Zn deposition. Wang et al. proposed a novel concentrated water-in-salt electrolyte containing 1 M $Zn(TFSI)_2$ and 21 M LiTFSI for AZIBs [54]. The Zn^{2+} solvation sheath is composed of $TFSI^-$ in this unique electrolyte, and Zn^{2+} is de-solvated from $TFSI^-$ sheath instead of water sheath during Zn deposition because of the strong interaction between Li^+ and water molecules (Figure 5.14e). The optimized energy barrier for the Zn^{2+} de-solvation endows the Zn anode with enhanced cyclic stability and uniform Zn deposition. Gel electrolytes can also reduce the activity of water through strong hydrogen-bonding interactions that restrict water molecules [138]. This mechanism to induce uniform Zn deposition and suppress water-related reactions is similar to the water-in-salt electrolyte. Gel electrolyte makes flexible AZIBs possible, which can be applied to emerging portable and wearable electronic products.

The enhanced electrochemical performances of AZIBs with dendrite-free zinc anodes using electrolyte engineering are summarized in Table 5.2. Highly concentrated salt aqueous electrolytes and gel electrolytes can greatly promote the uniformity of the Zn anode by reducing the solvation and de-solvation effects. However, their increased viscosity results in lower ionic conductivity, which is

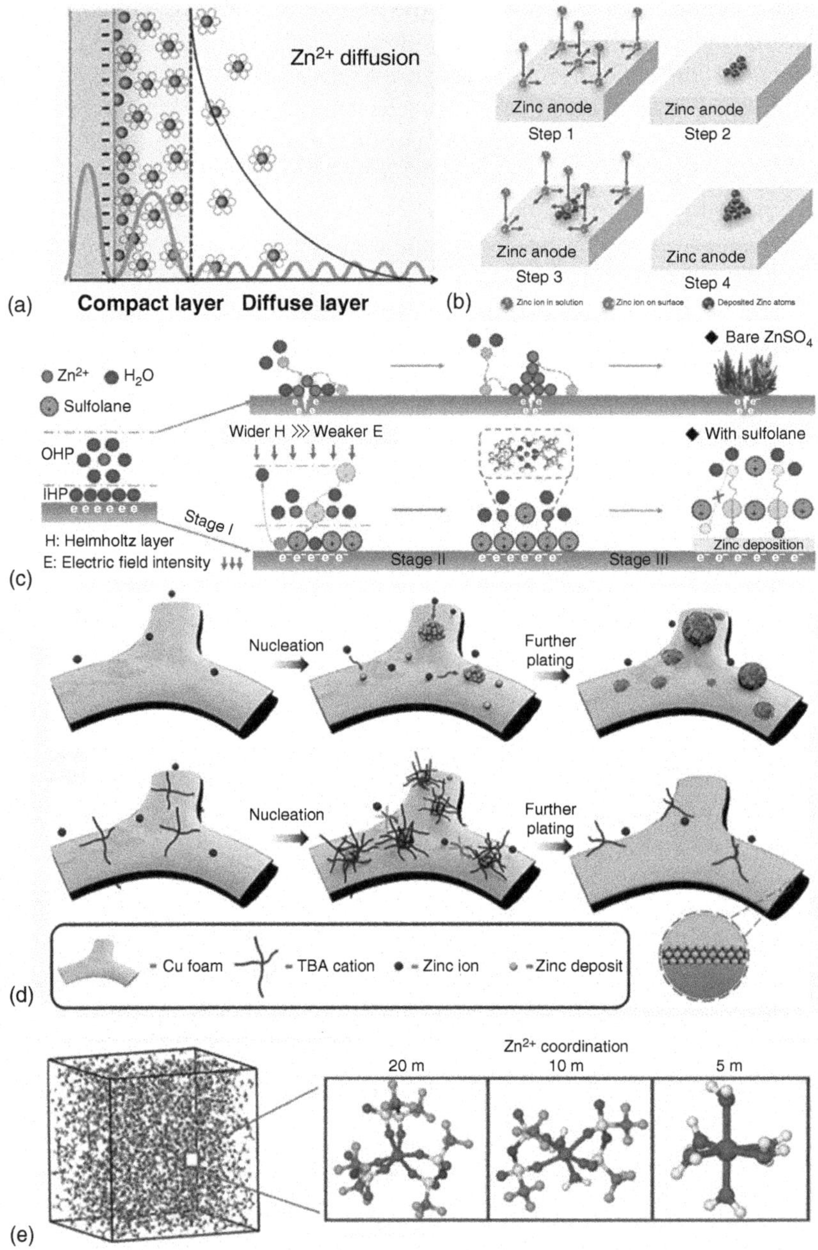

Figure 5.14 The schematic illustration of Zn deposition. (a) Electric double-layer structure of the interface between Zn anode and electrolyte. (b) Illustration for the Zn reduction and deposition process on Zn anode. (c) Schematic illustration of working principle of Zn deposition in the bare electrolyte and electrolyte with sulfolane. (d) Schematic illustration of Zn dendrite inhibition on Cu foam with TBA_2SO_4 as electrolyte additives. (e) Molecular dynamics simulations and Zn^{2+}-solvation structures in 1 M $Zn(TFSI)_2$ electrolytes with different concentrations of LiTFSI. Source: (a, b) Xie et al. [135]/Royal Society of Chemistry. (c, d) Adapted from Bayaguud et al. [41].

Table 5.2 Summary of recently reported electrolyte engineering for dendrite-zinc anode in AZIBs.

Anodes	Strategies	Operating condition	Lifespan	Refs.
Zinc-plated copper mesh	Polyacrylamide additive	0.2 mA cm^{-2}, 1 mAh cm^{-2}	350 h	[129a]
Zinc foil	Diethyl ether additive	0.2 mA cm^{-2}, 0.2 mAh cm^{-2}	250 h	[126]
Zinc foil	Hydrated eutectic electrolytes	0.05 mA cm^{-2}, 0.5 mAh cm^{-2}	800 h	[131]
Zinc-plated copper foam	Tetrabutylammonium sulfate additive	2 mA cm^{-2}, 2 mAh cm^{-2}	300 h	[134]
Zinc foil	Sodium dodecyl benzene sulfonate additive	0.5 mA cm^{-2}, 0.25 mAh cm^{-2}	1500 h	[129b]
Zinc foil	High concentration $Zn(TFSI)_2$ + Li(TFSI)	0.2 mA cm^{-2}, 0.032 mAh cm^{-2}	170 h	[136]
Zinc foil	PVA/$Zn(CF_3SO_3)_2$ hydrogels	0.1 mA cm^{-2}	800 h	[138]

detrimental to the battery's performance at high current density [139]. Besides, some additives can block grain growth to achieve uniform and denser Zn deposition by increasing Zn growth overpotential. High overpotential is not conducive to the development of high-energy-density AZIBs [140]. It is essential to develop new strategies to balance all related issues and enhance the overall performance of Zn anodes.

Inspired by the effects of SEI in Li-metal batteries, in situ SEI film has been reported to be able to successfully eliminate zinc dendrites. To construct this film on the zinc metal surface, the core is to change the decomposition object from water molecules to other substances [29]. It has been reported that $TFSI^-$ can form an in situ SEI layer compatible with zinc anodes by participating in the solvated structure of zinc ions (Figure 5.15a) [100]. DFT calculations show that the Zn^{2+}–$TFSI^-$ complex becomes unstable below 0.37 V (vs. Zn/Zn^{2+}) due to the close interaction between $TFSI^-$ and Zn^{2+}. In this case, $TFSI^-$ preferentially accepts electrons over Zn^{2+} to be reduced to form in situ SEI (Figure 5.15b). XPS spectra show that ZnF_2 mainly exists in the inner layer of SEI, and S/N-rich organic compounds mainly distribute in the outer layer of SEI. The high ZnF_2/Zn interface effectively suppresses dendrite growth by promoting lateral rather than vertical Zn^{2+} migration and deposition [143]. Since the SEI film can have a fast ion migration rate and high mechanical strength, the zinc metal anode in this electrolyte exhibits excellent reversibility. In addition, the addition of DMSO to $ZnCl_2$–H_2O can also achieve a similar effect [23]. During the charge–discharge process, the DMSO involved in the solvation will decompose to form an SEI enriched with $Zn_{12}(SO_4)_3Cl_3(OH)_{15}\cdot 5H_2O$, $ZnSO_3$, and ZnS, which avoids the formation of zinc dendrites and further inhibits water splitting.

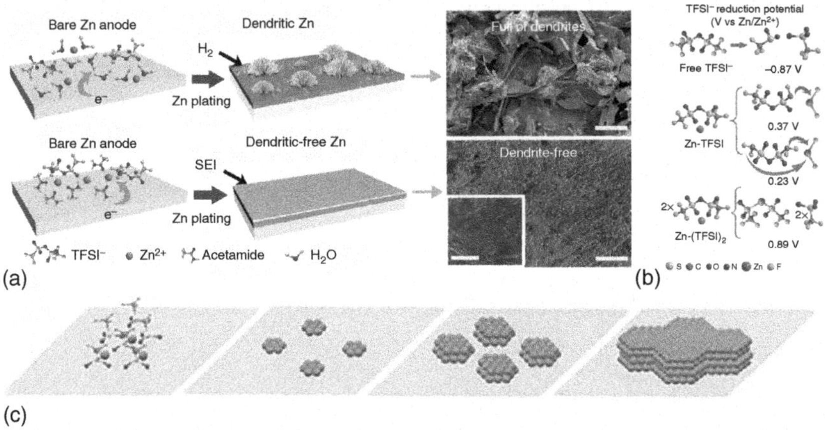

Figure 5.15 (a) Schematic illustration of zinc stripping/deposition in 1 M $Zn(TFSI)_2$ and ZES. (b) Predicted reduction potentials of $TFSI^-$. (c) Demonstration of OTf^- to induce Zn (002) texture-preferred growth. Source: (a, b) [141]/Springer Nature/Licensed under CC BY 4.0. (c) Yuan et al. [142]/John Wiley & Sons.

Concerning zinc growth behavior, studies have demonstrated that electrolyte optimization can achieve preferential crystal plane orientation for zinc growth and restrict the two-dimensional diffusion behavior of zinc ions on the surface, thereby driving uniform zinc deposition and avoiding dendrite formation. Chen et al. added organic additives [(hexadecyltrimethylammonium bromide (CTAB), SDS, polyethylene glycol (PEG-8000), and thiourea (TU)] into electrolytes. It was found that the orientation of the surface crystal planes changed during the zinc deposition process [144]. For example, CTAB can direct Zn^{2+} to deposit along the (101) crystal plane, generating zinc flakes perpendicular to the substrate that may induce short-circuiting of the cell. In contrast, SDS can easily promote zinc deposition along the (002) crystal plane, resulting in dense and uniform Zn deposition. Therefore, the capacity retention rates of Zn–SDS, Zn–PEG, and Zn–Tu anodes after more than 1000 cycles at 4 C are 79%, 76%, and 80%, respectively, while those of commercial zinc foil anodes are only 67%. Yuan et al. found that Zn^{2+} could be deposited in (002)-preferred crystal plane in the $Zn(OTF)_2$ electrolyte [142]. This is because the OTf^- can not only improve the stability of the adsorption of Zn^{2+} on the copper surface, but more importantly, it can promote the zinc nucleation behavior to proceed in a hexagonal arrangement. Since the basic hexagonal close packing has the lowest surface energy in the zinc lattice, the hexagonally arranged nuclei will grow in the (002) plane until the nuclei contact each other to form a continuous deposit (Figure 5.15c). This can greatly improve the stability of the battery.

5.2.5 Interaction Among HER, Corrosion, and Dendrite Growth

These three issues discussed above mutually interact with each other. Their interaction is summed up and displayed in Figure 5.16. In specific, dendrite formation can significantly proliferate the anode surface, which boosts hydrogen evolution [145].

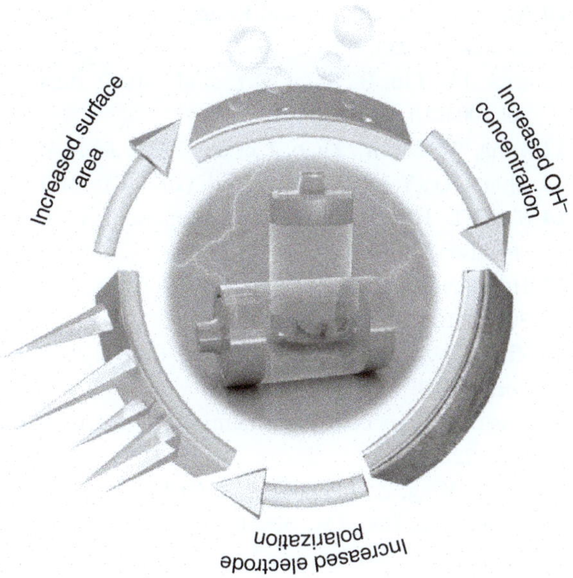

Figure 5.16 Schematic diagram of the interaction of dendrite formation, hydrogen evolution, and corrosion by-product formation.

As a result, the local pH in the proximity of the zinc anode changes owing to the concentration elevation of OH^-. Unfortunately, these generated OH^- will take part in side reactions to form electrochemically inert corrosion byproducts that can be deposited on the anode surface [146]. In this case, the growing degree of roughness of the anode surface will, in turn, promote dendrite formation [147]. Hence, in practice, when we are aiming to solve one of the aforementioned issues, we may indirectly alleviate one or more other issues at the same time. Conversely, the solution we adopted to suppress one issue may deteriorate another one in some cases.

References

1 Li, H., Ma, L., Han, C. et al. (2019). *Nano Energy* 62: 550.
2 Naveed, A., Yang, H., Shao, Y. et al. (2019). *Advanced Materials* 31: e1900668.
3 Kundu, D., Vajargah, S.H., Wan, L. et al. (2018). *Energy & Environmental Science* 11: 881.
4 Han, C., Li, W., Liu, H.K. et al. (2020). *Nano Energy* 74: 104880.
5 Song, M., Tan, H., Chao, D., and Fan, H.J. (2018). *Advanced Functional Materials* 28: 1802564.
6 Lee, J.-S., Kim, S.T., Cao, R. et al. (2011). *Advanced Energy Materials* 1: 34.
7 Mainar, A.R., Leonet, O., Bengoechea, M. et al. (2016). *International Journal of Energy Research* 40: 1032.
8 El-Sayed, A.-R., Mohran, H.S., and El-Lateef, H.M.A. (2010). *Journal of Power Sources* 195: 6924.
9 Zhao, Z., Fan, X., Ding, J. et al. (2019). *ACS Energy Letters* 4: 2259.

10 Xu, C., Li, B., Du, H., and Kang, F. (2012). *Angewandte Chemie, International Edition* 51: 933.
11 Kim, J.Y., Liu, G., Shim, G.Y. et al. (2020). *Advanced Functional Materials* 30: 2004210.
12 Yang, W., Du, X., Zhao, J. et al. (2020). *Joule* 4: 1557.
13 Zhang, Q., Yang, Z., Ji, H. et al. (2021). *SusMat* 1: 432.
14 Shoji, T., Hishinuma, M., and Yamamoto, T. (1988). *Journal of Applied Electrochemistry* 18: 521.
15 Hao, J.N., Li, X.L., Zhang, S.L. et al. (2020). *Advanced Functional Materials* 30: 2001263.
16 Zhang, N., Cheng, F., Liu, Y. et al. (2016). *Journal of the American Chemical Society* 138: 12894.
17 Kasiri, G., Trocoli, R., Hashemi, A.B., and La Mantia, F. (2016). *Electrochimica Acta* 222: 74.
18 Li, G., Yang, Z., Jiang, Y. et al. (2016). *Nano Energy* 25: 211.
19 Chae, M.S., Heo, J.W., Kwak, H.H. et al. (2017). *Journal of Power Sources* 337: 204.
20 Kim, S., Lee, J., Kim, S. et al. (2018). *Energy Technology* 6: 340.
21 Hao, J.N., Yuan, L.B., Ye, C. et al. (2021). *Angewandte Chemie, International Edition* 60: 7366.
22 Chang, N.N., Li, T.Y., Li, R. et al. (2020). *Energy & Environmental Science* 13: 3527.
23 Cao, L., Li, D., Hu, E. et al. (2020). *Journal of the American Chemical Society* 142: 21404.
24 Aguayo, A.T., Gayubo, A.G., Vivanco, R. et al. (2005). *Applied Catalysis A: General* 283: 197.
25 Qin, R.Z., Wang, Y.T., Zhang, M.Z. et al. (2021). *Nano Energy* 80: 105478.
26 Li, F.Y., Yu, L., Hu, Q.M. et al. (2021). *Science China Materials* 64: 1609.
27 Shi, J.Q., Xia, K.X., Liu, L.J. et al. (2020). *Electrochimica Acta* 358: 136937.
28 Dong, Y., Miao, L.C., Ma, G.Q. et al. (2021). *Chemical Science* 12: 5843.
29 Xu, W.N., Zhao, K.N., Huo, W.C. et al. (2019). *Nano Energy* 62: 275.
30 Feng, R.F., Chi, X.W., Qiu, Q.L. et al. (2021). *ACS Applied Materials & Interfaces* 13: 40638.
31 Nigatu, T.A., Bezabh, H.K., Taklu, B.W. et al. (2021). *Journal of Power Sources* 511: 230413.
32 Feng, D.D., Cao, F.Q., Hou, L. et al. (2021). *Small* 17: 2103195.
33 Hou, Z., Tan, H., Gao, Y. et al. (2020). *Journal of Materials Chemistry A* 8: 19367.
34 Nian, Q.S., Wang, J.Y., Liu, S. et al. (2019). *Angewandte Chemie, International Edition* 58: 16994.
35 Chao, D., Zhu, C., Song, M. et al. (2018). *Advanced Materials* 30: e1803181.
36 Huang, J., Chi, X., Han, Q. et al. (2019). *Journal of the Electrochemical Society* 166: A1211.
37 Wan, F., Zhang, L., Dai, X. et al. (2018). *Nature Communications* 9: 1656.

38 Sun, K.E.K., Hoang, T.K.A., The Nam Long, D. et al. (2018). *Chemistry—A European Journal* 24: 1667.
39 Huang, C., Zhao, X., Liu, S. et al. (2021). *Advanced Materials* 33: 2100445.
40 Lu, H.T., Zhang, X.L., Luo, M.H. et al. (2021). *Advanced Functional Materials* 31: 2103514.
41 Bayaguud, A., Luo, X., Fu, Y.P., and Zhu, C.B. (2020). *ACS Energy Letters* 5: 3012.
42 Cao, J., Zhang, D., Zhang, X. et al. (2022). *Energy & Environmental Science* 15: 499.
43 Yi, Z.H., Chen, G.Y., Hou, F. et al. (2021). *Advanced Energy Materials* 11: 2003065.
44 Xu, M.Y., Zhu, T., and Zhang, J.Z.H. (2019). *The Journal of Physical Chemistry. A* 123: 6587.
45 Suo, L.M., Borodin, O., Gao, T. et al. (2015). *Science* 350: 938.
46 Glatz, H., Tervoort, E., and Kundu, D. (2020). *ACS Applied Materials & Interfaces* 12: 3522.
47 Olbasa, B.W., Fenta, F.W., Chiu, S.F. et al. (2020). *ACS Applied Energy Materials* 3: 4499.
48 Chen, C.Y., Matsumoto, K., Kubota, K. et al. (2019). *Advanced Energy Materials* 9: 1900196.
49 Zhang, C., Holoubek, J., Wu, X.Y. et al. (2018). *Chemical Communications* 54: 14097.
50 Zhang, C., Shin, W., Zhu, L.D. et al. (2021). *Carbon Energy* 3: 339.
51 Blanc, L.E., Kundu, D., and Nazar, L.F. (2020). *Joule* 4: 771.
52 Liu, J., Hu, M., Wang, J. et al. (2019). *Nano Energy* 58: 338.
53 Liu, Z., Wang, D., Tang, Z. et al. (2019). *Energy Storage Materials* 23: 636.
54 Mo, F., Liang, G., Meng, Q. et al. (2019). *Energy & Environmental Science* 12: 706.
55 Wu, K., Huang, J., Yi, J. et al. (2020). *Advanced Energy Materials* 10: 1903977.
56 Liu, Z., Mo, F., Li, H. et al. (2018). *Small Methods* 2: 1800124.
57 Cui, Y., Zhao, Q., Wu, X. et al. (2020). *Energy Storage Materials* 27: 1.
58 Han, Q., Chi, X., Liu, Y. et al. (2019). *Journal of Materials Chemistry A* 7: 22287.
59 Tang, Y., Liu, C., Zhu, H. et al. (2020). *Energy Storage Materials* 27: 109.
60 Ghorai, S., Sarkar, A., Raoufi, M. et al. (2014). *ACS Applied Materials & Interfaces* 6: 4766.
61 Zhang, S., Yu, N., Zeng, S. et al. (2018). *Journal of Materials Chemistry A* 6: 12237.
62 Kaltenbrunner, M., Kettlgruber, G., Siket, C. et al. (2010). *Advanced Materials* 22: 2065.
63 Ma, L., Chen, S., Li, H. et al. (2018). *Energy & Environmental Science* 11: 2521.
64 Ma, L., Chen, S., Pei, Z. et al. (2018). *ACS Nano* 12: 8597.
65 Li, H., Liu, Z., Liang, G. et al. (2018). *ACS Nano* 12: 3140.
66 Tafur, J.P., Abad, J., Roman, E., and Fernandez Romero, A.J. (2015). *Electrochemistry Communications* 60: 190.
67 Kumar, G.G. and Sampath, S. (2003). *Solid State Ionics* 160: 289.

68 Rogers, R.D. and Seddon, K.R. (2003). *Science* 302: 792.

69 MacFarlane, D.R., Forsyth, M., Howlett, P.C. et al. (2016). *Nature Reviews Materials* 1: 15005.

70 Liu, C.X., Xie, X.S., Lu, B.G. et al. (2021). *ACS Energy Letters* 6: 1015.

71 Branco, L.C., Rosa, J.N., Ramos, J.J.M., and Afonso, C.A.M. (2002). *Chemistry—A European Journal* 8: 3671.

72 MacFarlane, D.R., Meakin, P., Sun, J. et al. (1999). *The Journal of Physical Chemistry B* 103: 4164.

73 Deng, M.J., Chen, P.Y., and Sun, I.W. (2007). *Electrochimica Acta* 53: 1931.

74 Serizawa, N., Katayama, Y., and Miura, T. (2010). *Electrochimica Acta* 56: 346.

75 Ma, L.T., Chen, S.M., Li, N. et al. (2020). *Advanced Materials* 32: e1908121.

76 Liu, Z., Cui, T., Lu, T.Q. et al. (2016). *Journal of Physical Chemistry C* 120: 20224.

77 Liu, Z., El Abedin, S.Z., and Endres, F. (2013). *Electrochimica Acta* 89: 635.

78 Liu, Z., Pulletikurthi, G., Lahiri, A. et al. (2016). *Dalton Transactions* 45: 8089.

79 Abbott, A.P., Capper, G., Davies, D.L. et al. (2001). *Chemical Communications* 19: 2010.

80 Tran, M.K., Rodrigues, M.T.F., Kato, K. et al. (2019). *Nature Energy* 4: 339.

81 Hammond, O.S., Bowron, D.T., and Edler, K.J. (2017). *Angewandte Chemie, International Edition* 56: 9782.

82 Zhang, C.K., Ding, Y., Zhang, L.Y. et al. (2017). *Angewandte Chemie, International Edition* 56: 7454.

83 Smith, E.L., Abbott, A.P., and Ryder, K.S. (2014). *Chemical Reviews* 114: 11060.

84 Wu, J.X., Liang, Q.H., Yu, X.L. et al. (2021). *Advanced Functional Materials* 31: 2011102.

85 Zhang, C.K., Zhang, L.Y., and Yu, G.H. (2020). *Accounts of Chemical Research* 53: 1648.

86 Du, C.L., Zhao, B.Y., Chen, X.B. et al. (2016). *Scientific Reports* 6: 29225.

87 Abbott, A.P., Barron, J.C., and Ryder, K.S. (2009). *Transactions of the Institute of Metal Finishing* 87: 201.

88 Zhao, J.W., Zhang, J., Yang, W.H. et al. (2019). *Nano Energy* 57: 625.

89 Qiu, H.Y., Du, X.F., Zhao, J.W. et al. (2019). *Nature Communications* 10: 5374.

90 Pan, H., Shao, Y., Yan, P. et al. (2016). *Nature Energy* 1: 16039.

91 Tang, B., Shan, L., Liang, S., and Zhou, J. (2019). *Energy & Environmental Science* 12: 3288.

92 Zhang, Q., Luan, J., Fu, L. et al. (2019). *Angewandte Chemie, International Edition* 58: 15841.

93 Wan, F. and Niu, Z. (2019). *Angewandte Chemie, International Edition* 58: 16358.

94 Cui, J., Liu, X., Xie, Y. et al. (2020). *Materials Today Energy* 18: 100563.

95 Chamoun, M., Brant, W.R., Tai, C.-W. et al. (2018). *Energy Storage Materials* 15: 351.

96 Shi, H.Y., Song, Y., Qin, Z. et al. (2019). *Angewandte Chemie, International Edition* 58: 16057.

97 Lu, Y., Zhu, T., van den Bergh, W. et al. (2020). *Angewandte Chemie, International Edition* 59: 17004.

98 He, P., Zhang, G., Liao, X. et al. (2018). *Advanced Energy Materials* 8: 1702463.

99 Wu, X., Xu, Y., Zhang, C. et al. (2019). *Journal of the American Chemical Society* 141: 6338.

100 Wang, F., Borodin, O., Gao, T. et al. (2018). *Nature Materials* 17: 543.

101 Huang, J., Chi, X., Wu, J. et al. (2022). *Chemical Engineering Journal* 430: 133058.

102 Shin, J., Choi, D.S., Lee, H.J. et al. (2019). *Advanced Energy Materials* 9: 1900083.

103 Hou, Z., Dong, M., Xiong, Y. et al. (2020). *Small* 16: e2001228.

104 Kühnel, R.-S., Reber, D., and Battaglia, C. (2017). A high-voltage aqueous electrolyte for sodium-ion batteries. *ACS Energy Letters* 2: 20052006.

105 Fic, K., Lota, G., Meller, M., and Frackowiak, E. (2012). *Energy & Environmental Science* 5: 5842.

106 Chao, D., Zhou, W., Ye, C. et al. (2019). *Angewandte Chemie, International Edition* 131: 7905.

107 Yu, F., Pang, L., Wang, X. et al. (2019). *Energy Storage Materials* 19: 56.

108 Zang, X., Shen, C., Sanghadasa, M., and Lin, L. (2019). *ChemElectroChem* 6: 976.

109 Borodin, O., Self, J., Persson, K.A. et al. (2020). *Joule* 4: 69.

110 Li, W., Wang, K., Zhou, M. et al. (2018). *ACS Applied Materials & Interfaces* 10: 22059.

111 Hou, Z., Zhang, X., Li, X. et al. (2017). *Journal of Materials Chemistry A* 5: 730.

112 Zheng, J., Tan, G., Shan, P. et al. (2018). *Chem* 4: 2872.

113 Yang, M., Zhu, J., Bi, S. et al. (2022). *Advanced Materials* 34: 2201744.

114 Zhou, J., Zhang, L., Peng, M. et al. (2022). *Advanced Materials* 34: 2200131.

115 Liu, Z., Huang, Y., Huang, Y. et al. (2020). *Chemical Society Reviews* 49: 180.

116 Dubouis, N., Lemaire, P., Mirvaux, B. et al. (2018). *Energy & Environmental Science* 11: 3491.

117 Shin, J., Lee, J., Park, Y., and Choi, J.W. (2020). *Chemical Science* 11: 2028.

118 Konarov, A., Voronina, N., Jo, J.H. et al. (2018). *ACS Energy Letters* 3: 2620.

119 Cai, Z., Ou, Y., Wang, J. et al. (2020). *Energy Storage Materials* 27: 205.

120 Su, L., Liu, L., Liu, B. et al. (2020). *iScience* 23: 100995.

121 Soundharrajan, V., Sambandam, B., Kim, S. et al. (2018). *ACS Energy Letters* 3: 1998.

122 Huang, J., Wang, Z., Hou, M. et al. (2018). *Nature Communications* 9: 1.

123 Zhao, S., Han, B., Zhang, D. et al. (2018). *Journal of Materials Chemistry A* 6: 5733.

124 Wang, L., Zhang, Y., Hu, H. et al. (2019). *ACS Applied Materials & Interfaces* 11: 42000.

125 Chen, M., Chen, J., Zhou, W. et al. (2021). *Advanced Materials* 33: 2007559.

126 Stankiewicz, A., Jagoda, Z., Zielińska, K., and Szczygieł, I. (2015). *Materials and Corrosion* 66: 1391.

127 Liu, Y., Shi, Q., Wu, Y. et al. (2021). *Chemical Engineering Journal* 407: 127189.

128 Zhang, Q., Luan, J., Tang, Y. et al. (2020). *Angewandte Chemie, International Edition* 59: 13180.

129 (a) Zhang, Q., Luan, J., Fu, L. et al. (2019). The three-dimensional dendritefree zinc anode on a copper mesh with a zinc-oriented polyacrylamide electrolyte additive. *Angewandte Chemie International Edition* 58: 1584115847. (b) Hao, J., Long, J., Li, B. et al. (2019). Toward high-performance hybrid Znbased batteries via deeply understanding their mechanism and using electrolyte additive. *Advanced Functional Materials* 29: 1903605.

130 (a) Zeng, Y., Zhang, X., Qin, R. et al. (2019). *Advanced Materials* 31: 1903675. (b) Yin, Y., Wang, S., Zhang, Q. et al. (2020). *Advanced Materials* 32: 1906803.

131 Naveed, A., Yang, H., Shao, Y. et al. (2019). *Advanced Materials* 31: 1900668.

132 Hao, J., Long, J., Li, B. et al. (2019). *Advanced Functional Materials* 29: 1903605.

133 Li, Y., Wu, P., Zhong, W. et al. (2021). *Energy & Environmental Science* 14: 5563.

134 (a) Zhang, Q., Luan, J., Fu, L. et al. (2019). *Angewandte Chemie, International Edition* 131: 15988. (b) Jin, Y., Han, K.S., Shao, Y. et al. (2020). *Advanced Functional Materials* 30: 2003932.

135 Xie, X., Liang, S., Gao, J. et al. (2020). *Energy & Environmental Science* 13: 503.

136 Mitha, A., Yazdi, A.Z., Ahmed, M., and Chen, P. (2018). *ChemElectroChem* 5: 2409.

137 Zhao, Z., Zhao, J., Hu, Z. et al. (2019). *Energy & Environmental Science* 12: 1938.

138 Ueno, K., Yoshida, K., Tsuchiya, M. et al. (2012). *The Journal of Physical Chemistry. B* 116: 11323.

139 Otani, T., Fukunaka, Y., and Homma, T. (2017). *Electrochimica Acta* 242: 364.

140 Liang, Y., Dong, H., Aurbach, D., and Yao, Y. (2020). *Nature Energy* 5: 646.

141 Qiu, H., Du, X., Zhao, J. et al. (2019). *Nature Communications* 10: 1.

142 Yuan, D., Zhao, J., Ren, H. et al. (2021). *Angewandte Chemie, International Edition* 133: 7289.

143 Cao, L., Li, D., Pollard, T. et al. (2021). *Nature Nanotechnology* 16: 902.

144 Sun, K.E., Hoang, T.K., Doan, T.N.L. et al. (2017). *ACS Applied Materials & Interfaces* 9: 9681.

145 Li, C., Xie, X., Liang, S., and Zhou, J. (2020). *Energy & Environmental Materials* 3: 146.

146 Liu, W., Dong, L., Jiang, B. et al. (2019). *Electrochimica Acta* 320: 134565.

147 Lu, W., Xie, C., Zhang, H., and Li, X. (2018). *ChemSusChem* 11: 3996.

6

Separators for Aqueous Zinc-Ion Batteries

Currently, research on aqueous zinc-ion batteries (AZIBs) is mainly focused on the design of cathode materials, the optimization of electrolytes, and the improvement of zinc anodes. However, as one of the key components of batteries, the study of separators has not attracted enough attention. The essential function of a separator is to prevent the physical contact of electrodes while serving as the electrolyte reservoir to enable ion transport [1]. Although the separator is not directly involved in the electrochemical reaction, it affects the battery's performance by directly influencing the interfacial structure and electrochemical dynamics process of batteries [2]. In this section, the main performance requirements of the separator for AZIBs will be listed, and the relevant separator properties that meet the good performance of the separator will be described in detail. Then, several separator design strategies will be summarized based on the most recent findings to construct high-performance separators. In addition, separator-free AZIBs will be introduced.

6.1 Performance Requirements and Properties of Separator

6.1.1 Performance Requirements of Separator

6.1.1.1 Chemical and Electrochemical Stability

The chemical stability of the separator mainly refers to the corrosion resistance and dimensional stability of the electrolyte. The separator is an inactive substance in batteries, which should not react with the electrolyte and electrode materials to avoid degradation and mechanical strength loss. At the same time, the separator is required to have good dimensional stability, so that there is no expansion, contraction, or deformation during battery operation. The expansion or contraction of the separator in either direction may cause misalignment or electrode contact at the edges during battery assembly, resulting in compromised performance or even short circuits. In addition, it should be electrochemically stable under the reduction and oxidation reactions encountered during the charge/discharge process to ensure long service life. The common cyclic voltammetry (CV) method can be used to determine whether the separator material has reacted with the electrode/electrolyte

Aqueous Zinc Ion Batteries: Fundamentals, Materials, and Design, First Edition.
Haiyan Wang, Qi Zhang, Yixin Li, and Yougen Tang.

within the operational voltage range. The stability and electrochemical reactivity of the separator can be judged from the redox peaks presented in the CV scans [3].

6.1.1.2 Wettability, Electrolyte Uptake, and Electrolyte Retention

The separator can accommodate liquid electrolytes, and its compatibility with electrolytes directly affects ion transport, which has a significant impact on the voltage hysteresis and rate capability of the batteries. During the charging/discharging process, the separator should be able to adsorb the electrolyte quickly and retain it to achieve high ionic conductivity. The ability of the separator to absorb and retain electrolytes is associated with many factors, such as composition, pore size, porosity, and surface chemical characteristics of the separator [4]. The contact angle is an important parameter that measures the wettability of the electrolyte on the separator. It is obtained by placing an electrolyte drop on a dry separator and measuring the shape change over time. The smaller the contact angle, the better the wettability of the separator for the electrolyte. When the separator has a rough surface, uniform composition, texture, or rapid fluid uptake, it is usually difficult to obtain a precise measurement since the angle changes rapidly as the electrolyte is absorbed by the separator. At this time, the actual situation of separator wetting can be better reflected by measuring the dynamic contact angle. Electrolyte uptake and electrolyte retention are also widely used to assess the wettability of separators, which can be calculated by Eqs. (6.1) and (6.2), respectively.

$$\text{Electrolyte uptake}(\%) = \frac{W_{\text{wet}} - W_{\text{dry}}}{W_{\text{dry}}} \times 100 \tag{6.1}$$

$$\text{Electrolyte retention}(\%) = \frac{W_{\text{wet}} - W_{\text{dry}}}{W_{\text{wet}}} \times 100 \tag{6.2}$$

where W_{dry} and W_{wet} are the weights of dry and electrolyte-soaked separators, respectively.

6.1.1.3 Mechanical Strength

The mechanical strength of the separator refers to its ability to resist deformation and damage, which is crucial for battery assembly and throughout its life. The separator with low mechanical strength is easily damaged during the battery assembly, which increases the difficulty of battery assembly, especially for flexible AZIBs. During operation and under abuse conditions, the separator is subject to mechanical deformation caused by external compression or electrode expansion. In addition, the separator is required to inhibit the piercing of zinc dendrites to avoid an internal short circuit [5]. The mechanical strength of the separator is usually measured in terms of tensile strength and puncture resistance, which can be characterized by an electronic universal testing machine.

6.1.2 Properties Requirements of Separator

6.1.2.1 Pore Size

The pore size of the separator should be larger than the diameter of solvent zinc ions to allow the transport of zinc ions through the absorbed liquid electrolyte. However, it must be smaller than the size of electrode material particles to avoid particles or zinc dendrites penetrating the separator, resulting in a short circuit inside the battery. That is to say, the large aperture separator is beneficial to improve its wettability and liquid absorption, while the small aperture separator is conducive to enhancing its anti-penetration ability. In addition, the pore size of the separator is typically obtained by Brunauer Emmett Teller (BET) specific surface area method.

6.1.2.2 Pore Distribution

An ideal separator should have uniform morphology and tight pore size distribution. The uniform pore distribution facilitates the separator being more evenly and fully wetted by the electrolyte, preventing dry spots in the separator during cycling, resulting in adjacent electrode active materials not being fully utilized. In addition, the uniform pore distribution is conducive to the uniform distribution of the battery current. It can avoid continuous reactions in the local area of the electrode, which will promote the growth of zinc dendrites [6]. The pore size distribution of the separator is usually characterized using microscopy techniques, e.g. scanning electron microscopy images.

6.1.2.3 Porosity

The porosity of a separator indicates whether it has abundant pore density to accommodate liquid electrolytes for free ion transport [7]. Low porosity will increase internal resistance, while excessive porosity may decrease mechanical strength. Therefore, proper porosity should be chosen when separators are designed and fabricated. The porosity is often estimated by weighing separators before and after absorbing the electrolyte as the following equation [8]:

$$\text{porosity}\ (\%) = \frac{W - W_0}{\rho_L V_0} \times 100 \tag{6.3}$$

where W_0 and W indicate the weights of the separator before and after immersion in the liquid electrolyte, respectively, and ρ_L and V_0 represent the density of the electrolyte and the geometric volume of the separator before immersion, respectively.

6.1.2.4 Thickness

Separators need to be in an appropriate range of thickness to strike balance between good wettability and adequate mechanical strength. On the one hand, the thicker separator will affect the wetting and absorption of the electrolyte, which will increase the overall resistance of the battery and thus reduce the energy and power density. On the other hand, the thinner separators lack sufficient mechanical strength to withstand rough burrs on electrode surfaces and zinc dendrite punctures during charge/discharge. Moreover, the thickness of the separator should be uniform to ensure even current distribution during battery operation. Therefore, it is ideal to have enough mechanical strength and a relatively thin battery separator, which puts forward higher requirements for the preparation process of the separator. The separator of commercial lithium-ion batteries (LIBs) generally requires less than 25 μm [9], but the ideal thickness range of AZIB separator has not been reported in the literature.

To sum up, the different characteristics of the separator are interrelated, and their influence on the battery's performance is complex. For example, a thin and porous separator with a large mean pore size is conducive to wetting the electrolyte, minimizing the ionic resistance to obtain a high-power density for batteries. However, it ineluctably reduces the mechanical strength and increases the risk of inner battery electrical shorting. Therefore, a variety of factors should be considered when choosing an appropriate separator for ZIBs. The separator may not be able to meet the above requirements at the same time, but when developing and designing a separator, it is often necessary to make trade-offs according to the characteristics of the working environment and application scenarios. At present, there is no unified test standard for battery separators, and many test projects refer to the product standards of plastics, films, paper products, textiles, and other industries. Unlike mature LIBs, there are no ideal separator specifications for AZIBs. Hence, there is an urgent need to establish a general evaluation method for various basic requirements of the AZIB separator, including chemical and electrochemical stability, wettability, and mechanical strength. Similarly, electrolyte uptake, electrolyte retention, and other interface parameters also need to be considered.

6.2 Commercial Separators

Compared with the commercialized separators for LIBs, the research on AZIB separators is still in the preliminary stage [10]. Currently, available separators for AZIBs are only able to satisfy one or two of the requirements mentioned in the previous section at the expense of others [11]. In this section, we will introduce some common and commercial separators used in AZIBs.

6.2.1 Polyolefin Separator

Polyolefin separators (e.g. polyethylene (PE) or polypropylene (PP) porous membranes) have dominated the commercial separator market for LIBs [12]. No matter whether fabricated by dry or wet processes, there are several uniformly distributed pores throughout the entire separator with good mechanical strength and chemical stability [13] (Figure 6.1a). However, they cannot be utilized directly in aqueous AZIBs because of their poor wettability and hydrophobic nature, leading to large interfacial contact resistance [11]. Up to now, some efforts have been dedicated to coating the hydrophilic interface layer on polyolefin separators to improve wettability and increase the water retention capacity to reduce electrolyte loss [15].

6.2.2 Glass Fiber Separator

Glass fiber (GF) is the most widely used separator material for AZIBs in experimental practices. As shown in Figure 6.1b, this separator is made from randomly distributed or weaved GFs with 0.2–2 μm diameters, leaving plenty of interconnected micron pores for electrolyte permeation [10c]. With the advantages of high ionic

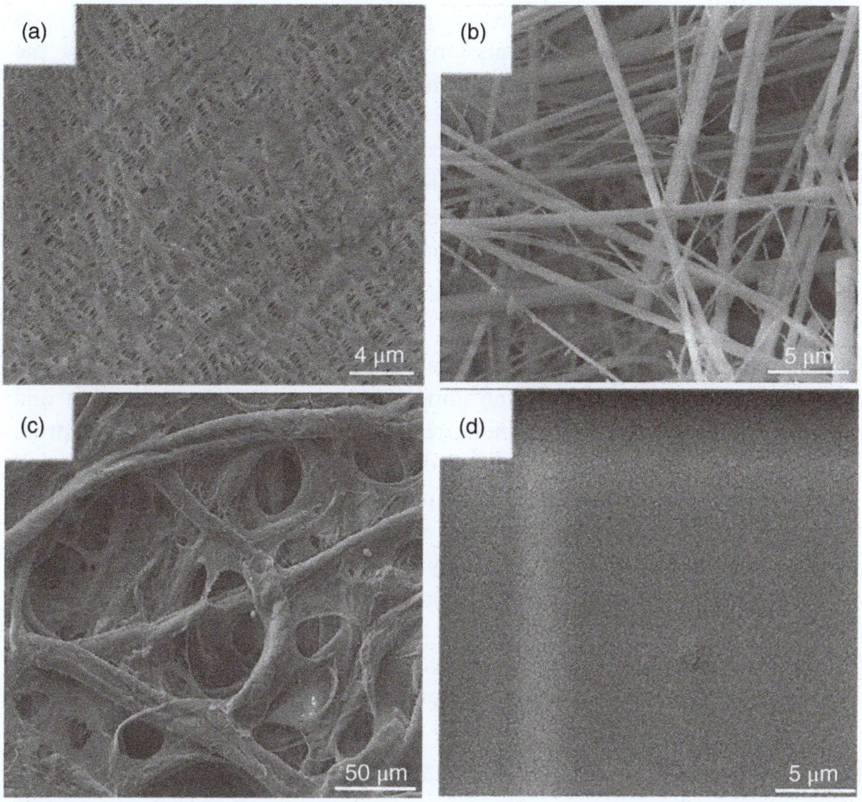

Figure 6.1 SEM images of (a) PP separator, (b) GF separator, (c) filter paper, and (d) Nafion separator. Source: Ghosh et al. [14]/John Wiley & Sons.

conductivity, high electrolyte uptake, and high chemical stability, GF separator is universally used in aqueous batteries. However, GF separators are so fragile that they can be pierced easily by zinc dendrites, resulting in battery failure. Although the performance of GF separator could be optimized by compositing with other functional materials, due to the large thickness (a few hundred microns) and heavy areal weight, it is difficult to assemble the entire battery to achieve high energy density and meet industrial requirements.

6.2.3 Cellulose-Based Separator

Cellulose is the most abundant natural organic polymer, which has multiple merits, such as good hydrophilicity, high mechanical strength, and low cost. Compared with other separator materials, cellulose-based separators possess the unique advantage of being environmentally friendly, as they can decompose slowly and return to nature. Filter paper (mainly composed of cellulose) made from wood pulp is an attractive separator material for AZIBs [16]. The wood fabric and inhomogeneous pore structure of filter paper can be seen from the typical scanning electron microscope (SEM) image in Figure 6.1c. There are tens-micron pores in some parts of the filter paper, while the rest parts are much denser. In AZIBs, the dense parts of filter papers can limit the transmission of local electrolytes and increase the overpotentials of Zn plating/stripping. On the other hand, the pore parts kinetically accelerate the growth and detachment of Zn dendrites on the local anode surface. In conclusion, separators with uneven pores could cause microscopic pore deficiency and kinetically depress the electrochemical performance of Zn anodes by limiting local electrolyte transport. Hence, there are a lot of reported papers putting effort into constructing advanced separators.

6.2.4 Nafion Separator

Separators mentioned above cannot regulate ion transport, which could cause a concentration gradient near the zinc anode, inducing dendrite formation and growth. On the other hand, they cannot arrest the diffusion of undesired side products from one electrode to the other [17]. Nafion separator (Figure 6.1d) has entered the perspective of researchers because of its single ion conduction characteristics and rich ionic functional groups. Meena Ghosh first proved that the Zn^{2+}-integrated Nafion ionomer separator could effectively suppress the growth of irregular zinc deposits [14]. The coordination of Zn^{2+} ions with SO_3^{2-} moieties in Nafion results in a high transfer number of Zn^{2+} cations and high ionic conductivity. Although Nafion separator shows good electrochemical performance when used in AZIBs, it is not a suitable choice in terms of its high price of about US$1200 per m^{-2}.

6.3 Constructing High-Performance Separators

In general, apart from blocking electrons and isolating the direct contact of cathode and anode in the battery, the most important function of a separator is to transport ions [18]. The former can be relatively easily achieved by using insulating and high-mechanical-strength materials to prepare the separator, and all battery separators are universal. For the latter, there is still much scope for improvement in the separator of AZIBs. Specifically, when ions are transmitted through the pores of the separator, they accumulate in the pores resulting in a nonuniform ion distribution of Zn^{2+} flux across the separators [19]. This phenomenon is particularly serious when the pores of the separator are unevenly distributed. The thick separator and the tortuous ion channels in the separator can lead to sluggish ion transport. In addition, the routine separators do not have a regulatory effect on ion diffusion behavior, resulting in the deposition of zinc in any direction, which is manifested as random crystallographic orientation zinc deposition and vertical growth of zinc. Therefore, it is necessary to find ways to homogenize Zn^{2+} distribution, accelerate Zn^{2+} transport, and manipulate Zn growth direction to construct high-performance separators for AZIBs.

6.3.1 Promoting Homogeneous Ion Distribution

In AZIBs, it is easy for the inhomogeneous distribution of zinc ions to stimulate the uneven zinc nucleation sites at the anode interface, and these nucleation sites are easy to cause the subsequent random growth of zinc to become uncontrollable zinc dendrites due to the tip effect [20]. It is well known that zinc dendrites are fatal to AZIBs, which not only affect the cycle performance and coulombic efficiency of the battery but also directly pierce the separator and lead to short circuits in severe cases [21]. Thus, it is necessary to promote uniform ion distribution. To achieve this purpose, in addition to the development and modification of current collectors, zinc anodes, and electrolytes, an increasing number of researchers have also focused on the separator that provides ion transport channels [1, 22]. As described in the first subsection above, the pores in the separator act as ion channels, and zinc ions will reach the anode surface through the separator for electrochemical deposition. It is not difficult to imagine that the purpose of uniform ion flux can be achieved by optimizing the pore structure of the separator to standardize the ion path. In addition, during the battery cycle, the random distribution of zinc ions will affect the relatively balanced electric field between the separator and electrode, and the disturbed electric field will reversely affect the distribution of ions [23]. Therefore, to ensure homogeneous ion distribution through the separator, a conductive layer can be introduced on the separator to uniformize the interface electric field.

6.3.1.1 Constructing Ordered Pore Structure

The separator with regular pores can not only homogenize the distribution of zinc ions at the anode interface, but also the current density inside the separator shows an even distribution, indicating that the migration rate of zinc ions in all directions is consistent, that is, zinc ions in different regions can almost migrate to the surface of zinc anode at the same time, which will be beneficial to the homogeneous deposition process of zinc ions on the anode surface (Figure 6.2a,c). In contrast, when the separator has a nonhomogeneous pore distribution, the zinc ion concentration and current density in different pore regions will be different. Specifically, more zinc ions are faster concentrated on the surface of the anode corresponding to a large pore size to achieve electrochemical deposition, while the zinc ion concentration is lower, and the migration is relatively slow in the area with a small pore size (Figure 6.2b,d). This inevitably leads to the growth of zinc dendrites in local areas. Therefore, modifying and developing new separators possessing an ordered pore structure (uniform pore distribution, narrow pore size, and suitable porosity) are desirable for constructing high-performance AZIBs.

Commercial GF separators and filter paper have been widely used in AZIBs owing to their excellent electrolyte compatibility and stability. However, their large and disordered pore structure often leads to a random distribution of zinc ions, triggering inhomogeneous zinc plating. The unsatisfactory mechanical properties caused by brittle GF materials and large pores make it easy to be pierced by zinc dendrites, resulting in the failure of AZIBs [24]. To overcome these shortcomings, Zhang et al. utilized a facile vacuum filtration method to uniformly anchor $BaTiO_3$ (BTO) nanoparticles on GFs while filling gaps of different sizes on the surface of GF separators [25]. BTO anchored on the fibers can accelerate the transport of ions in the separator–electrolyte interface due to its high dielectric constant and high zinc affinity. And the BTO-filled interfacial layer can ensure the uniform redistribution

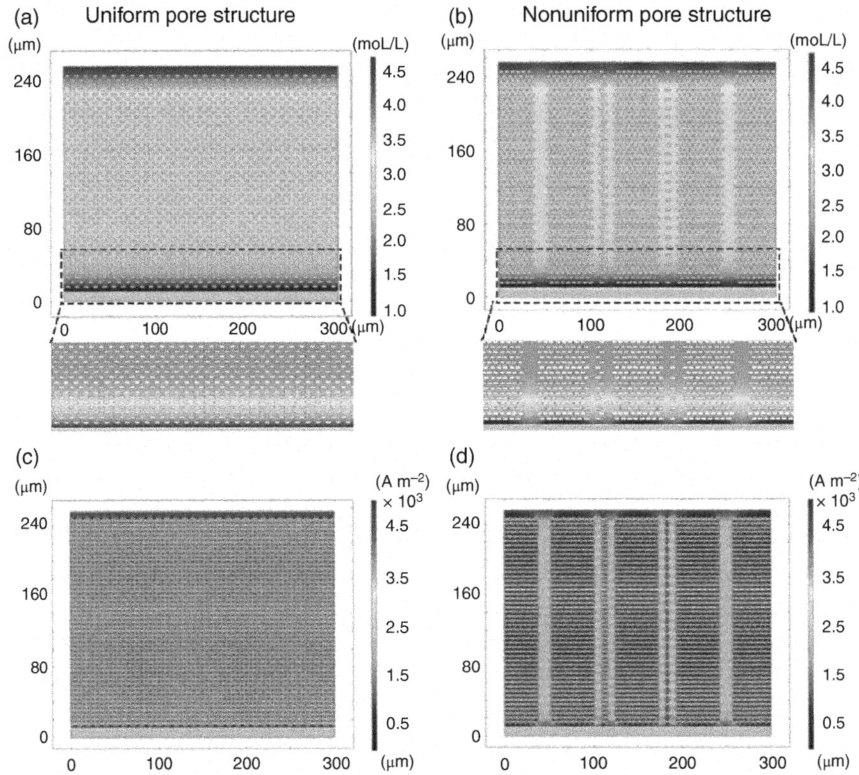

Figure 6.2 The effect of pore structure on the distribution of zinc-ion concentration and current density. (a, c) separator with uniform pore structure (filter membrane), (b, d) separator with nonuniform pore structure (GF separator and filter paper). The dotted rectangle is the corresponding partial analysis. Source: (a, b) Adapted from Qin et al. [1]. (c, d) [1]/John Wiley & Sons.

of zinc ions in the separator–anode interface (Figure 6.3a). This dual-interface engineering requires optimizing the loading quality of BTO nanoparticles modified on GF. In detail, as the loading mass increases, BTO nanoparticles are first anchored on the fiber, and the excess will tend to gather and fill the surface interspace of the separator, while excessive BTO will aggregate to form large caking (Figure 6.3b). Therefore, the purpose of uniform ion flux cannot be achieved by filling deficient or excessive BTO on the separator surface. Li et al. used a similar interface engineering strategy to anchor persimmon branch carbon (PBC) material onto the cellulose-filter paper separator. The PBC particles filled the uneven macropores in the cellulose-filter paper separator and the curved channel that was not conducive to ion transport, which standardized and shortened the ion path and ensured the homogeneous plating/stripping of the zinc anode [27]. In addition, metal–organic frameworks (MOFs) with high chemical robustness and high porosity were also applied to the separator modification of AZIBs [28]. An appropriate amount of UiO-66 (UiO, University of Oslo) can uniformly fill the uneven voids in GF to target the redistribution and acceleration of zinc ion flux. Moreover, large specific surface

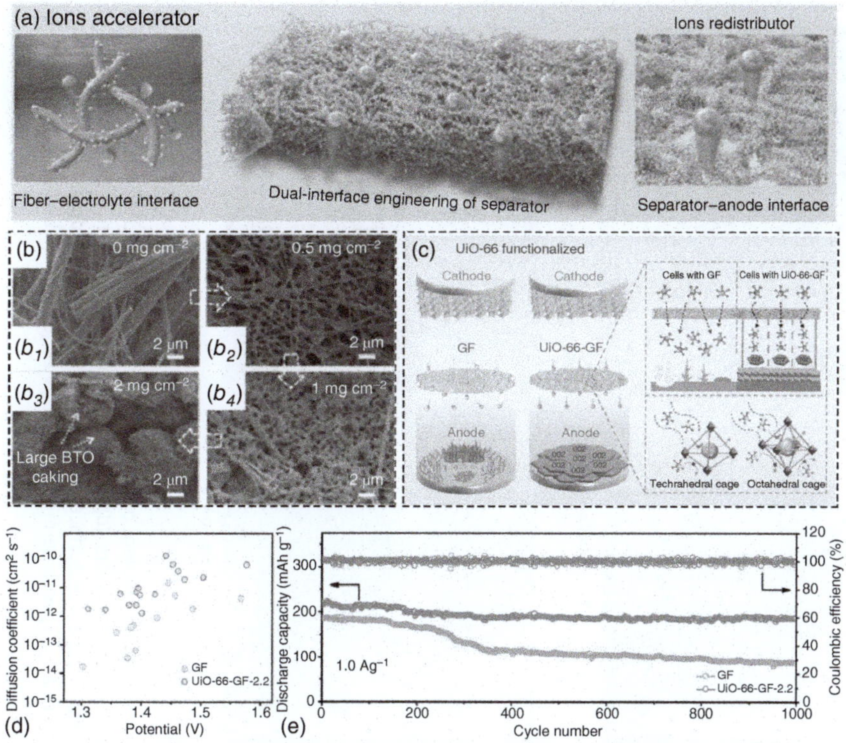

Figure 6.3 (a) Dual-interface engineering diagram of BTO-modified GF separator. (b) FESEM images of GF separators loaded with 0 (b1), 0.5 (b2), 1 (b3), and 2 (b4) mg cm^{-2} BTO dispersions. (c) The comparison diagram of the action mechanism of GF and UiO-66-GF-2.2 separators. (d) Zinc-ion diffusion coefficients during discharging and (e) cycling performances at 1.0 A g^{-1} of Zn||GF||MnO_2 and Zn||UiO-66-GF-2.2||MnO_2 full cells. Source: (a, b) Liang et al. [25]/John Wiley & Sons. (c–e) [26]/Springer Nature/Licensed under CC BY 4.0.

areas and porous structures in UiO-66 also provide more transport channels for the migration and diffusion of zinc ions (Figure 6.3c). Therefore, compared with the unmodified pure GF separator, UiO-66-GF is applied to the zinc–manganese full cell, showing a superior zinc-ion diffusion coefficient (1.46465×10^{-11} cm^2 s^{-1} vs. 1.30906×10^{-10} cm^2 s^{-1}, Figure 6.3d) and long cycle stability (the capacity remains 85% after 1000 cycles at 1.0 A g^{-1}, Figure 6.3e) [26].

In addition to filling the uneven interspace of the separator, directly introducing an interface layer with a uniform pore structure on the separator surface to regulate the ion distribution is also an efficient and easy-to-operate method. It is a very typical example to introduce porous two-dimensional (2D) g-C_3N_4 nanosheets as an additional interface layer to modify commercial cellulose fiber separators [29]. The bare cellulose fiber separator with a large and disordered pore structure guides the anisotropic migration of zinc ions (Figure 6.4a), resulting in a large zinc-ion concentration gradient near the anode to form zinc dendrites (Figure 6.4b,c). In stark contrast, g-C_3N_4 nanosheets have a large number of small

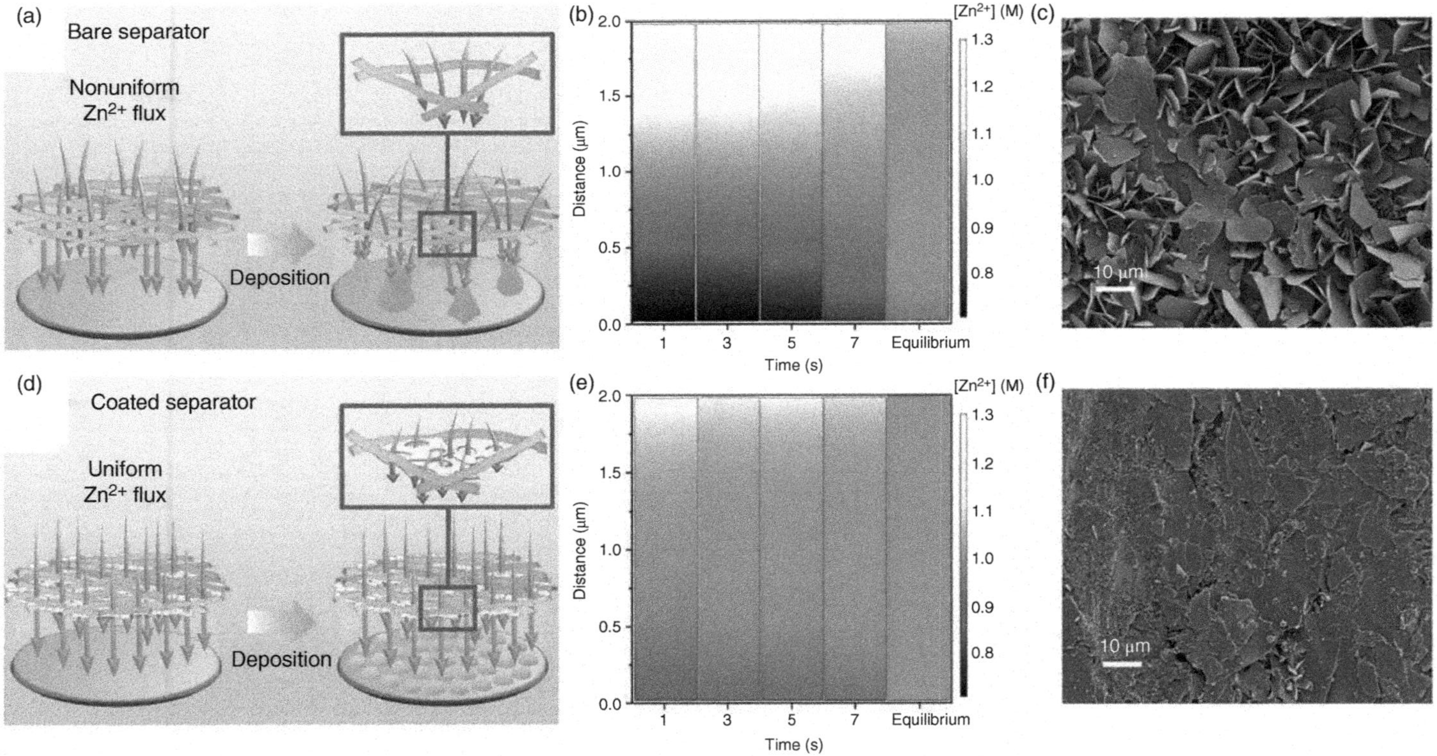

Figure 6.4 (a) The zinc deposition schematic diagram, (b) simulation of the concentration gradient of zinc ions with time, and (c) SEM images of the zinc surface after 10 minutes of deposition at a constant overpotential of −50 mV of the bare separator. (d) The zinc deposition schematic diagram, (e) simulation of the concentration gradient of zinc ions with time, and (f) SEM images of the zinc surface after 10 minutes of deposition at a constant overpotential of −50 mV of the g-C_3N_4-coated separator. Source: (a, b, d, e) Yang et al. [29]/with permission of Elsevier. (c, f) Yang et al. [29]/Royal Society of Chemistry.

and regular pores. Therefore, g-C_3N_4-coated separator manipulates a directional ion flux and reduces the concentration gradient of zinc ions at the electrode interface, ultimately inducing uniform zinc deposition (Figure 6.4d–f). Gan et al. reported a thin-to-250 nm hydroxyl-modified silica nanosheet (IS/SiO_2–OH) ion sieve with a micro-mesoporous (1.5 and 3.4 nm) ion channel, which is adhered to the GF surface by a simple vacuum filtration method rather than dispersed into the GF separator [30]. Different from single pore distribution (1, 3, and 50 nm), the IS/SiO_2–OH ion sieve can uniformly redistribute the electric field as well as effectively accelerate the migration of Zn ions at the separator/electrode interface due to the capillary effect exerted by the unique pore structure. Therefore, the Zn//Zn cell equipped with IS/SiO_2–OH@GF separator has a cycle life of more than 3400 hours at 1 mA cm^{-2}, which is about 25 times that of the bare cell. Even at a high current density of 10 mA cm^{-2}, it delivers an impressively long cycle life of 2500 hours and a cumulative capacity of 12 500 mAh cm^{-2}. Recently, the strategy of introducing an ordered pore structure interface layer on the separator in AZIBs has been increasingly applied and achieved very good results [31], such as sulfonated polysulfone-modified poly(*m*-phenylenedicarboxamide) nanofiber separator with micro-nanometer-sized pores [32], porous graphene oxide loaded on hydrophilic polytetrafluoroethylene (PTFE) membrane [33], and MXene/nanoporous NiO-modified GF separator [34].

The introduction of additional materials to modify the disordered pore structure of common separators can effectively regulate the migration of zinc ions to improve battery performance. However, these separators still have some inherent shortcomings, such as high price, excessive thickness, and increased weight, and the additional materials also complicate the separator manufacturing process. Therefore, it is necessary to explore a special separator with an ordered pore structure for AZIBs. Due to the rich and homogeneous pore structure of the filter membrane that is widely used in food, medicine, chemistry, and other fields, our group has applied a water-based filter membrane directly to AZIBs as the separator and achieved better electrochemical performance than GF and filter paper [1]. As shown in Figure 6.5a, the Zn||$NaV_3O_8 \cdot 1.5\ H_2O$ cell assembled by filter membrane has a capacity retention rate of 83.8% after 5000 cycles at 5 A g^{-1}, while the full cell based on GF and filter paper has only 15.9% and 59.8%, respectively. Based on the work of our group, Zhang et al. prepared nanocellulose separator by a simple papermaking process to further study the effect of pore size on zinc deposition [35]. Compared with the filter membrane with uniform pores of 1–3 μm, the nanocellulose separator has the same uniform but narrower pore structure (15 nm), which enables it to obtain more homogeneous zinc ion flux and form a dense zinc nucleation in the initial cycle. After multiple cycles, a compact dendrite-less Zn deposit layer can also be obtained (Figure 6.5b,d). The Zn||Zn cell based on the nanocellulose separator has a voltage hysteresis of as low as 29 mV and a cycle stability of more than 2000 hours at a current density of 2 mA cm^{-2} (Figure 6.5e). Recently, more and more porous AZIB separators have been developed [37], such as cellulose-based separators with uniform nanoporous structures prepared from various biomass sources like cotton [38], bamboo [39], bacteria [40], and organic/inorganic composite separators

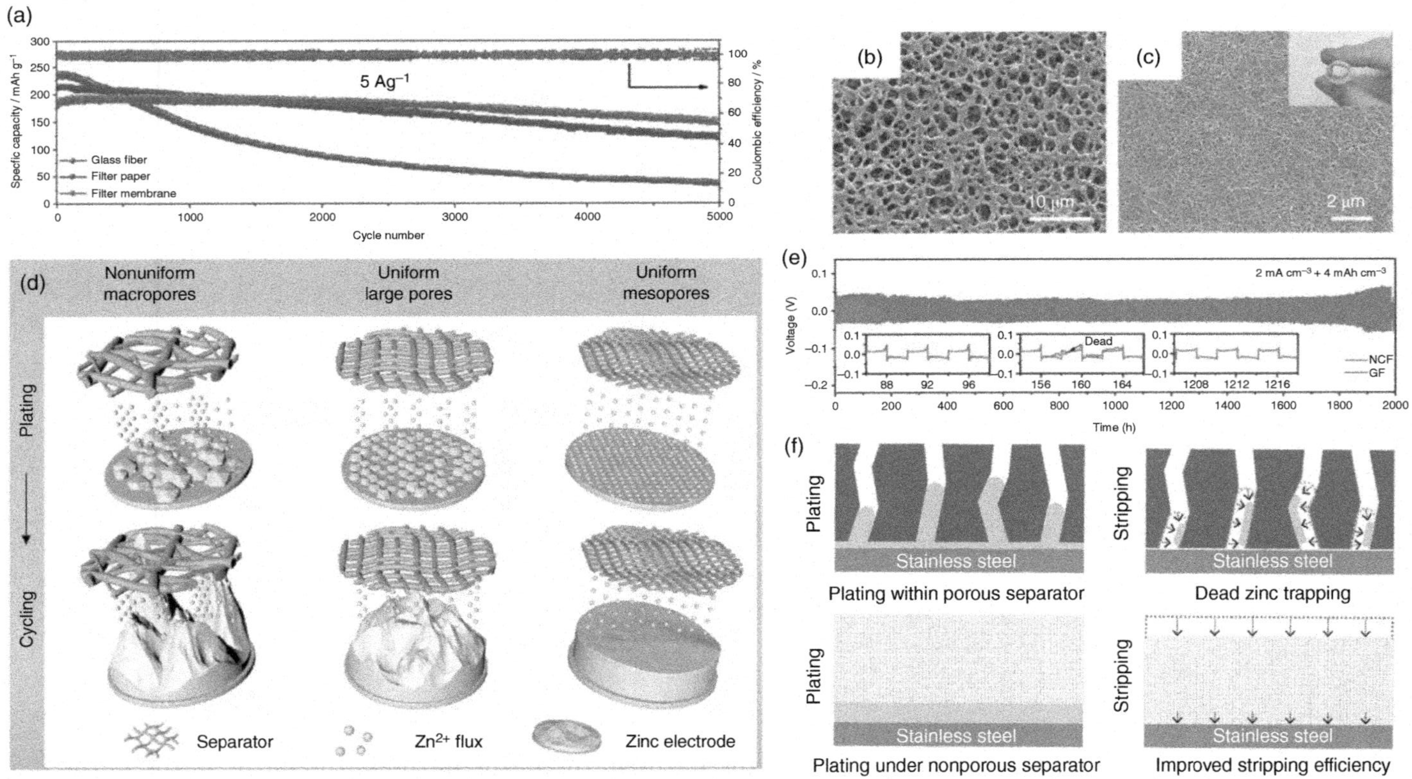

Figure 6.5 (a) The cycling performance of Zn||$NaV_3O_8 \cdot 1.5\ H_2O$ full cells with different separators. SEM images of (b) filter membrane and (c) nanocellulose separator. (d) Schematic illustration of Zn deposition in separators with different pore structures. (e) Cycling performance of Zn||Zn cells with GF and nanocellulose separators. (f) Schematic illustration of the electrochemical behavior of zinc in porous and non-porous separators. Source: Qin et al. [1]/John Wiley & Sons. (b, c) Zhang et al. [35]/with permission from Elsevier. (d, e) Zhang et al. [35]/with permission of Elsevier. (f) Zhang et al. [36]/American Chemical Society.

with the porous structure of inorganic particles [41]. In fact, the porous structure of the separator has its own drawbacks, which will guide zinc to fill in the tortuous and rich pores and leave dead zinc upon stripping. Hence, the more uniform and smaller the pores of the separator, the more effective it is to alleviate this problem. Small to nonporous is an optimal solution to this problem, so that zinc deposition can be limited underneath the separator to prevent zinc penetration (Figure 6.5f) [36]. However, the ionic conductivity of the nonporous separator will be poor. Therefore, the research on the pore structure of the AZIB separator still needs to be strengthened, and it is necessary to figure out the specific pore structure requirements of an ideal AZIB separator as soon as possible.

6.3.1.2 Introducing Conductive Layer

As mentioned in the previous section, an inhomogeneous electric field distribution inevitably forms at the pores of the separator during zinc deposition. This is because the electric field is inclined to concentrate at the tips, and the electric field intensity on the pores of the separator is significantly higher than that of the skeleton (Figure 6.6a). In addition, zinc ions preferentially aggregate at high electric field intensity regions to obtain electrons for zinc nucleation and growth. Such preferential zinc deposition further evolves into zinc dendrites through self-amplification mechanism, ultimately leading to a short circuit inside the cell (Figure 6.6b). Thus, introducing a conductive layer on the separator is a feasible approach to uniformizing the electric and ion fields between the separator and the anode. In this case, some functional materials with excellent electrical conductivity, such as metal and carbon-based materials, have been used for AZIB separator modification.

Inherently, conductive metals can take advantage of their equipotential characteristics to homogenize electric field distribution (Figure 6.6c) [43]. However, not all metals are suitable as conductive layers to be introduced into the separator. It should be noted that the superior conductive layer should have a higher redox potential than zinc metal to avoid electrochemical oxidation and higher hydrogen evolution overpotential to avoid severe hydrogen evolution reactions. Zhao et al. coated a multifunctional conductive layer of tens of microns thick on the PP film by roller pressing method (specifically composed of metallic Bi, metallic zinc with high conductivity, Al_2O_3 with high zincophilicity, and potassium polyacrylate with good hydrophilicity) [15]. Such a conductive layer can not only ensure uniform zinc deposition but also effectively mitigate side reactions and electrolyte loss. However, the additional metal material itself is relatively dense, which reduces the overall energy density of the battery. Controlling the thickness of the conductive metal layer by appropriate methods can minimize this effect. Hou et al. constructed nano-thick Sn coating on filter paper by magnetron sputtering [42]. In areas where the Sn coating is not in contact with the electrode, the good conductivity of the Sn layer makes the electric field distributed evenly, which contributes to the uniform nucleation and growth of zinc on the anode surface. In the contact regions, zinc metal would be preferentially deposited on modified separator due to its higher zinc affinity than anode, which avoids the short circuit caused by zinc growth at anode tips. After multiple cycles, the continuous deposition of zinc leads to the connection of the

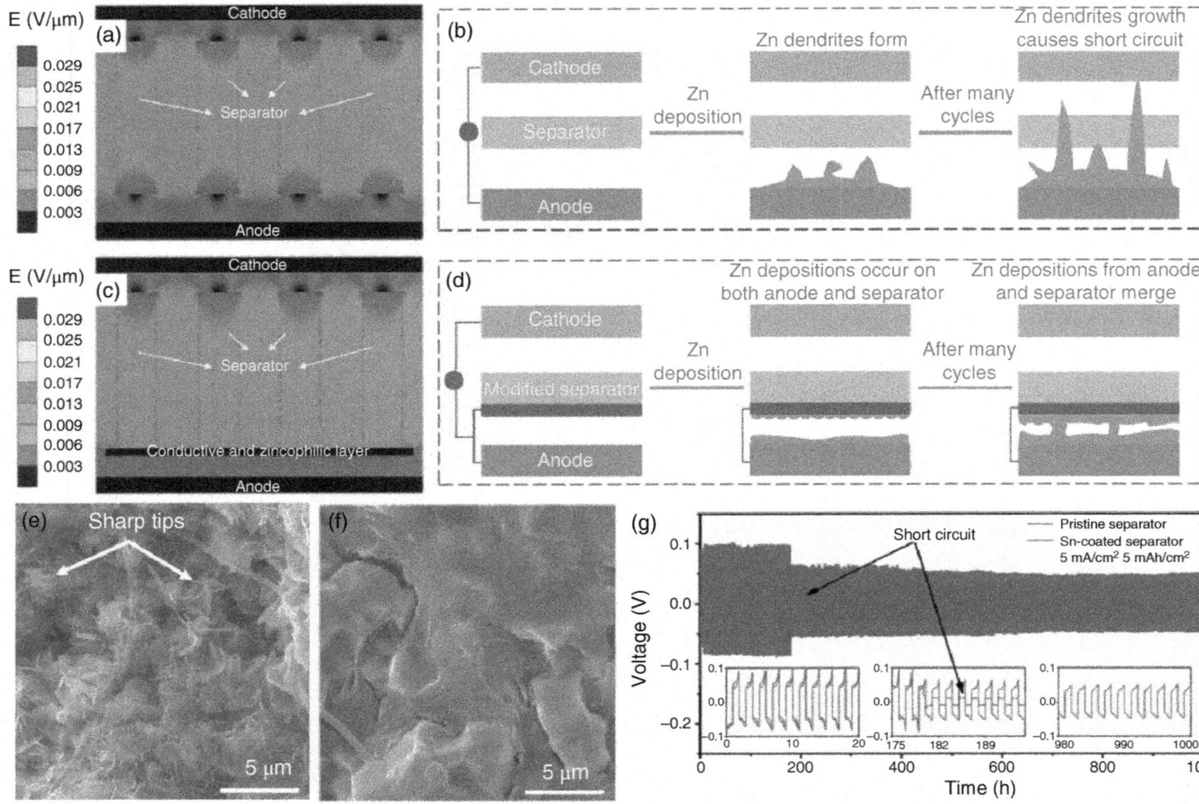

Figure 6.6 (a) Electric field distribution and (b) schematic illustration of zinc deposition with filter paper. (c) Electric field distribution in the non-contact region of the Sn-coated separator and the anode. (d) Schematic illustration of zinc deposition in the contact region of the Sn-coated separator and the anode. SEM images of the Ti current collector using (e) pristine separator and (f) Sn-coated separator after Zn deposition at 1 mA cm^{-2} for 4 mAh cm^{-2}. (g) Cycling performance of the Zn||Zn symmetric cells at 5 mA cm^{-2} for 5 mAh cm^{-2}. Source: (a–g) Hou et al. [42]/Springer Nature/Licensed under CC BY 4.0.

conductive layer and anode to form a compact and dense deposit (Figure 6.6d). At a high deposition capacity of 4 mAh cm^{-2}, the Sn layer can also induce a dense and uniform zinc coating rather than sharp tips (Figure 6.6e,f). Even under the rigorous testing conditions of 5 mA cm^{-2} and 5 mAh cm^{-2}, the Zn||Zn symmetric cell based on Sn-coated separator can operate stably for 1000 hours (Figure 6.6g). These all prove that the conductive and zincophilic layers on separator at high current and high cycle capacity can still effectively induce stable zinc plating/stripping.

The lighter carbon-based material is another ideal conductive material for modifying the separator. Carbon-based materials have been widely adopted in AZIBs, such as current collectors to accommodate the Zn metal [44], cathodes for adsorbing Zn ions [45], and interlayers for regulating Zn deposition [46]. According to the experience of these works, adjusting the microstructure of carbon-based materials and surface functionalization of carbon materials can regulate the zinc deposition behavior. Li et al. devised a Janus separator via in situ growth of vertical 3D graphene on the GF separator near the anode throughout plasma-enhanced chemical vapor deposition [47]. As shown in the simulation of the electric field distribution in Figure 6.7a, the 3D conductive graphene skeleton can significantly increase the surface area, and the current density of the interface reaction is reduced to 0.24 mA cm^{-2}, which is better than the 2D planar graphene without microstructure adjustment (0.38 mA cm^{-2}). Reducing the local current density is beneficial to delay the initial formation point of the dendrite, and this 3D scaffold can provide a uniformly distributed electric field, allowing zinc to be deposited uniformly inside the framework.

In contrast, the pristine commercial GF separator with a planar configuration prefers to trigger individual nucleation sites and further evolves into an inhomogeneous electric field, inducing the formation of zinc dendrites. Most carbon materials are hydrophobic and zincophobic, and they are introduced as a conductive layer onto the separator to detrimentally affect the wetting of the electrolyte and the transport of ions. The property of the separator can be improved by surface functionalization. For instance, hydroxyl groups are introduced to functionalize carbon nanotubes, increasing hydrophilicity and improving electrolyte permeability [48].

Another example is the high nitrogen-doped carbon-modified GF separator with high zincophilicity, which can have abundant zinc-affinity sites, thereby inducing uniform zinc-ion distribution [49]. In addition, introducing a conductive layer on the separator cannot only reduce the Zn^{2+}/Zn redox energy barrier (Figure 6.7b), but also provide additional electronic pathways to suppress "dead zinc," thus improving the reversibility of the zinc anode (Figure 6.7c). A potential discrepancy of different regions on the heterogeneous surface between anode creates numerous corrosion cells [50]. The corrosion process not only continuously consumes electrolytes to generate H_2 and passivation byproducts (ZnO, $Zn(OH)_2$, and zincates) but also leads to severe self-discharge of AZIBs [51]. The conductive rGO layer on the separator can make the corrosion cells short-circuit, reduce the microscopic potential discrepancy, and alleviate the corrosion problem (Figure 6.7d) [28a]. Therefore, the introduction of a conductive layer on the separator is worthy of further research.

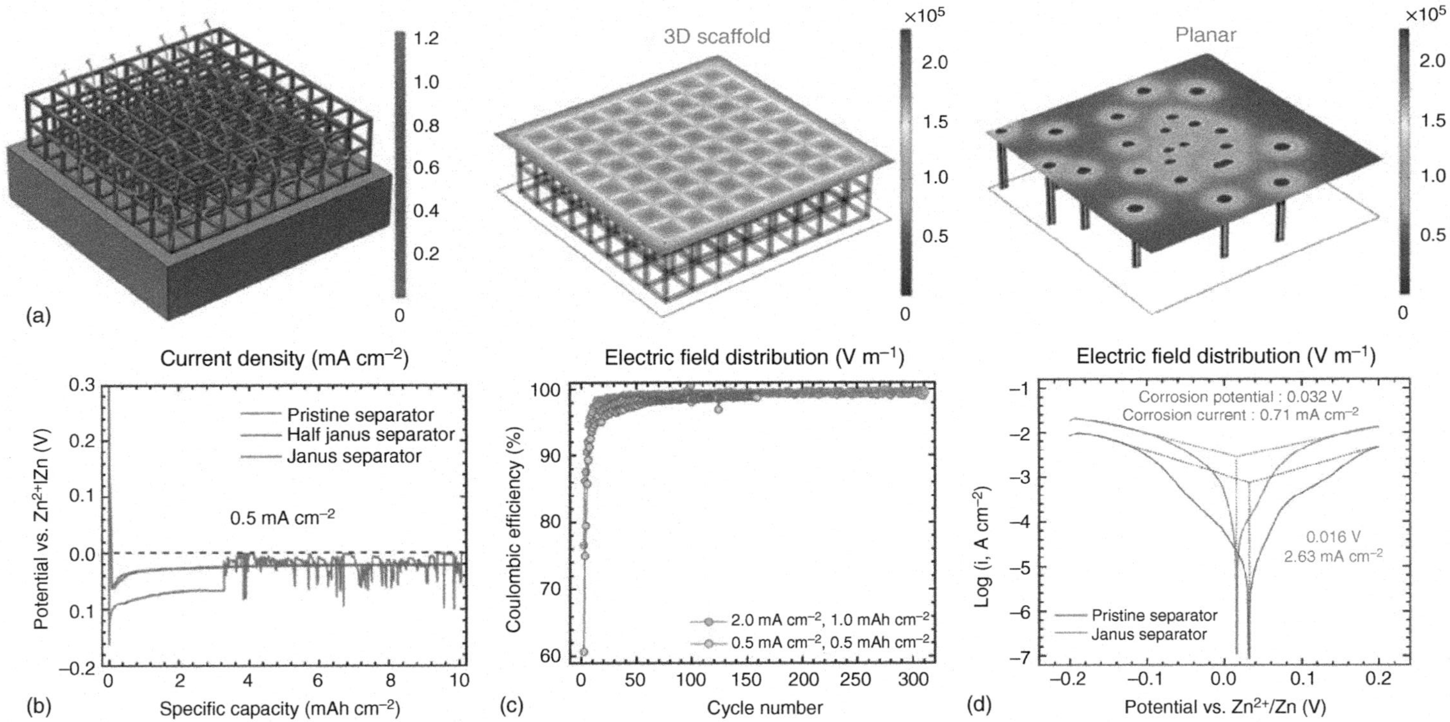

Figure 6.7 (a) Electric field distribution simulation of 3D conductive scaffold and 2D planar structure. (b) Zn-plating profiles of Zn||Cu cells with different separators. (c) CE of the Zn||Cu cells with a Janus separator. (d) Linear polarization curves of the Zn||Zn cells with Janus and pristine separators at a scan speed of 1 mV s^{-1}. Source: (a) [47] / John Wiley & Sons. (b–d) Wang et al. [28a]/Springer Nature/Licensed under CC BY 4.0.

6.3.2 Accelerating Zn^{2+} Transport

Sluggish Zn^{2+} diffusion kinetics based on common separators could cause a concentration gradient near the zinc anode, which is a vital factor responsible for the fast dendrite growth [10a, 25]. Thus, accelerating Zn^{2+} transport is significant for developing dendrite-free AZIBs. In addition, the uncontrolled transport of anions could induce other problems [52]. For instance, in $ZnSO_4$ electrolyte, $SO_4{}^{2-}$ would participate in side reactions and form side products like $Zn_4SO_4(OH)_6{\cdot}4H_2O$, resulting in shortened battery cycle life. The Zn^{2+} transference number ($\boldsymbol{t}_{\mathbf{Zn}^{2+}}$) could be used to describe diffusion rate of Zn^{2+}. The separator with high $\boldsymbol{t}_{\mathbf{Zn}^{2+}}$ could not only boost Zn^{2+} transport effectively but also repel anions, and it is expected to inhibit dendrites effectively. In recent years, several advances have been made in the development of separators with high $\boldsymbol{t}_{\mathbf{Zn}^{2+}}$ to accelerate Zn^{2+} transport, and the typical design strategies include zincophilicity, electrostatic interaction, and Maxwell–Wagner polarization. Table 6.1 summarizes some typical examples of high-performance separators for accelerating Zn^{2+} transport.

6.3.2.1 Zincophilicity

Zincophilicity refers to the ability of material to absorb and bond zinc [23a]. A zincophilic separator could validly capture the Zn^{2+} between the fiber–electrolyte interface and lower zinc nucleation overpotential. Inducing zincophilic materials or groups to modify separators is the typical method to improve the zincophilicity of separators. The adsorption energy (E_{ads}) is widely accepted to illustrate the affinity between Zn and base structure and is calculated by density functional theory (DFT) calculation (Eq. (6.4)):

$$E_{ads} = E_{total} - E_{base} - E_{Zn} \tag{6.4}$$

where E_{total} is the total energy of the Zn-adsorbed system, and E_{base} and E_{Zn} are the energies of the pure base structure and the isolated Zn atom, respectively.

Polar organic molecules are often used as the zincophilic layer to regulate zinc deposition behavior. For example, Liu et al. decorated GF separator with functional supramolecules (Figure 6.8a) [53]. According to DFT calculation, the binding energies of zinc atoms with the selected sites in supramolecules, O1 (carbonyl), N2 (imino groups), N3 (amino groups), and N4 (triazine) are −2.37, −1.96, −2.36, and −1.89 eV, respectively (Figure 6.8b,c). Thus, compared to GF, the modified separator (GF@SM) shows strong attraction with Zn^{2+} and prevents the local aggregation of Zn^{2+} thanks to the abundant zincophilic groups in supramolecules. Such a functional separator demonstrated excellent cycling performance during the Zn-plating/stripping process, as Zn||Zn symmetrical battery adopting GF@SM separator delivered an ultralong cycle life (2000 hours). Similarly, chitosan-coated filter paper separator showed a larger $\boldsymbol{t}_{\mathbf{Zn}^{2+}}$ (0.62) than that of bare filter paper (0.16) [55]. This enhancement could be attributed to preferential adsorption of H^+, Zn^{2+}, and $SO_4{}^{2-}$ with functional groups on chitosan. The larger $\boldsymbol{t}_{\mathbf{Zn}^{2+}}$ helps to reduce Zn^{2+} concentration gradient and facilitate uniform ion distribution. Besides, in Guo's research, a lab-widely used weighing paper (WP) was inserted as an interlayer

Table 6.1 Several examples of separators with high $t_{Zn^{2+}}$.

Materials	Strategies	Electrolyte	$t_{Zn^{2+}}$	Performance (Zn‖Zn cells)	Reference
Supramolecules/glass microfiber	Zincophilicity	2 M $ZnSO_4$	–	2000 h (1 mA cm^{-2}, 1 mAh cm^{-2})	[53]
Weighing paper/filter paper	Zincophilicity	1 M $ZnSO_4$ (30 μL)	0.45	>2000 h (1 mA cm^{-2}, 1 mAh cm^{-2})	[54]
N-doped carbon-modified GF separator	Zincophilicity	2 M $ZnSO_4$	0.57	1100 h (1 mA cm^{-2}, 1 mAh cm^{-2})	[49]
Chitosan/filter paper	Zincophilicity	1 M $ZnSO_4$	0.62	>2900 h (0.5 mA cm^{-2}, 0.5 mAh cm^{-2})	[55]
Polyacrylonitrile (PAN) separator	Zincophilicity	2 M $ZnSO_4$	0.85	800 h (0.283 mA cm^{-2})	[56]
Polyacrylonitrile/graphene oxide (PG) separator	Zincophilicity and electrostatic interaction	3 M $Zn(OTf)_2$	0.74	13 000 h (1 mA cm^{-2}, 1 mAh cm^{-2})	[57]
Collagen hydrolysate-coated absorbed glass mat (CH@AGM)	Electrostatic interaction	1 M $ZnSO_4$ + 2 M Li_2SO_4	–	>3300 h (1 mA cm^{-2}, 1 mAh cm^{-2})	[58]
Negatively charged covalent organic framework (COF)-coated GF separator	Zincophilicity and electrostatic interaction	2 M $ZnSO_4$	0.76	1200 h (0.5 mA cm^{-2}, 1 mAh cm^{-2})	[59]
Cellulose nanowhisker–graphene (CNG) membrane	Electrostatic interaction	3 M $ZnSO_4$	–	5500 h (0.25 mA cm^{-2}, 0.5 mAh cm^{-2})	[60]
Parallelly grown graphene sheets modified with sulfonic cellulose on one side of the commercial glass fiber separator	Electrostatic interaction	3 M $ZnSO_4$	0.826	1400 h (10 mA cm^{-2}, 10 mAh cm^{-2})	[61]
Single-ion-functionalized CNF separator	Electrostatic interaction	0.5 M $ZnSO_4$	0.70 ± 0.12	>500 cycles (1 mA cm^{-2}, 0.5 mAh cm^{-2})	[11]
Sulfonated polysulfone @ poly(*m*-phenylene dicarboxamide) (PMIA) separator	Electrostatic interaction	2 M $Zn(OTf)_2$	0.74	>1000 h (1 mA cm^{-2}, 1 mAh cm^{-2})	[32]
$Ti_3C_2T_x$ MXene/glass fiber	Maxwell–Wagner polarization	2 M $ZnSO_4$	0.35	1180 h (1 mA cm^{-2}, 1 mAh cm^{-2})	[62]
ZrO_2/cellulose	Maxwell–Wagner polarization	2 M $ZnSO_4$	0.69	2000 h (0.5 mA cm^{-2}, 0.25 mAh cm^{-2})	[63]
$BaTiO_3$/glass fiber	Maxwell–Wagner polarization and zincophilicity	2 M $ZnSO_4$	0.69	1600 h (10 mA cm^{-2}, 2.5 mAh cm^{-2})	[25]

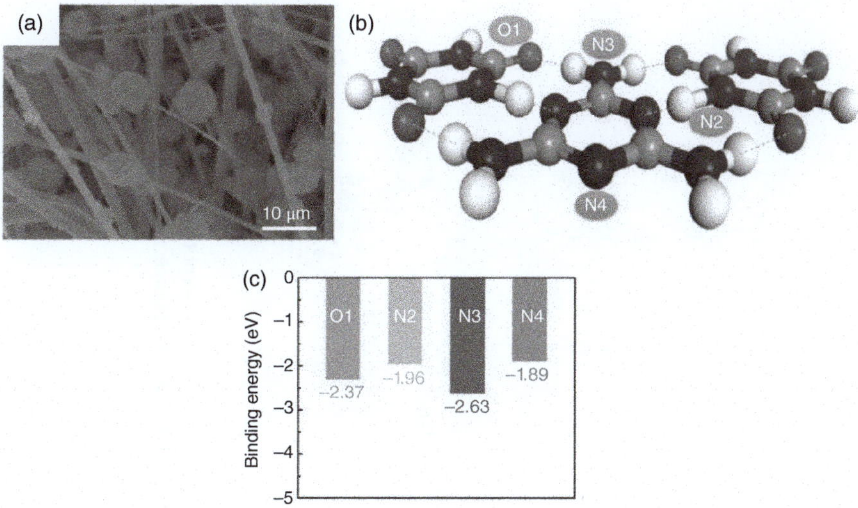

Figure 6.8 (a) SEM images of composite separator decorated with supramolecules. (b) Simulation of the interaction between Zn atoms and possible adsorption sites in the structure of supramolecules, and (c) the corresponding calculated binding energies. Source: (a–c) Liu et al. [53]/with permission from Elsevier.

between zinc anode and separator [54]. In order to increase water resistance during the production process, WP needs to be impregnated with glycerol. Owing to zincophilic —O groups, the adsorption energy between Zn^{2+} and glycerol is stronger than that between Zn^{2+} and water. After the insertion of WP, $\boldsymbol{t}_{Zn^{2+}}$ rose from 0.38 to 0.45, and charge transfer resistance (R_{ct}) was always lower than that of control group at various temperatures, indicating faster ion transport behavior at the interface. Theoretical simulation also proved that WP interlayer could lead to a peaceful Zn^{2+} flux and a gentle concentration gradient, which are beneficial for dendrite-free growth.

Due to the strong coordination between Zn^{2+} and CN^-, CN^--containing baths are commonly used in industrial zinc plating, which helps to obtain a smooth zinc coating [64]. Therefore, the —CN in polyacrylonitrile (PAN) can provide an interfacial interaction for Zn^{2+} and repel anions. To understand the regulation mechanism of Zn^{2+} by PAN separator, whose transference number can reach 0.85, Fang et al. analyzed the XPS spectrum of PAN immersed in 2 M $ZnSO_4$ solution [56]. As shown in Figure 6.9a,b, XPS peaks of Zn $2p_{3/2}$ and $2p_{1/2}$ are monotonic-positive shifts, and the peak value of N 1s showed a negative shift, which could be attributed to the existence of electron transfer between Zn and N, leading to the trend of bonding between Zn and N. Yao and co-authors developed a PAN/graphene oxide (PG) separator and conducted the mechanism of Zn^{2+} migration behavior through a series of theoretical calculations [57]. The PAN chains can capture Zn^{2+} from the aqueous solution due to the functional groups. Subsequently, the Zn^{2+} ions are rapidly transported along the directed migration paths of the fibers, reaching a state where the distribution of Zn is spread across the PG separator. Finally, PG separator can act as a template skeleton to achieve uniform zinc deposition (Figure 6.9c–e).

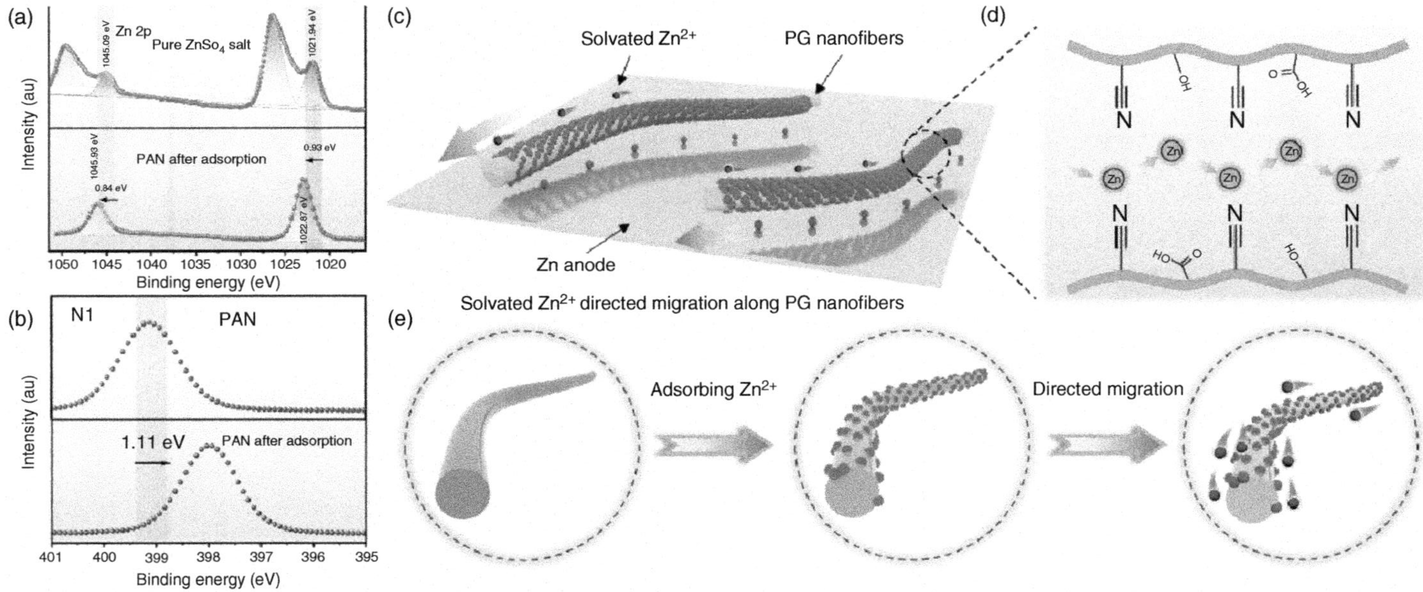

Figure 6.9 (a) XPS spectra of Zn 2p spectrum after adsorption for pure $ZnSO_4$ salt. (b) XPS spectra of N 1 spectrum and adsorption energy of N–Zn. (c–e) Schematic of solvated Zn^{2+} migrating along the directed migration paths of PG nanofibers targeting stabilized Zn anode. Source: (a) Adapted from Fang et al. [56]. (b) Fang et al. [56]/John Wiley & Sons. (c-e) Yao et al. [57]/John Wiley & Sons.

Numerous similar researches have been reported on the induction of zincophilic materials on separator to suppress dendrite growth, for instance, in situ-grown N-doped carbon-modified GF separator [49], polydopamine-functionalized PVDF separator [65], and PBC-modified cellulose-filter paper separator [27], all of which achieved good performance.

6.3.2.2 Electrostatic Interaction

The charge property of the separator skeleton also has a great influence on Zn^{2+} transport behavior. According to space–charge theory, during metal electrodeposition, an ion depletion region exits near the anode because cations are continually and rapidly reduced, which follows an amplified axial electric field to sustain the current [66]. Recently, several researchers have reported that the surface of a bare GF separator is positively charged according to Zeta potential experiment [58, 59]. Thus, owing to the electrostatic interaction, anions can be absorbed on the surface of the fiber and move away from the anode, further limiting the transport of Zn^{2+} (Figure 6.10a). Under this situation, Zn^{2+} ions deposit at a relatively slow rate. Zinc metal tends to grow, avoiding the pore walls of the separator, and forms irregular dendrites (Figure 6.10b). To cope with this phenomenon, researchers designed negatively charged separators.

Zhi et al. developed a collagen hydrolysate (CH)-coated absorbed glass mat (CH@AGM), which showed a negative surface charge, as evidenced by Zeta potential experiment, as CH backbones have abundant oxygen functional groups [58]. Under this condition, as shown in Figure 6.10c, Zn^{2+} ions provide surface conduction to maintain electrodeposition instead of anions providing surface conduction when the separator skeleton is positively charged. Surface conduction finally dominates bulk diffusion in carrying the current through the depleted region, causing a significantly increased limiting current. At the same time, SO_4^{2-} ions are repelled by the amplified electric field, further enhancing bulk depletion and resulting in "shock electrodeposition" in the wake (Figure 6.10d). As a result, fast conduction of Zn^{2+} is carried out to suppress dendritic growth. Similarly, a negatively charged covalent organic framework (COF)-coated GF separator (named "COF@GF") was developed to regulate uniform zinc deposition through a synergistic method combining zincophilicity and electrostatic interaction. The negatively charged separator enables Zn^{2+} ions to migrate along the pore wall boundary, increasing the limiting current. Meanwhile, the high adsorption energy between COF and Zn effectively overcomes the "tip effect." In electrolyte-wetted COF@GF separators, $\boldsymbol{t}_{Zn^{2+}}$ (=0.76) is significantly higher than that of GF separator (=0.22), demonstrating the accelerated Zn^{2+} transport. Similarly, as shown in Figure 6.10e, Zhang et al. developed a cellulose nanowhisker (CNW)–graphene (CNG) membrane, assembled with graphene (GN) and CNWs by CH–π interaction [60]. Zeta potential experiment shows that CNG membrane presents a deanionization shock due to the negative surface charge generated by oxygen-containing groups in the backbone of cellulose nanowhiskers, which promotes cation transport but shields anions to inhibit side reactions. Moreover, compared with hydrophobic GN, CNWs have a strong interaction with H_2O molecules and $[Zn(H_2O)_6]^{2+}$, which was

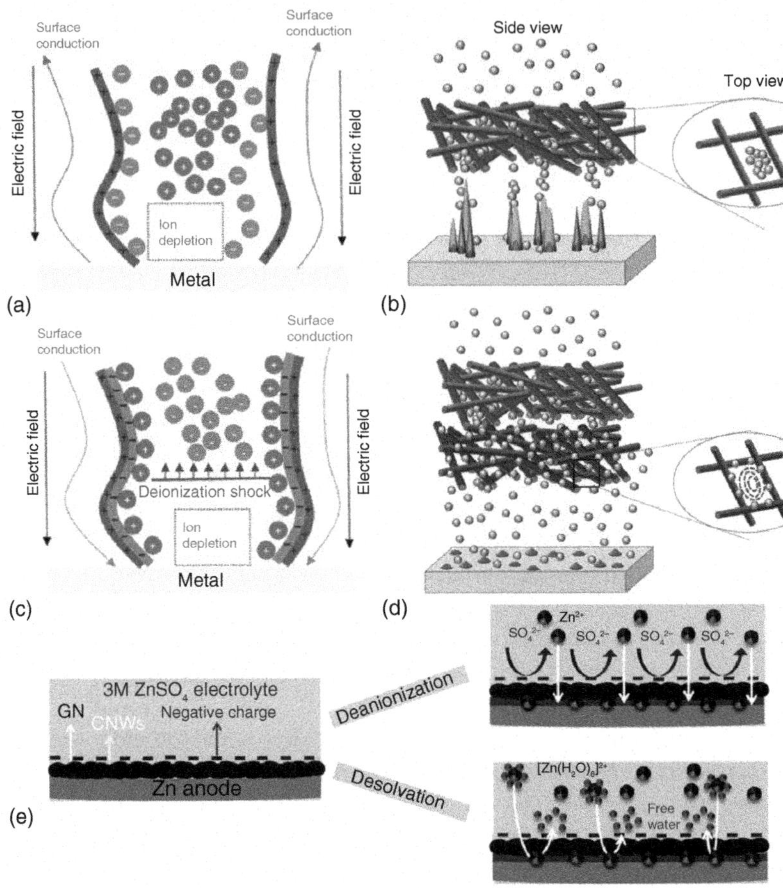

Figure 6.10 Mechanistic sketches of the effect of surface conduction on metal electrodeposition in a (a) positively and (c) negatively charged pore. Schematic diagrams of Zn^{2+} movements in (b) the absorbed glass mat (AGM) and (d) collagen hydrolysate (CH)-coated AGM (CH@AGM). (e) Schematic illustration of Zn plating/stripping behavior with CNG membrane and combination mode of artificial CNG layer. Source: (a–d) Zhi et al. [58]/American Association for the Advancement of Science. (e) Zhang et al. [60]/Royal Society of Chemistry.

confirmed by DFT calculations. As a result, the adsorption effect of hydrophilic CNWs and the repulsive effect of hydrophobic GN can prevent H_2O molecules from reaching the Zn anode, inhibiting water-induced side reactions around zinc anode. Beyond these researches, in fact, compounds with negative charge groups, such as MOFs [67], graphdiyne oxide [33], graphene oxide [68], and polymers containing sulfonate anions [11, 61] , could effectively regulate Zn^{2+} transport behavior.

6.3.2.3 Maxwell–Wagner Polarization

Materials with high dielectric constant (ε) can contrast an enhanced ordered built-in electrical field under an external electric field, and this phenomenon is called Maxwell–Wagner polarization (also called interfacial polarization or space charge polarization) [52]. Such an electrical field can motivate fast ion diffusion kinetics and diminish the dynamic disparity between redox reaction and mass transfer [62].

ZrO_2 is a typical ceramic insulating material with high dielectric constant ($\varepsilon = 25$) [67]. In Cao's research, a ZrO_2/cellulose separator was developed by the solution casting method [63]. The electric dipoles could induce a polarized electric field on the surface around the ZrO_2 particles, rendering uniform ion pathways and motivating ion diffusion kinetics. Zn||ZrO_2/cellulose||Zn symmetrical battery presents a $\boldsymbol{t}_{\mathbf{Zn}^{2+}}$ of 0.69, which is much higher than that of Zn||cellulose||Zn (0.47) and Zn||GF||Zn (0.39), indicating the accelerated Zn^{2+} transport behavior. At the same time, the nucleation overpotential is also greatly reduced by Maxwell–Wagner polarization of ZrO_2, facilitating uniform zinc deposition. The authors also replaced ZrO_2 with other ceramics, including TiO_2, Sb_2O_3, Al_2O_3, SiO_2, and mixed ceramics like $ZrO_2 + TiO_2$ to form ceramics–cellulose separators. The overpotential of Zn||Zn symmetrical batteries drops gradually with the increasing ε value of the ceramics in the separators, representing that ceramics with high dielectric constant can induce a strong polarized electric field, which enables high ionic conductivity and fast ion diffusion, thus reducing the overpotential. Su et al. reported a $Ti_3C_2T_x$ MXene-decorated GF (MXene-GF) separator with a high dielectric constant of 53.3, offering a directional built-in electric field with 50% enhanced intensity to expedite Zn^{2+}-ion transfer (Figure 6.11a). Accordingly, Zn||Zn cells equipped with MXene-GF could cycle stably for 1180 h at 1 mA cm^{-2} (Figure 6.11b).

6.3.3 Manipulating Zn Growth Direction

Common separators are not ionically conductive, and the only pathway for Zn^{2+} ions through them is through the pores immersed with electrolytes [19]. That means common separators cannot regulate ions' diffusion behavior after ions transport across the separators, resulting in a random direction of zinc deposition, which is manifested as random crystallographic orientation and vertical growth of zinc. Therefore, many researchers have been dedicated to manipulating the crystallographic orientation or lateral growth of zinc.

6.3.3.1 Manipulating Crystallographic Orientation

The crystal structure of zinc is hexagonal close-packed (HCP) with space group of $P6_3/mmc$. Therefore, zinc metal has low symmetry and shows a high degree of crystal anisotropy. In addition, the c/a value of the Zn is 1.85, which is larger than the ideal value (1.63), i.e. the face spacing along the c-axis is higher than the theoretical value, which further increases the crystal anisotropy of zinc metal [69]. The Zn (002) plane has the highest atomic packing density, which is preferentially exposed in the electrodeposition process to minimize the surface free energy, and tends to grow parallel to the substrate, as the angle between the (002) plane and the substrate is 0–30°. The angles between the substrate and the (100) and (110) planes are 70–90°, which tends to form rough and uneven surfaces, thus increasing the risk of short circuit of the battery [68a, 70]. Therefore, regulating Zn (002) plane

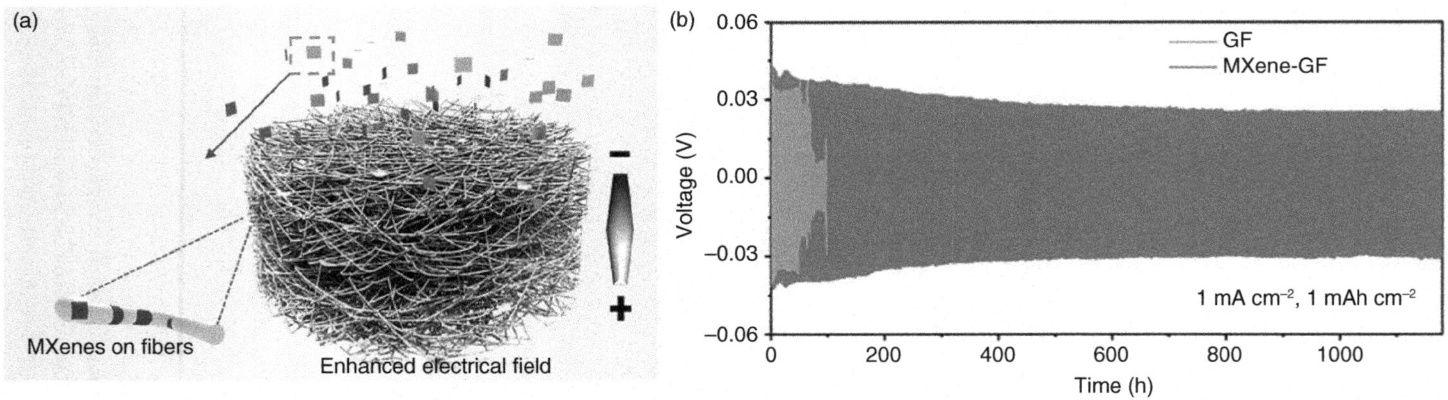

Figure 6.11 (a) Schematic illustration depicting the function of MXene-GF separator. (b) Cycling performances of symmetric cells based on MXene-GF and GF at 1 mA cm^{-2} with a capacity of 1 mAh cm^{-2} [62].

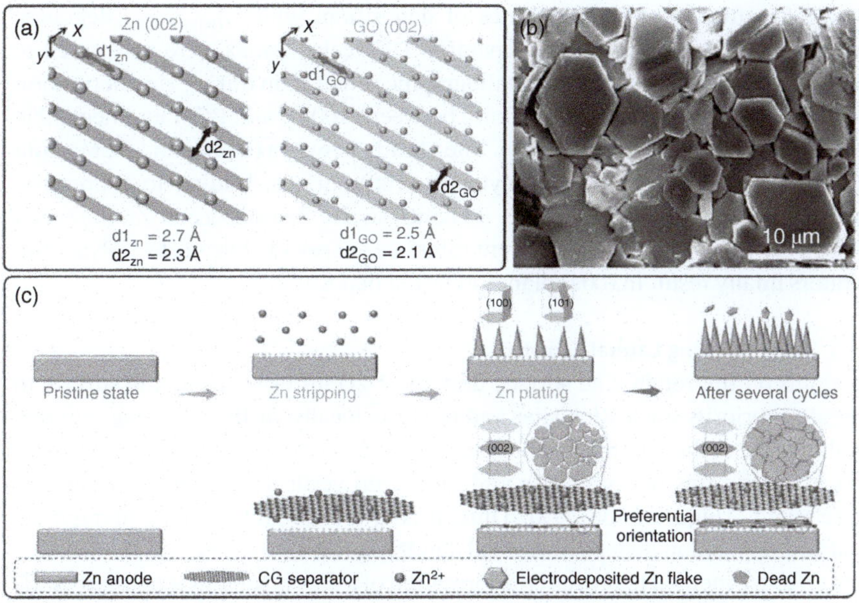

Figure 6.12 (a) Atomic arrangements of (002) plane of Zn and (002) plane of GO. (b) Hexagonal platelet morphology of zinc anode surface after cycling at 2 mA cm^{-2}. (c) Schematic illustration of zinc deposition behavior with bare or GO-modified separator. Source: Cao et al. [68a]/John Wiley & Sons.

as the dominant orientation of electrodeposition is of great significance to inhibit dendritic growth and increase battery cycle life.

Zheng et al. first reported that the Zn nuclei as crystals have a strong tendency to form coherent interfaces with a textured, low lattice-mismatched graphene layer as an epitaxial substrate [71]. Recent researches have shown that lattice-matched material-modified separator could also play a role in inducing Zn (002) plane preferential orientation deposition. Graphene oxide is one of the materials with a hexagonal lattice structure in atomic arrangement. And the lattice mismatch degree between $(002)_{Zn}$ and $(002)_{GO}$ is about 7.4% (Figure 6.12a), which can effectively manipulate the initial Zn^{2+} ions to preferentially deposit via the regular hexagonal (002) plane rather than the protruding and dendritic (100) or (101) planes [68]. These zinc deposits with preferential orientation prefer the parallel direction to the anode surface during the initial cycling process and induce subsequent deposition to grow with the same preferred orientation, thus facilitating nondendritic electrodeposition of Zn (Figure 6.12b,d).

Besides the lattice match principle, some materials could induce zinc deposition with preferential orientation owing to their special affinity to specific zinc crystal plane. Zhang and co-authors constructed ultrathin and high-toughness bacterial cellulose membranes (BCM) for substituting GF separators [40]. According to ex situ X-ray diffraction (XRD), during zinc plating process, the deposited Zn with BCM exhibited increasing diffraction intensity of (002) plane. The peak intensity ratio between (002) and (100), denoted as $I_{(002)}/I_{(100)}$, showed an increase from

2.68 to 3.85 after 10 and 120 minutes of deposition, while $I_{(002)}/I_{(100)}$ degraded from 1.75 to 1.15 when using GF separator. DFT calculations were performed to reveal why BCM could regulate crystallographic orientation during zinc deposition process, suggesting a remarkable affinity between $(100)_{Zn}$ and BCM and negligible interaction between $(002)_{Zn}$ and BCM. The affinity between $(002)_{Zn}$ and BCM could hinder Zn^{2+} transport from electrolytes to (100) surface, leading to suppressed crystal growth along the (100) plane. By contrast, the free surface of Zn (002) can replenish Zn^{2+} over time. Such different interface energies between electrolytes and separators finally result in (002) plane-preferred deposition.

6.3.3.2 Manipulating Lateral Growth

Zinc atoms on the rough zinc surface undergo a 2D diffusion along the surface to aggregate at priority nucleation sites and nucleate locally, further evolving into tiny crystallite. Due to the inhomogeneous electric field, Zn^{2+} ions in the electrolyte tend to absorb at the tips of crystallites and grow into vertical dendrites to minimize surface energy (Figure 6.13a) [72]. Thus, it is of great significance to manipulate lateral zinc deposition instead of vertical dendrite growth.

One effective method to prevent the formation of local crystallites is to improve 2D diffusion barriers to induce spacious atomic clusters and inhibit dendrite formation. Cao et al. directly developed a commercial cotton towel (CT) as a separator [73]. As shown in Figure 6.13b, chronoamperometry (CA) experiment demonstrated that the 2D diffusion process occurred within only 18 s and then underwent a constant 3D diffusion when equipped with CT separator, while 300 s for the GF separator. This is because the strong interaction between hydrate Zn^{2+} ions and —OH inside the CT separator leads to an increased energy barrier for the 2D diffusion barrier, resulting in a reduced 2D diffusion current (<4 mV). In other words, Zn^{2+} ions are forced to deposit close to their initial absorbed sites instead of rediffusing to locations with lower surface energy.

Our group first reported that zinc plating reaction can be well defined under faceted titanium dioxide protective layer with relatively low zinc affinity, preventing dendrites from growing vertically [74]. Similarly, zincophobic separators could force Zn to deposit preferentially away from the separator rather than at the tips of crystallites. Gan et al. designed an ion sieve consisting of silica nanosheets with micro-mesoporous ion channels and hydroxyl groups (IS/SiO_2–OH) to modify commercial GF separator [30]. Compared to bare silica, IS/SiO_2–OH showed a much lower binding energy due to the introduction of hydroxyl groups, suggesting a low zinc affinity or even zincophobicity (Figure 6.13c). The zincophobic layer facilitates zinc deposition preferentially away from the ion sieve, preventing the failure of the ion sieve and the puncture of the separator (Figure 6.13d). Besides, our group implanted a commercial zincophobic cellulose paper between GF separator and zinc anode to improve deposition progress [75]. Compared to the zinc anode with high zincophilic GF separator, the zinc with highly zincophobic cellulose paper grows toward the edges of initially deposited islands rather than the tops, which is due to the generated repulsive field on the anode surface. This lateral growth mechanism effectively induced smooth zinc deposition.

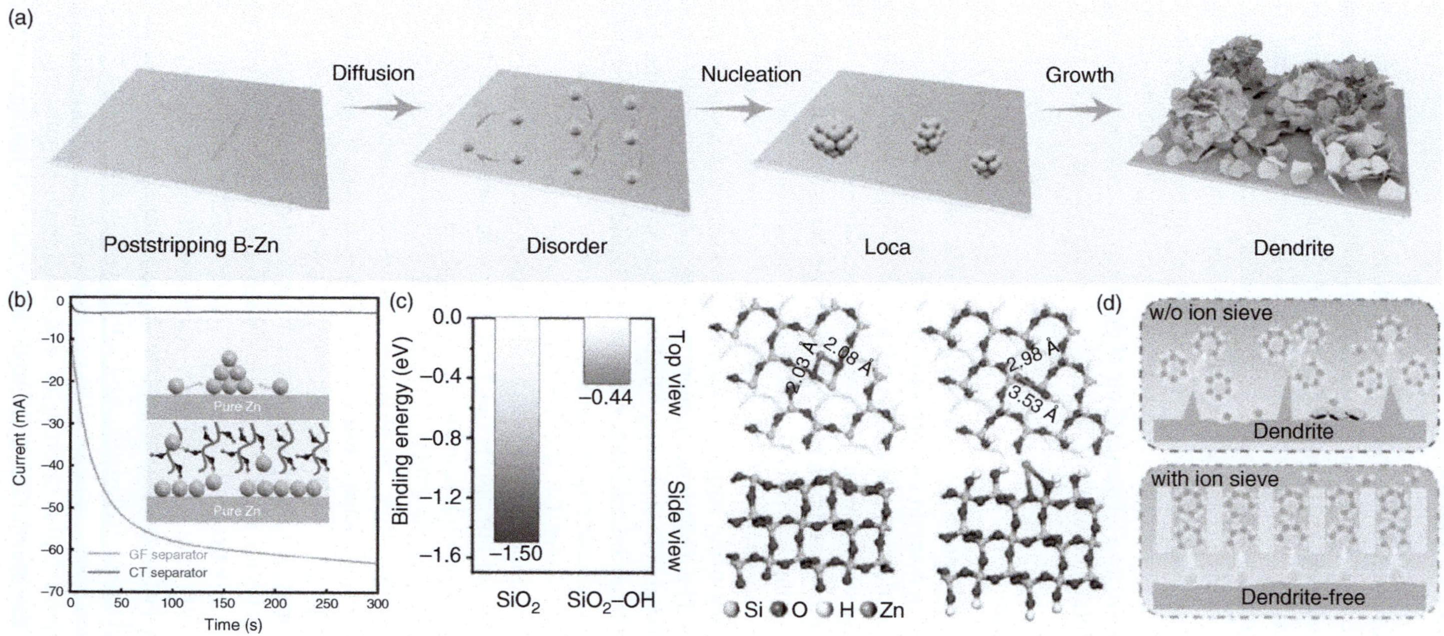

Figure 6.13 (a) Schematic illustration of zinc deposition on post-stripping bare Zn. (b) Chronoamperometry (CA) test of the cell with glass fiber separator and cotton towel separator under −150 mV. (c) DFT calculation for the binding energy of SiO_2 and hydroxylated SiO_2 with Zn atom. (d) Schematic illustration of zinc deposition with and without ion sieve. Source: (a) Adapted from Yang et al. [72]. (b) Cao et al. [73]/American Chemical Society. (c, d) Gan et al. [30]/With permission of Elsevier.

6.4 Separator-Free AZIBs

Gel electrolytes (GEs) and solid electrolytes (SEs) can be regarded as a combination of separator and electrolyte with both ion transport and mechanical barrier functions, which we classify as separator-free type. With the advantages of minimum risk of electrolyte leakage, high security, easy processing, and high thermal stability, separator-free batteries have attracted great interest. In addition, considering the reduction of free water, GE and SE can alleviate the uncontrollable solid–liquid interface side reactions (including dissolution of cathode materials, corrosion, and passivation of zinc anodes) caused by free water during cycling. Therefore, the development of high-performance GEs and SEs is of great significance in promoting the development and application of AZIBs.

6.4.1 Gel Electrolyte

The vigorous development of portable and wearable electronic products has generated an increasing demand for corresponding electrochemical energy-storage systems [76]. One of the key challenges identified in this field is fabricating flexible energy-storage devices with high performance, safety, and strong ability to function under different conditions, especially extreme conditions such as cutting, bending, or extreme temperature [77]. Thus, the development of flexible GE has become a hotspot for researchers. GE is composed of polymeric frameworks and Zn-containing salts, featuring liquid-like ionic conductivity and solid-like dimensional stability [78]. The polymeric frameworks are formed by crosslinking by monomers with a 3D network structure, which not only provides high mechanical integrity but also a large internal space for better liquid uptake and retention. Therefore, GE meets the basic requirements of a separator (high strength and high ionic conductivity). Moreover, GE with limited free water can also play a positive role in the cathode and anode, which not only effectively alleviates the dissolution of cathode-active material but also reduces the side reactions (such as hydrogen evolution reaction and corrosion reaction) of zinc anode [79].

Niu et al. [80] developed the PVA/$Zn(CF_3SO_3)_2$ hydrogel electrolyte through a freezing/thawing strategy, which can autonomously self-heal and recover electrochemical performance completely after cutting (Figure 6.14a). Huang et al. [82], designed PANa GE to prepare NiCo//Zn batteries, which could increase the capacity under 400% tensile and 50% compression conditions. Zhou et al. [83], prepared a GG/SA/EG hydrogel with excellent freezing resistance, which still had a high ionic conductivity of 6.19 $Ms\,cm^{-1}$ at −20 °C. Huang and Zhi et al. [84] reported an intelligent thermally responsive polymer electrolyte that can prevent thermal runaway through a fast-responsive reversible sol–gel transition. Niu et al. [83] developed a self-adhesive, degradable, and edible sustainable dough-based GE (Figure 6.14b,c).

Compared with pure liquid electrolytes and traditional separators, GE can effectively alleviate the dissolution of electrode materials and maintain relatively stable electrochemical performance [85]. It not only avoids the leakage and

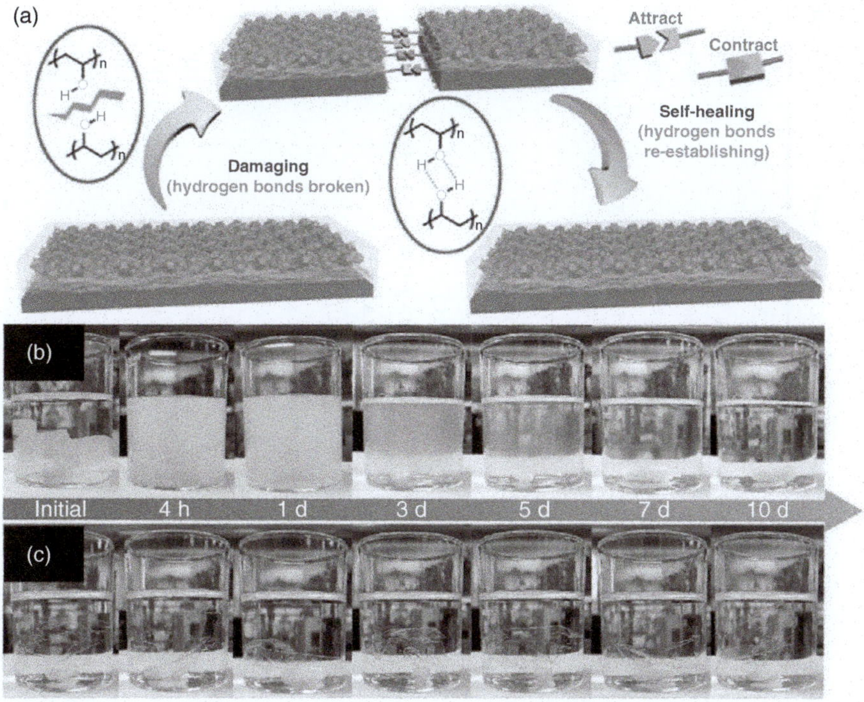

Figure 6.14 (a) Illustration of the structure of self-healing integrated all-in-one AZIBs. The degradation processes of (b) dough electrolyte and (c) PVA/H_2SO_4 electrolyte in simulated gastric fluid. Source: (a) Huang et al. [80]/John Wiley & Sons. (b, c) Wang et al. [81]/John Wiley & Sons.

volatilization of conventional aqueous electrolytes but also simplifies the battery assembly. Table 6.2 summarizes the recent reports on the properties of GEs used in AZIBs. Most of the reported GEs still have poor ionic conductivity, relatively narrow voltage windows, and poor cycle stability. More researchers must vigorously develop novel materials and study mechanisms to achieve the commercialization goal. In addition, we should pay attention to the compatibility between GE and electrode, consider the cost, power consumption, environment, and other issues, simplify the battery manufacturing process, and optimize the battery packaging technology, to realize the practical application of GE in AZIBs as soon as possible.

6.4.2 Solid Electrolyte

GEs can only inhibit the water-induced side effects to a certain extent because active water still remains. SE, with the absence of solvent, can avoid some side reactions fundamentally. Moreover, SE can maintain high rigidity in the case of thinner than GEs, so SE can effectively inhibit dendrite generation and puncture, preventing short circuits inside the battery. Therefore, the study of SE is conducive to the further development of AZIBs. However, the development of SE is still in its infancy. The strong electrostatic bonding nature of the divalent Zn^{2+} ions leads to slow diffusion

Table 6.2 Summary of the basic properties of GEs, matching cathode materials, and electrochemical performance for recently reported GEs in AZIBs.

Electrolyte	Ionic conductivity ($S\,m^{-1}$)	Tensile strength (kPa)	Stretchability (%)	Cathode-active material	Capacity ($mAh\,g^{-1}$)	Capacity retention/cycle number	Reference
PVA/$Zn(CF_3SO_3)_2$	1260	N/A	N/A	PANI	123 (0.1 $A\,g^{-1}$) 94 (3 $A\,g^{-1}$)	97.1%/1000 (1 $A\,g^{-1}$)	[80]
PVA/$ZnCl_2$/LiCl/$MnCl_2$	N/A	N/A	N/A	MnO_2	290 (0.1 $A\,g^{-1}$) 51 (2 $A\,g^{-1}$)	75%/300 (1 $A\,g^{-1}$)	[86]
PVA/$ZnCl_2$/LiCl/$MnSO_4$	N/A	N/A	N/A	MnO_2	366.6 (0.74 $A\,g^{-1}$)	77.7%/300 (1.86 $A\,g^{-1}$)	[87]
Gelatin /SA/$ZnSO_4$	3.7	2140	~140	V_2O_5	212 (2 $A\,g^{-1}$)	79%/400 (2 $A\,g^{-1}$)	[88]
Gelatin /Li_2SO_4/$ZnSO_4$	0.61	~50	~50	$LiMn_2O_4$	110.2 (0.025 $A\,g^{-1}$)	90%/100 (0.025 $A\,g^{-1}$) 89%/200 (0.1 $A\,g^{-1}$)	[89]
Gelatin/$ZnSO_4$/$MnSO_4$	10	N/A	N/A	MnO_2	93 (20 C)	91%/1000 (1 C)	[90]
Gelatin/PAM/$ZnSO_4$/$MnSO_4$	1.76	7760	~175	MnO_2	306 (0.0616 $A\,g^{-1}$)	97%/1000 (2.772 $A\,g^{-1}$)	[91]
PAM/$ZnSO_4$/$MnSO_4$	1.73	N/A	300	MnO_2	302.1 (0.06 $A\,g^{-1}$)	98.5%/500 (2 $A\,g^{-1}$)	[92]
PAM/$ZnSO_4$/$CoSO_4$	12	N/A	N/A	Co_3O_4	205 (0.5 $A\,g^{-1}$)	92%/5000 (4 $A\,g^{-1}$)	[93]
PAM/NFC/$ZnSO_4$/$MnSO_4$	2.28	158	1400	MnO_2	200 (4 C)	88.3%/1000 (4 C)	[94]
PAM/CNFs/$Zn(CF_3SO_3)_2$	0.68	200	730	$Mg_{0.23}V_2O_5$	216 (5 $A\,g^{-1}$) 367 (0.5 $A\,g^{-1}$)	81.7%/180 (0.5 $A\,g^{-1}$) 98.6%/2000 (5 $A\,g^{-1}$)	[95]
SA/$ZnSO_4$/$MnSO_4$	1.83	N/A	N/A	MnO_2	300 (0.2 $A\,g^{-1}$)	~77%/100 (0.2 $A\,g^{-1}$)	[96]
PHE/Li_2SO_4/$ZnSO_4$	0.633	N/A	N/A	$LiFePO_4$	169 (0.1 C)	90%/300 (0.5 C)	[97]
Dough (Na, K)/$Zn(CF_3SO_3)_2$	0.25	65.1	~225%	PANI	137.8 (0.1 $A\,g^{-1}$)	N/A/1000 (1 $A\,g^{-1}$)	[81]
Cotton/TEOS/glycerol/$ZnSO_4$/$MnSO_4$	3.23	2110	846.50	MnO_2	278.1 (0.2 $A\,g^{-1}$)	99.2%/2000 (3 $A\,g^{-1}$, 20 °C)	[98]
PAAm/DMSO/$Zn(CF_3SO_3)_2$	~3.8	48.8	1122	$Zn_3V_2O_8$	265.2 (0.2 $A\,g^{-1}$) 185.1 (10 $A\,g^{-1}$)	95.27%/3000 (1 $A\,g^{-1}$)	[99]
GG/SA/EG/$ZnSO_4$/$MnSO_4$	2.537	N/A	N/A	MnO_2	135.0 (3 $A\,g^{-1}$) 90.3 (6 $A\,g^{-1}$)	78.39%/1000 (6 $A\,g^{-1}$)	[83]
SA/glycerol/$Zn(CF_3SO_3)_2$	3.28	N/A	N/A	I_2	214.7 (0.2 $A\,g^{-1}$) 120.8 (2 $A\,g^{-1}$)	66.8%/2000 (2 $A\,g^{-1}$) 97.6%/200 (0.2 $A\,g^{-1}$)	[100]
PAM/ChNF/$Zn(CF_3SO_3)_2$	1.52	114.62	~500	VO_2	263.9 (0.2 $A\,g^{-1}$)	93.3%/100 (0.2 $A\,g^{-1}$)	[101]
PANa/$Zn(CH_3COO)_2$/KOH	20	80%	1700%	NiCo	80 (56.5 C)	86%/340 (56.5 C)	[82]
PNA/$ZnSO_4$/$MnSO_4$	N/A	N/A	N/A	MnO_2	145 (0.1 $A\,g^{-1}$)	167.7%/550 (0.5 $A\,g^{-1}$)	[84]

Abbreviations: PVA, poly(vinyl alcohol) [80]; SA, sodium alginate [88]; PAM, polyacrylamide [91]; NFC, nanofibrillated cellulose [94]; CNF, cellulose nanofiber [95]; PHE, pluronic hydrogel electrolyte [97]; TEOS, tetraethyl orthosilicate [98]; PAAm, polyacrylamide [99]; GG, guar gum [83]; EG, ethylene glycol [83]; ChNF, chitin nanofiber [101]; PANa, sodium polyacrylate [82]; PNA, poly(*N*-isopropylacrylamide-*co*-acrylic acid) [84].
All the above data are measured at room temperature.
The ionic conductivity of PAM/$ZnSO_4$/$CoSO_4$ is measured at 300% swelling ratio [93].

kinetics, resulting in low ionic conductivity (10^{-8}–10^{-6} S cm^{-1}), which is difficult to meet the practical needs of the battery (10^{-4} S cm^{-1}) [102].

The most common SE is formed by the appropriate complexation of metal salts within a polymer matrix [103]. At present, it is generally believed that the polar group of the polymer can effectively dissolve the metal salt, resulting in the formation of polymer–metal salt complex in which the metal salt ions are mainly conducted through the chain segments of the polymer in the amorphous regions of the polymer. Therefore, the ionic conductivity can be improved by adding inorganic fillers or regulating zinc salts to destroy the crystallinity of the polymer matrix and increase the proportion of amorphous regions. Karan et al. synthesized SE [90PEO:10Zn(CF_3SO_3)$_2$] (the ionic conductivity is 1.09×10^{-6} S cm^{-1}) by compounding large anionic zinc salt Zn(CF_3SO_3)$_2$ based on PEO (the ionic conductivity is 3.20×10^{-9} S cm^{-1}) [104]. Then ZnO nanoparticles were dispersed therein, and the ionic conductivity was raised to 1.84×10^{-5} S cm^{-1} [105]. However, their ion conductivity is still far from meeting the needs of practical applications. In order to fully excavate the conductivity of the amorphous region, an amorphous ion-conductive polymer is synthesized by controlling the synthesis method. Zhi et al. [106] developed a liquid-free amorphous SE containing a high concentration of Zn(BF_4)$_2$ by in situ performing ring-opening polymerization of a precursor of 1,3-dioxolane inside an electrochemical cell (1.96×10^{-2} S cm^{-1} at room temperature, Figure 6.15a,b). The problems of high interface resistance, low ionic conductivity of conventional SEs, and high liquid containment of GEs were successfully addressed by in situ polymerization method. In addition to using a polymer matrix, SE can take into account multiple excellent properties through a plurality of polymer grafting, blending, cross-linking, etc. [108]. The abundant groups on the surface of MXene were grafted with poly(methyl acrylate) and added to the PVHF/Zn(OTF)$_2$ matrix to form PVHF/MXene-g-PMA (Figure 6.15c) [107]. It delivers remarkable ionic conductivity (2.69×10^{-4} S cm^{-1}) at room temperature with a high Zn^{2+} transference number. The full cells based on the SEs exhibited obvious HER-free features and excellent cycling performance of 10 000 cycles at 2 °C and can operate normally within a wide temperature range from −35 to 100 °C.

Overall, in order to promote the practical application of AZIBs, there is an urgent need to develop relatively thin separators with high electrochemical and structural stability, good electrolyte wettability, high electrolyte uptake and retention, uniform pore structure, and strong mechanical strength. In addition, the manufacturing process should be as green and low-cost as possible. At present, separators for AZIBs mostly refer to commercialized LIBs separators and have their own inherent defects. As a result, it is of great significance to develop novel and low-cost separators for AZIBs. Homogenizing Zn^{2+} distribution, accelerating Zn^{2+} transport, and manipulating Zn growth direction are three typical design strategies for modifying commercial separators or developing novel separators to construct high-performance AZIBs. For the vigorous development of flexible batteries and solid-state batteries, developing high-performance GEs and solid-state electrolytes is also the future development direction.

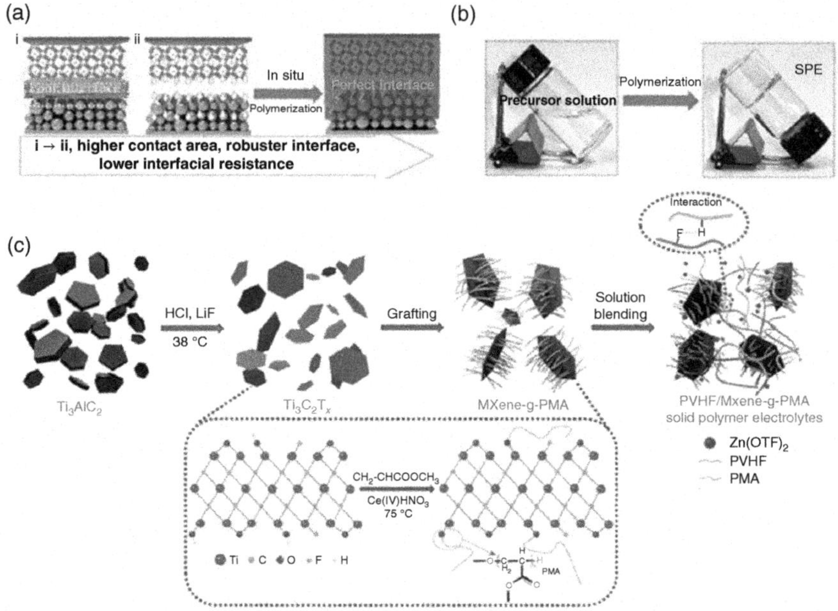

Figure 6.15 (a) Schematic illustration of ex situ and in situ-formed SPEs for Zn batteries. (b) Digital photograph depicting the liquid 4 M $Zn(BF_4)_2$/DOL solution and SPEs formed spontaneously in an electrolyte containing 2 mM $Al(OTf)_3$ salt. (c) Schematic illustration of the overall preparation process of the SPEs. Source: (a) Ma et al. [106]/John Wiley & Sons. (b) Ma et al. [106]/John Wiley & Sons. (c) Chen et al. [107]/Royal Society of Chemistry. (d) Wang et al. [81]/With permission of Elsevier.

References

1 Qin, Y., Liu, P., Zhang, Q. et al. (2020). *Small* 16: e2003106.

2 Arora, P. and Zhang, Z. (2004). *Chemical Reviews* 104: 4419–4462.

3 Luo, W., Cheng, S., Wu, M. et al. (2021). *Journal of Power Sources* 509: 230372.

4 Francis, C.F., Kyratzis, I.L., and Best, A.S. (2020). *Advanced Materials* 32: 1904205.

5 Liu, F. and Chuan, X. (2021). *RSC Advances* 11: 16633.

6 Yang, J., Wang, C.-Y., Wang, C.-C. et al. (2020). *Journal of Materials Chemistry A* 8: 5095.

7 Li, J., Cai, Y., Wu, H. et al. (2021). *Advanced Energy Materials* 11: 2003239.

8 Lee, H., Yanilmaz, M., Toprakci, O. et al. (2014). Energy & Environmental. *Science* 7: 3857.

9 Chen, J., Liu, Q., Wang, B. et al. (2017). *Journal of the Electrochemical Society* 164: A1526.

10 (a) Du, W., Ang, E.H., Yang, Y. et al. (2020). *Energy & Environmental Science* 13: 3330. (b) Li, Q., Zhao, Y., Mo, F. et al. (2020). *EcoMat* 2: e12035. (c) Kang, L., Cui, M., Zhang, Z., and Jiang, F. (2020). *Batteries & Supercaps* 3: 966.

11 Ge, X., Zhang, W., Song, F. et al. (2022). *Advanced Functional Materials* 32: 2200429.

12 Mun, S.C. and Won, J.H. (2021). *Crystals* 11: 1013.

13 Kim, P.J. (2021). *Nanomaterials* 11: 2275.

14 Ghosh, M., Vijayakumar, V., and Kurungot, S. (2019). *Energy Technology* 7: 1900442.

15 Zhao, Z., Xu, L., Wang, C. et al. (2022). *Chemical Engineering Journal* 136079.

16 (a) Han, X., Chen, L., Yanilmaz, M. et al. (2023). *Chemical Engineering Journal* 454: 140311. (b) Isik, M., Sardon, H., and Mecerreyes, D. (2014). *International Journal of Molecular Sciences* 15: 11922.

17 Shang, W., Li, Q., Jiang, F. et al. (2022). *Nano-Micro Letters* 14: 82.

18 (a) Ni, Q., Kim, B., Wu, C., and Kang, K. (2022). *Advanced Materials* 34: 2108206. (b) Song, Y., Sheng, L., Wang, L. et al. (2021). *Electrochemistry Communications* 124: 106948. (c) Huang, X., He, R., Li, M. et al. (2020). *Materials Today* 41: 143.

19 Zhao, C.-Z., Chen, P.-Y., Zhang, R. et al. (2018). *Science Advances* 4: eaat3446.

20 (a) Cao, P., Tang, J., Wei, A. et al. (2021). *ACS Applied Materials & Interfaces* 13: 48855. (b) Liu, H., Zhang, Y., Wang, C. et al. (2021). *ACS Applied Materials & Interfaces* 13: 32930.

21 (a) Qin, R., Wang, Y., Yao, L. et al. (2022). *Nano Energy* 98: 107333. (b) Guo, L., Guo, H., Huang, H. et al. (2020). *Frontiers in Chemistry* 8: 557. (c) Chen, S., Wang, H., Zhu, M. et al. (2023). *Nanoscale Horizons* 8: 29.

22 (a) Liu, H., Zhou, Q., Xia, Q. et al. (2023). *Journal of Energy Chemistry* 77: 642. (b) Li, Z., Ye, L., Zhou, G. et al. (2023). *Chemical Engineering Journal* 457: 141160.

23 (a) Zhang, Q., Luan, J., Tang, Y. et al. (2020). *Angewandte Chemie (International Ed. in English)* 59: 13180. (b) Zhang, Q., Luan, J., Huang, X. et al. (2020). *Small* 16: e2000929.

24 (a) Liu, X., Wei, W., Yang, Y. et al. (2022). *Chemical Engineering Journal* 437: 135409. (b) Xu, S., Nie, W., Sun, P. et al. (2023). *Materials Letters* 331: 133546.

25 Liang, Y., Ma, D., Zhao, N. et al. (2022). *Advanced Functional Materials* 32: 2112936.

26 Song, Y., Ruan, P., Mao, C. et al. (2022). *Nano-Micro Letters* 14: 218.

27 Li, L., Peng, J., Jia, X. et al. (2022). *Electrochimica Acta* 430: 141129.

28 (a) Wang, Z., Dong, L., Huang, W. et al. (2021). *Nano-Micro Letters* 13: 73. (b) Maeboonruan, N., Lohitkarn, J., Poochai, C. et al. (2022). *Journal of Science* 7: 100467. (c) Yang, H., Qiao, Y., Chang, Z. et al. (2020). *Advanced Materials* 32: 2004240. (d) Gopalakrishnan, M., Ganesan, S., Nguyen, M.T. et al. (2023). *Chemical Engineering Journal* 457: 141334.

29 Yang, Y., Chen, T., Yu, B.-X. et al. (2022). *Chemical Engineering Journal* 433: 134077.

30 Gan, H., Wu, J., Zhang, F. et al. (2023). *Energy Storage Materials* 55: 264.

31 (a) Guo, R., Liu, X., Xia, F. et al. (2022). *Advanced Materials* 34: 2202188. (b) Wu, C., Guo, F., Zhuang, L. et al. (2020). *ACS Energy Letters* 5: 1644.

32 Hu, W., Ju, J., Zhang, Y. et al. (2022). *Journal of Materials Chemistry A* 10: 24761.

33 Li, J., Chen, Y., Guo, J. et al. (2020). *Advanced Functional Materials* 30: 2004115.

34 An, Y., Tian, Y., Man, Q. et al. (2022). *ACS Nano* 16: 6755.

35 Zhang, X., Yang, S., Huang, Z. et al. (2022). *Electrochimica Acta* 430: 141081.

36 Zhang, Y., Yang, G., Lehmann, M.L. et al. (2021). *Nano Letters* 21: 10446.

37 Zhang, Y., Zhao, T., Yang, S. et al. (2022). *Journal of Energy Chemistry* 75: 310.

38 Zhou, W., Chen, M., Tian, Q. et al. (2022). *Energy Storage Materials* 44: 57.

39 Fu, J., Wang, H., Xiao, P. et al. (2022). *Energy Storage Materials* 48: 191.

40 Zhang, Y., Li, X., Fan, L. et al. (2022). *Cell Reports Physical Science* 3: 100824.

41 Ogawa, S., Takazawa, Y., Harada, H., and Nogami, M. (2022). *Electrochemistry* 90: 017001.

42 Hou, Z., Gao, Y., Tan, H., and Zhang, B. (2021). *Nature Communications* 12: 3083.

43 (a) Zhang, G., Zhang, X., Liu, H. et al. (2021). *Advanced Energy Materials* 11: 2003927. (b) Li, C., Shi, X., Liang, S. et al. (2020). *Chemical Engineering Journal* 379: 122248.

44 Zhou, Y., Wang, X., Shen, X. et al. (2020). *Journal of Materials Chemistry A* 8: 11719.

45 (a) Zhu, X., Guo, F., Yang, Q. et al. (2021). *Journal of Power Sources* 506: 230224. (b) Zhang, H., Chen, Z., Zhang, Y. et al. (2021). *Journal of Materials Chemistry A* 9: 16565.

46 (a) Qiu, M., Jia, H., Liu, H. et al. (2022). *Journal of Alloys and Compounds* 891: 161886. (b) Dong, L., Yang, W., Yang, W. et al. (2020). *Chemical Engineering Journal* 384: 123355.

47 Li, C., Sun, Z., Yang, T. et al. (2020). *Advanced Materials* 32: e2003425.

48 Huang, J.-Q., Hou, Z., Gao, P. et al. (2022). *Materials Today Communications* 32: 103939.

49 Yang, X., Li, W., Lv, J. et al. (2022). *Nano Research* 15: 9785.

50 (a) Jia, H., Wang, Z., Tawiah, B. et al. (2020). *Nano Energy* 70: 104523. (b) Chen, L.N., Yan, M.Y., Mei, Z.W., and Mai, L.Q. (2017). *Journal of Inorganic Materials* 32: 225.

51 Chen, C.-Y., Matsumoto, K., Kubota, K. et al. (2019). *Advanced Energy Materials* 9: 1900196.

52 Hao, Z.D., Zhao, Q., Tang, J.D. et al. (2021). *Materials Horizons* 8: 12–32.

53 Liu, T., Hong, J., Wang, J. et al. (2022). *Energy Storage Materials* 45: 1074.

54 Guo, Y., Cai, W., Lin, Y. et al. (2022). *Energy Storage Materials* 50: 580.

55 Yang, X., Wu, W., Liu, Y. et al. (2022). *Chemical Engineering Journal* 450: 137902.

56 Fang, Y., Xie, X.S., Zhang, B.Y. et al. (2022). *Advanced Functional Materials* 32: 2109671.

57 Yao, L., Hou, C.X., Liu, M.Q. et al. (2023). *Advanced Functional Materials* 33: 2209301.

58 Zhi, J., Li, S.K., Han, M., and Chen, P. (2020). *Science Advances* 6: eabb1342.
59 Yang, D., Wu, X., He, L. et al. (2023). *Nano Letters* 23: 336.
60 Zhang, X.T., Li, J.X., Liu, D.Y. et al. (2021). *Energy & Environmental Science* 14: 3120.
61 Zhang, X., Li, J., Qi, K. et al. (2022). *Advanced Materials* 34: 2205175.
62 Su, Y., Liu, B., Zhang, Q. et al. (2022). *Advanced Functional Materials* 32: 2204306.
63 Cao, J., Zhang, D.D., Gu, C. et al. (2021). *Nano Energy* 89: 106322.
64 Fang, Y., Xie, X., Zhang, B. et al. (2021). *Advanced Functional Materials* 2109671.
65 Liu, Y., Liu, S., Xie, X. et al. (2022). *InfoMat* 5: e12374.
66 (a) Li, G., Liu, Z., Huang, Q. et al. (2018). *Nature Energy* 3: 1076. (b) Chazalviel, J.N. (1990). *Physical Review A* 42: 7355.
67 Liang, P., Yi, J., Liu, X. et al. (2020). *Advanced Functional Materials* 30: 1908528.
68 (a) Cao, J., Zhang, D., Gu, C. et al. (2021). *Advanced Energy Materials* 11: 2101299. (b) Cao, J., Zhang, D., Zhang, X. et al. (2020). *Journal of Materials Chemistry A* 8: 9331. (c) Luo, Y., Yang, Y., Tao, Y. et al. (2021). *ACS Applied Energy Materials* 4: 14599.
69 Du, W., Yan, J., Cao, C., and Li, C.C. (2022). *Energy Storage Materials* 52: 329.
70 Zhao, Z., Wang, R., Peng, C. et al. (2021). *Nature Communications* 12: 6606.
71 Zheng, J.X., Zhao, Q., Tang, T. et al. (2019). *Science* 366: 645.
72 Yang, Z., Lv, C., Li, W. et al. (2022). *Small* 18: 2104148.
73 Cao, P., Zhou, H., Zhou, X. et al. (2022). *ACS Sustainable Chemistry & Engineering* 10: 8350.
74 Zhang, Q., Luan, J., Huang, X. et al. (2020). *Nature Communications* 11: 3961.
75 Yang, Z., Li, W., Zhang, Q. et al. (2022). *Materials Today Energy* 28: 101076.
76 Wen, L., Chen, J., Liang, J. et al. (2016). *National Science Review* 4: 20.
77 (a) Lu, X., Yu, M., Wang, G. et al. (2014). *Energy & Environmental Science* 7: 2160. (b) Zhou, G., Li, F., and Cheng, H.-M. (2014). *Energy & Environmental Science* 7: 1307.
78 Gwon, H., Hong, J., Kim, H. et al. (2014). *Energy & Environmental Science* 7: 538.
79 Yu, P., Zeng, Y., Zhang, H. et al. (2019). *Small* 15: e1804760.
80 Huang, S., Wan, F., Bi, S. et al. (2019). *Angewandte Chemie (International Ed. in English)* 58: 4313.
81 Wang, R., Yao, M., Huang, S. et al. (2021). *Advanced Functional Materials* 31: 202009209.
82 Liu, J., Hu, M., Wang, J. et al. (2019). *Nano Energy* 58: 338.
83 Wang, J., Huang, Y., Liu, B. et al. (2021). *Energy Storage Materials* 41: 599.
84 Mo, F., Li, H., Pei, Z. et al. (2018). *Science Bulletin* 63: 1077.
85 Cheng, X., Pan, J., Zhao, Y. et al. (2018). *Advanced Energy Materials* 8: 1702184.
86 Wang, K., Zhang, X., Han, J. et al. (2018). *ACS Applied Materials & Interfaces* 10: 24573.
87 Zeng, Y., Zhang, X., Meng, Y. et al. (2017). *Advanced Materials* 29: 1700274.

88 Lu, Y., Zhu, T., Xu, N., and Huang, K. (2019). *ACS Applied Energy Materials* 2: 6904.
89 Han, Q., Chi, X., Zhang, S. et al. (2018). *Journal of Materials Chemistry A* 6: 23046.
90 Zhao, T., Zhang, G., Zhou, F. et al. (2018). *Small* 14: e1802320.
91 Hongfei Li, C., Han, Y., Huang, Y. et al. (2018). *Energy & Environmental Science* 11: 941.
92 Li, H., Liu, Z., Liang, G. et al. (2018). *ACS Nano* 12: 3140.
93 Ma, L., Chen, S., Li, H. et al. (2018). *Energy & Environmental Science* 11: 2521.
94 Wang, D., Li, H., Liu, Z. et al. (2018). *Small* 14: e1803978.
95 Xu, W., Liu, C., Wu, Q. et al. (2020). *Journal of Materials Chemistry A* 8: 18327.
96 Tang, Y., Liu, C., Zhu, H. et al. (2020). *Energy Storage Materials* 27: 109.
97 Zhao, J., Sonigara, K.K., Li, J. et al. (2017). *Angewandte Chemie (International Ed. in English)* 56: 7871.
98 Chen, M., Chen, J., Zhou, W. et al. (2021). *Advanced Materials* 33: e2007559.
99 Lu, H., Hu, J., Wang, L. et al. (2022). *Advanced Functional Materials* 32: 2112540.
100 Shang, W., Zhu, J., Liu, Y. et al. (2021). *ACS Applied Materials & Interfaces* 13: 24756.
101 Liu, C., Xu, W., Mei, C. et al. (2021). *Advanced Energy Materials* 11: 2003902.
102 Prasanna, C.M.S. and Suthanthiraraj, S.A. (2015). *Ionics* 22: 389.
103 Liu, C., Xie, X., Lu, B. et al. (2021). *ACS Energy Letters* 6: 1015.
104 Karan, S., Sahu, T.B., Sahu, M. et al. (2017). *Ionics* 23: 2721.
105 Karan, S., Sahu, M., Sahu, T.B. et al. (2017). *Materials Today Communications* 13: 269.
106 Ma, L., Chen, S., Li, X. et al. (2020). *Angewandte Chemie (International Ed. in English)* 59: 23836.
107 Chen, Z., Li, X., Wang, D. et al. (2021). *Energy & Environmental Science* 14: 3492.
108 Zhu, J., Zhang, Z., Zhao, S. et al. (2021). *Advanced Energy Materials* 11: 2003836.

7

Development of Full Zinc-Ion Batteries

7.1 Types of AZIBs

The full aqueous zinc-ion batteries (AZIBs) consist of cathode, anode, electrolyte, separator, and package. AZIBs can be categorized into various types based on shape and packaging, such as button batteries, coin batteries, and soft-packed batteries. This section primarily discusses the structure and unique features of different AZIB types.

7.1.1 Initial Test Molds

Figure 7.1a demonstrates the development history of AZIBs. After the successful use of the cylindrical structure in 1960, the development of alkaline primary zinc–manganese batteries rapidly progressed, demonstrating great potential for further advancement [2]. However, the corrosive potassium hydroxide electrolyte can cause severe dendrite growth on the zinc anode and bring about environmental pollution, which hinders further applications of alkaline primary zinc–manganese batteries.

In order to reduce the toxicity of alkaline zinc–manganese battery, Yamamoto reported the first aqueous zinc ion secondary battery in 1986, which was prepared using neutral and weakly acidic zinc salt solutions (ZnX_n, X = SO_4, Cl, NO_3, BF_4, SiF_6, etc.) as electrolytes instead of alkaline potassium hydroxide solution, providing a new direction for the development of AZIBs [1].

Figure 7.1b depicts the zinc ion full cell assembled by Shoji and Yamamoto in the early stage. Two glass epoxy resin end plates are part of the battery shell, and three silicone resin interlayers are used to seal the battery to avoid water evaporation. The battery reaction is carried out in the hole of silicone resin in the middle layer. This kind of model with a similar structure is still widely used in aqueous batteries. Also, materials for making models have been enriched with the rapid development of material chemistry. In addition, many new molds have been created based on this model to meet the requirements of diverse testing environments. However, the ZIB with this structure mode is not the battery in the full sense of the word. It serves primarily as a means to test the electrochemical performance of the full battery system and is not typically employed as a chemical power source.

Aqueous Zinc Ion Batteries: Fundamentals, Materials, and Design, First Edition.
Haiyan Wang, Qi Zhang, Yixin Li, and Yougen Tang.

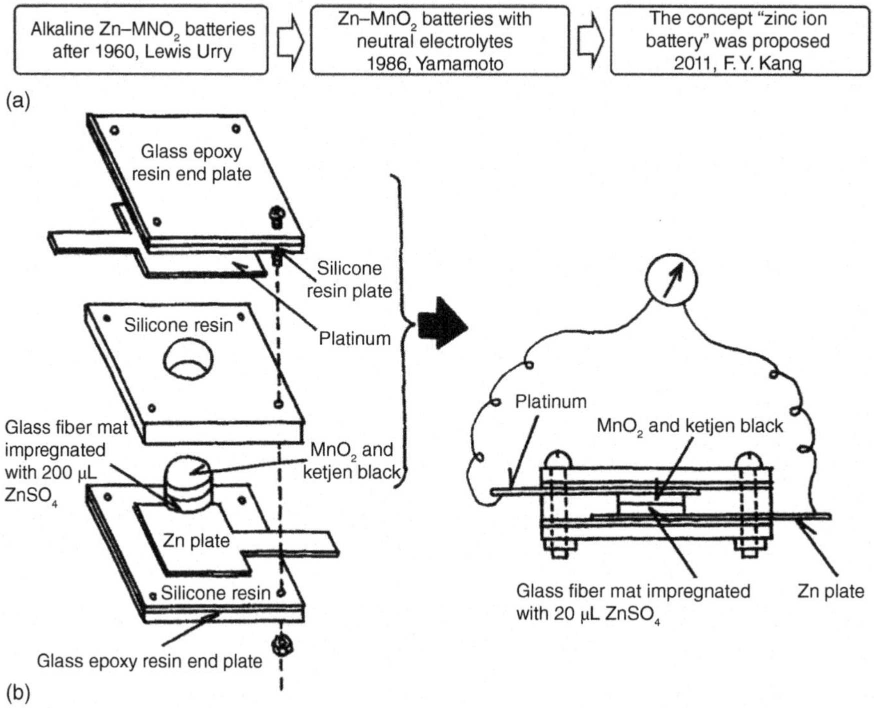

Figure 7.1 (a) Development history of Zn-ion batteries. The first reported Zn–MnO_2 secondary batteries with $ZnSO_4$ electrolyte (b). Source: (b) Adapted from Yamamoto and Shoji [1].

At present, in addition to electrochemical characterizations (such as open-circuit voltage and charge–discharge curves), such models can be used in combination with other characterization methods.

7.1.2 Coin Cell

Although AZIBs made leap-type progress as early as 1986 by replacing alkaline electrolytes with neutral or weakly acidic electrolytes, the reaction mechanism of ZIBs was not explored in the subsequent decades. Plagued by the high-speed development of lithium-ion batteries, AZIBs developed slowly for more than two decades after 1987. Meanwhile, in the 1980s, other Zn-based batteries, such as alkali–Mn, Ni–Zn, Ag–Zn, and Zn–air batteries, entered the commercial development stage, and a large number of Zn-based coin cells were put into the market and developed rapidly [3].

After the concept of AZIBs was formally proposed in 2011, this field began to develop rapidly, and certain breakthroughs have been made in recent years. However, there are still many key technical issues that limit industrialization. Nowadays, AZIBs are generally assembled as coin cells for rapid detection of the electrochemical performance of materials.

Coin cells, also known as button cells, refer to cells with external shapes that look like small buttons. It is one of the common battery types that has gained popularity in daily life and experimental tests. A wide variety of coin cells are now being sold, most of which are classified according to their electrochemical system by the rules of the International Electrotechnical Commission (IEC). For example, the letter C represents the lithium–manganese dioxide electrochemical system, and the letter R represents the cylindrical battery.

The main component building blocks of coin-type AZIBs generally include positive shell, negative shell, (positive/negative) pole piece, separator, other parts (gasket and warhead), and electrolyte. At present, the coin-type AZIBs reported in the literature generally adopt the CR#### (i.e. CR2016 and CR2032) model. The first two digits following the "CR" letter are diameter parameters (in mm), and the last two digits are the thickness (in 0.1 mm). For example, a battery model CR2016 has a diameter of 20 mm and a thickness of 1.6 mm. The electrochemical performance of coin cells can be measured rapidly and conveniently to evaluate the electrode materials.

However, developing ZIBs based on coin cell types cannot satisfy researchers' requirements. Coin cells are small in size with limited active substance loading, belong to miniature energy-storage devices, and have limited applications in the field of large energy storage.

7.1.3 Soft-Packed Cell

Compared to coin cells, soft-packed or prismatic cells can satisfy the application requirements of scaled-up energy-storage systems. Soft-packed batteries, especially lithium-ion batteries with aluminum–plastic membrane as an outer shell, have been widely used in large-scale electrochemical energy-storage systems, power trams, small- and medium-sized mobile electronics, portable electronics, and electric toys [4]. In the same way, the ultimate development path of AZIBs is not limited to micro energy-storage devices such as coin batteries. Therefore, in recent years, the performance of soft-packed ZIBs has attracted more attention from researchers. At present, there are three forms of electrolyte for soft-packed zinc ion cells: liquid [5], solid [6], and gel [7].

The technology of soft pack battery core comes in two types: spiral wound and lamination, which consists of packed-forming lamination, Z-lamination, and coiling lamination. The three ways of lamination are given in Figure 7.2e. Packed-forming lamination is also known as coated lamination. The positive or negative electrode plates are fitted into a pocket made of the separator, and then the packed and unpacked poles are stacked sequentially. The Z-lamination is a z-word fold in separator, and the negative and positive electrode plates are spaced to insert into the diaphragm sandwich. Coiling lamination combines the advantages of spiral wounds and lamination. The procedures are roughly as follows: the positive electrode sheet, the film, and the negative electrode sheet are stacked into spaced units, and then these units are coiled by separators. The greater the target number of coiling layers, the greater the spacing of the later monoliths. In contrast, the coiled core is

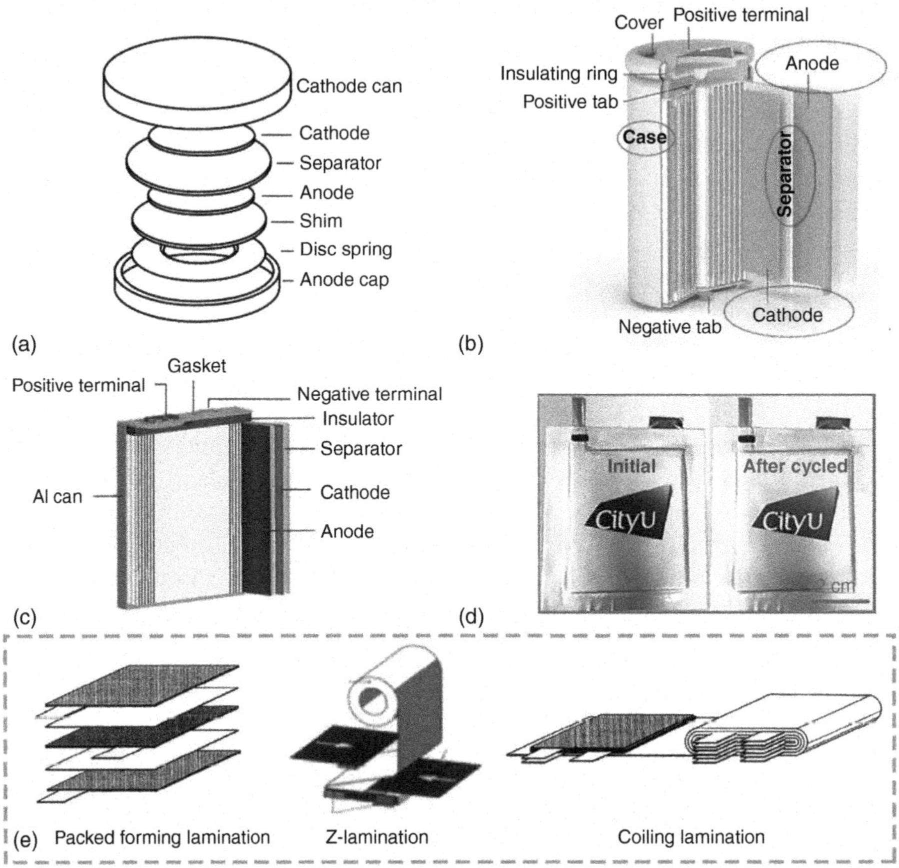

Figure 7.2 (a) Coin cell, (b) cylinder cell, (c) prismatic cell, and (d) soft-packed cell. Three types of lamination (e). Source: (a-c, e) Adapted from Ma et al. [6]. (d) Ma et al. [6]/John Wiley & Sons.

formed by severing and pasting the tape after the entire coil of the positive sheet, the diaphragm, the negative sheet, and the diaphragm are stacked sequentially to terminate the formation of several layers. In combination with the different electrolyte morphologies, positive and negative electrodes, separator materials, etc., different preparation processes for the electric core should be chosen.

7.1.4 Cylinder Cell

As one of the three major battery forms, cylinder battery still occupies a large market share, including MH–Ni batteries, nickel–cadmium batteries, lithium-ion batteries, alkali manganese batteries, zinc–air batteries, and other commercial battery systems. The basic structures of a cylinder battery include a cover, positive terminal, negative plate (anode), positive plate (cathode), separator, positive tab, negative tab, insulating ring, and case. Compared with square battery and soft-packed battery, cylinder battery has a mature production process, high pass rate, low cost, and good heat dissipation performance. Moreover, according to specific needs, it can be combined in various forms, which is conducive to the full layout of the internal space of the equipment. However, the cylinder battery is usually encapsulated in a steel case or an aluminum case, which is weighty, resulting in a relatively low specific energy density of the battery.

AZIBs are also in accordance with the conditions for assembling cylinder batteries. Moreover, AZIBs with aqueous electrolyte is more secure than batteries using organic electrolyte. Although there are no commercial AZIBs available on the market yet, many enterprises have applied for patents related to cylinder AZIBs, including positive and negative electrode processes, winding processes, packaging equipment, etc. It is believed that in the future, with the advancement of technology, cylindrical AZIBs will make good development.

7.1.5 Prismatic Cell

In addition to cylindrical batteries, prismatic batteries entered the automotive field very early. At present, the main types of prismatic batteries are nickel metal hydride batteries and lithium-ion batteries. The main components of a classic prismatic battery include a cap plate cover, can, lamination or winding core composed of the positive plate, negative plate, and separator, insulator, gasket, etc. The can is mostly made of aluminum alloy, stainless steel, and other materials, which have a better protective effect on the battery cell than the soft-packed battery and are conducive to the closer rolling of the internal materials of the battery. Limited by the high hardness of the hard can, the battery is not easy to expand with higher packaging reliability. In addition, the prismatic battery has the advantages of convenient capacity expansion, large single-cell capacity, relatively simple system composition, and good stability. Therefore, it is also one of the good choices for the development direction of AZIBs.

7.2 Performance Parameters of AZIB

7.2.1 Electromotive Force (EMF)

A battery is a device that converts electrical energy into chemical energy, allowing it to be stored and then converted back into electrical energy when required. When two dissimilar conductive materials come into contact, a potential difference is created at the phase interface, resulting in electromotive force (EMF). The EMF is the algebraic sum of the potential difference at the interface of each phase within the battery and can be described using Eq. (7.1).

$$E = \varepsilon_+ + \varepsilon_- + \varepsilon_{\mathrm{CPD}} + \varepsilon_{\mathrm{LJPD}} \tag{7.1}$$

Here, $\boldsymbol{\varepsilon_+}$ and $\boldsymbol{\varepsilon_-}$ represent the potential difference between the positive and negative electrodes and the solution interface, respectively. $\varepsilon_{\mathrm{CPD}}$ is the contact potential difference (CPD), and $\boldsymbol{\varepsilon_{LJPD}}$ is the liquid junction potential difference (LJPD).

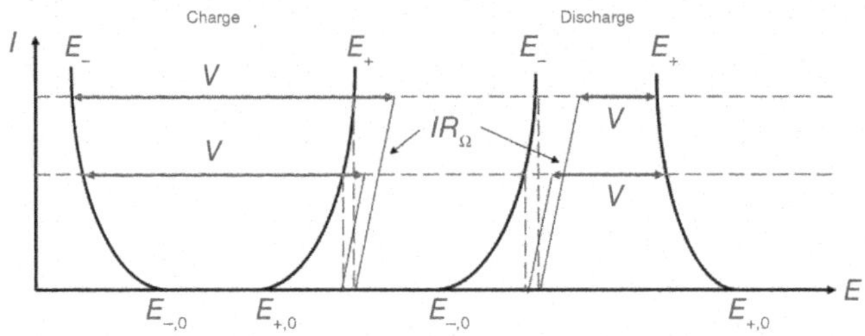

Figure 7.3 Composition of battery voltage and its relationship with working current [8].

7.2.2 Battery Internal Resistance (R_i)

The internal resistance of the battery includes the ohmic internal resistance (R_Ω) and polarization resistance (R_f) of the electrode during the electrochemical reaction. Ohmic resistance (R_Ω) consists of electrode material, electrolyte, diaphragm resistance, and contact resistance of other parts. Polarization resistance also includes electrochemical polarization resistance (R_e) and concentration polarization resistance (R_C). Electrochemical impedance spectroscopy (EIS) can effectively help researchers analyze the internal resistance of batteries.

7.2.3 Open-Circuit Voltage (V_{OC}) and Working Voltage (V)

Open-circuit voltage (V_{OC}) refers to the potential difference between electrodes when there is no current flowing through the external circuit. Generally, the open-circuit voltage is less than the battery EMF (E).

Working voltage (V), also known as discharge voltage, refers to the potential difference between the two electrodes of the battery when current passes through the external circuit.

When a battery is discharged and current flows through it, the resistance caused by its internal resistance must be overcome, resulting in ohmic voltage drop and electrode polarization. This leads to a working voltage that is lower than the open-circuit voltage. Conversely, during charging, the charging voltage (V_C) is always higher than the open-circuit voltage due to polarization's contribution. The following equation shows the relationship between these factors (Figure 7.3):

$$V = E - IR_i = E - I(R_\Omega + R_\mathrm{f}) \tag{7.2}$$

$$\text{Discharge}: V = E_+ - E_- - IR_\Omega = (E_{+,0} - \eta_+) - (E_{-,0} - \eta_-) - IR_\Omega$$
$$= E - \eta_+ - \eta_- - IR_\Omega$$

$$\text{Charge}: V_\mathrm{C} = E_+ - E_- + IR_\Omega = (E_{+,0} + \eta_+) - (E_{-,0} - \eta_-) + IR_\Omega$$
$$= E + \eta_+ + \eta_- + IR_\Omega$$

7.2.4 Capacity (C) and Theoretical Capacity (C_0)

Battery capacity (C) measures the amount of electrical energy that a battery can store and deliver over a period of time under certain conditions (discharge rate, temperature, termination voltage, etc.). It is typically measured in ampere-hours (Ah; 1 Ah = 3600 C) or milliampere-hours (mAh). Battery capacity includes theoretical capacity (C_0), actual capacity, and rated capacity. Formulas (7.3) and (7.4) are used to calculate the theoretical capacity of active substances.

$$K = \frac{M}{nF} = \frac{M}{n96485}\ \mathrm{g \cdot As^{-1}} = \frac{M}{n26.8}\ \mathrm{g \cdot Ah^{-1}} \tag{7.3}$$

$$C_0 = \frac{m_0}{K} = \frac{m_0 n26.8}{M}\ \mathrm{Ah} \tag{7.4}$$

C_0: theoretical capacity, m_0: quality of complete reaction of active substances, M: molar amount of active substance, n: number of electrons gained and lost in current forming reaction, and K: electrochemical equivalent of active substance.

For example, the theoretical capacity of metal zinc negative electrode is 53.6 Ah.

$$C_{0,\mathrm{Zn}} = \frac{m_0 n26.8}{M} = \frac{65.39 \times 2 \times 26.8}{65.39} = 53.6\ \mathrm{Ah} \tag{7.5}$$

Actual capacity refers to the amount of electricity released by current under specific discharge conditions. There are two different situations. One is constant current discharge, and the other is constant resistance discharge.

During the constant current discharge, the voltage of the battery gradually decreases as its energy is used up until it reaches a predetermined cutoff voltage. The time it takes for the battery to reach the cutoff voltage is used to calculate the battery's capacity.

During a constant resistance discharge test, the resistance in the circuit remains constant while the voltage across the battery decreases as it discharges. As a result, the current flowing through the circuit also decreases over time. The rate of decrease in the current depends on the capacity of the battery and the resistance of the circuit.

$$\text{Constant current discharge}: C = It \tag{7.6}$$

$$\text{Constant resistance discharge}: C = \int_0^t Idt = \frac{1}{R}\int_0^t Udt \tag{7.7}$$

R: discharge resistance, t: time from discharge to termination voltage, and U: average discharge voltage of battery.

7.2.5 Depth of Discharge (DOD)

Depth of discharge (DoD) is a term used to describe the amount of energy removed from a battery compared to the total amount of energy it can provide when fully charged. DoD is an important parameter for battery performance and longevity. Generally, deeper discharge levels can result in shorter battery life because batteries are subjected to more stress and chemical changes as they are discharged. In terms of zinc anodes, their DOD refers to the percentage of capacity used really (C_{real}) to the theoretically available capacity ($C_{\mathrm{theoretical}}$).

$$C_{\mathrm{theoretical}} = C_{\mathrm{specific,theory}} m_{Zn} \tag{7.8}$$

$$\mathrm{DOD} = \frac{C_{\mathrm{real}}}{C_{\mathrm{theoretical}}} \times 100\% \tag{7.9}$$

$C_{\text{specific, theory}}$: theoretical specific capacity of zinc anode (820 mAh g^{-1}) and m_{Zn}: mass of zinc electrode.

Here, the DOD calculation of 30 μm zinc foil electrode with a diameter of 12 mm is taken as an example.

$$\rho_{\text{Zn}} = 7.14\,\text{g} \times \text{cm}^{-3}$$

$$h_{\text{Zn}} = 0.003\,\text{cm}$$

$$m_{\text{Zn}} = \rho_{\text{Zn}} V_{\text{Zn}} = \rho_{\text{Zn}} \pi r^2 h_{\text{Zn}} = 7.14 \times 3.14 \times 0.6^2 \times 0.003\,\text{g} = 0.02421\,\text{g}$$

$$C_{\text{theoretical}} = C_{\text{specific,theory}} m_{\text{Zn}} = 820 \times 0.02421 = 19.8522\,\text{mAh}$$

If the capacity density is 1 mAh cm^{-2}, then:

$$C_{\text{real}} = 1\,\pi r^2\,\text{mAh} = 1.1304\,\text{mAh}$$

$$\text{DOD} = \frac{C_{\text{real}}}{C_{\text{theoretical}}} \times 100\% = \frac{1.1304}{19.8522} \times 100\% = 5.69\%$$

7.2.6 Energy Density

Energy density refers to the amount of energy stored in a certain unit of space or mass material, which is determined by the material characteristics of the battery. The energy density of a battery is the electric energy released by the average unit volume or mass of the battery. The energy density of a battery is generally divided into two dimensions: weight energy density and volume energy density.

Therefore, under the condition that the same energy consumption remains unchanged and the battery pack volume and weight remain unchanged are strictly limited, the maximum single driving mileage of new energy vehicles mainly depends on the energy density of the battery. The greater the energy density of the battery, the more electricity is stored per unit volume or weight.

7.2.7 Power Density

Power density refers to the amount of power delivered per unit of volume (W l^{-1}) or weight (W kg^{-1}) of a system or device. It measures of how much power can be generated or transmitted within a given space or weight constraint.

In the context of energy storage, power density is a critical factor for many applications, such as electric vehicles, portable electronic devices, and renewable energy systems. Higher power density batteries or energy-storage systems can deliver more power in a smaller or lighter package. This is especially important in applications where weight or size constraints are significant factors.

The power density, like energy density, is determined by the characteristics of the material, but there is no direct correlation between the two. A higher energy density does not necessarily mean a higher power density. Energy density is a measure of battery capacity, while power density is an indicator of the instantaneous dynamic characteristics and torque of the battery. Power density is crucial for battery and electric vehicle development as it describes the rate at which the battery can discharge, indicating how much current it can deliver. A high power density is especially important for electric vehicles, enabling fast acceleration.

7.3 Assembly Process of Full Battery

7.3.1 Cathode Flake

The cathode materials of ZIBs include manganese-based materials, vanadium-based materials, Prussian blue derivatives (PBA), organics, and other special materials. The cathode preparation processes include coating [9], rolling [10], in situ synthesis [11], and slurry method.

7.3.1.1 Coating

As the battery field has evolved, the coating method has become widely adopted in the production of electrodes, and it is not surprising that this method has also found a place in the preparation technology of pole pieces for AZIBs. The coating process in the laboratory can be briefly summarized as the following steps: slurry mixing, coating, drying, and cutting, as Figure 7.4a describes.

(a) Dissolve binder.
(b) Add the positive active material and conductive agent to the solution-contenting binder.
(c) The solvent with a small amount of binder is added and stirred to form a slurry with moderate viscosity (similar to ointment with good fluidity). Use simple experimental devices such as scraper or slot extrusion to evenly coat the slurry on the collector to obtain the electrode flake.
(d) Put the electrode flake into the vacuum-drying oven to dry. The temperature shall be set according to the actual situation.
(e) *Cut*: Some popular positive current collectors used in batteries include graphite paper, stainless steel foil, titanium foil, and more. While stainless steel foil is advantageous due to its smooth surface, low cost, and resistance to scratching, it is not always firmly bonded to the positive electrode, leading to electrode material detachment and reduced capacity. Additionally, stainless steel has a higher density, which can lower the energy density of the battery. Graphite paper can bond well with positive material, but it is susceptible to issues such as powder shedding, breaking, and uneven collector thickness. On the other hand, Titanium foil has good conductivity and adhesion, making it a common choice in experimental research; however, it is expensive and not well-suited for commercial applications. Therefore, rolling method and in situ growth method have gradually attracted the attention of researchers.

7.3.1.2 Rolling

With the deepening understanding of AZIBs, the advantages of the network-conductive skeleton are gradually revealed. If mesh-type collectors (e.g. stainless

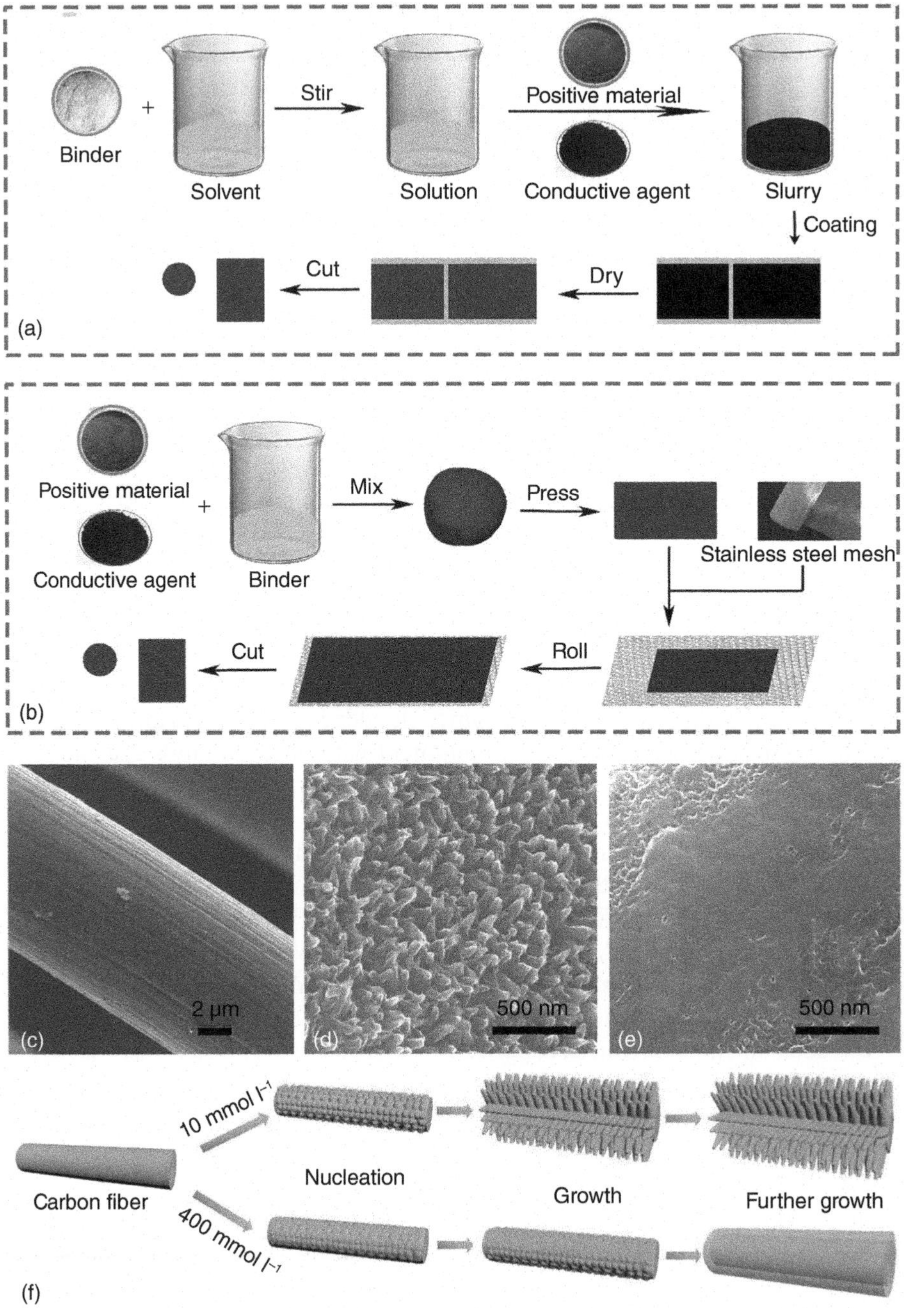

Figure 7.4 (a) Coating, (b) rolling and in situ synthesis (c–f). Source: (a) Deng et al. [9]/John Wiley & Sons, Inc. (b) Xie et al. [10]/John Wiley & Sons, Inc. (c–e) Oldham and Myland [8]/American Chemical Society. (f) Fu et al. [11]/John Wiley & Sons.

steel mesh) are used, it is more difficult to load active substances by coating method due to the general existence of holes in the network collector. Therefore, the rolling method is undoubtedly a good choice. Rolling technology has been available since the 1970s. Alkaline manganese batteries, zinc–air batteries, silver–zinc batteries, and other batteries containing mesh-conductive skeleton electrodes are using this technology to prepare electrodes.

The steps of rolling method include mixing, cutting, rolling, and drying (shown in Figure 7.4b). Among these, mixing is the most important step. The specific steps are as follows:

(a) Mix the active material, conductive agent, and binder (polytetrafluoroethylene, PTFE dispersion is generally used) evenly in proportion to form a "dough." This step requires the materials to be well mixed. The dough should be soft, have good ductility, have moderate humidity with no small pieces.
(b) Spread the "dough" evenly, press it to a certain thickness according to the required load, lay it on the collector and use the roller press to tightly embed the positive electrode into the collector grid.
(c) Put the positive electrode into the vacuum drying oven to dry. The temperature shall be set according to the actual situation.

The mesh current collector has a larger specific surface area, which can be more effectively bonded with the cathode material and play the role of a fast electron transmission channel. At the same time, the positive material is embedded in the grid, which can effectively improve the bonding type of the positive material and reduce the content of the binder, thus further improving the energy density of the battery. In addition, the uniform distribution of the electric field in the three-dimensional mesh current collector can effectively reduce the internal resistance of the battery and facilitate the electrochemical reaction, which brings better charge and discharge performances. Stainless steel mesh is generally selected as the collector, and the appropriate mesh count is selected according to the particle size of the cathode material.

7.3.1.3 In Situ Synthesis

In situ synthesis is a synthetic method that takes a certain material as the matrix and directly generates some functional groups on the surface or inside the matrix. It has been gradually applied to AZIBs in recent years. Carbon cloth or stainless steel mesh is generally used as the substrate in the preparation of cathode plates, and the cathode material is directly grown through electrochemical synthesis or hydrothermal reaction. The advantages of this preparation method are that the cathode with fine particles and stable performance can be obtained without interface pollution. In addition, the in situ synthesis electrode enhances the binding between the collector and the cathode material, integrates them, and improves the electrochemical performance. Polyaniline [11], manganese dioxide [12], vanadium-based materials [13], and other cathode materials can be synthesized in situ on carbon cloth collectors by electrochemical synthesis.

7.3.1.4 Slurry Method

The slurry method is one of the earlier battery electrode preparation processes. The process of the slurry method includes mixing, slurry coating, drying, and cutting. The specific methods are as follows:

(a) Thoroughly mix the positive active material, conductive agent, and binder in a certain proportion.
(b) Add solvent to prepare positive slurry.
(c) Place the slurry in a special slurry-pulling tank and pull it through with a very thin fluid current collector to make it stick to the current collector.
(d) Drying and cutting.

In the manufacture of positive slurry, adjusting the ratio of substances and adopting a suitable binder can make the stickiness and fluidity of the positive slurry reach the requirement of the process simultaneously. So, the positive slurry can be automatically applied to the foil of the current collector. The slurry method is suitable for large-scale production because of its simple and fast process and low cost.

7.3.2 Anode Flake

The negative electrode of AZIBs generally adopts zinc-based materials, including zinc metal negative electrode (zinc plate, zinc powder, galvanized metal, and three-dimensional zinc metal) and zinc alloy negative electrode. In addition, zinc-metal-free negative electrode has also been reported in some studies [14–16]. Zinc metals with simple structures, such as zinc powder and zinc plate, were used earlier in AZIBs because they are readily available commercial materials due to their simple preparation process. The zinc negative electrode used by Shoji in 1986 is a zinc plate, while the earlier alkaline manganese battery uses the zinc gel negative electrode made of mercury-homogenized zinc powder. With the deepening understanding of zinc metal negative electrode, zinc electrode has been continuously innovated.

7.3.2.1 Zinc Powder

Powdered negative electrode materials such as zinc particles and zinc fibers are generally pasted on the current collector by being modulated into gel or made into electrodes by molding and sintering.

(a) *Zinc gel negative electrode:* Zinc gel negative electrode is one of the negative electrode materials with a long history. It is widely used in alkaline manganese battery, zinc–mercuric oxide battery, silver–zinc battery, and zinc–air battery. The zinc gel negative electrode adopts the paste-coating process, which is to mix the zinc powder with the conductive agent, additive, and binder in proportion. Then it is prepared into a paste and loaded on the current collector to be negative electrode. Last, cut it into negative electrode plate according to the required size.

The conductive agent can use activated carbon (AC), superconducting carbon black (SP), carbon nanotube (CNT), and acetylene black (AB). The conductive agent does not participate in the electrode reaction directly but can collect micro-current among each component of the electrode. With the contact resistance of the electrode reduced, the electron movement rate accelerated, and the migration rate of negative metal ions in the electrode material improved. As a result, the electrochemical performance of the electrode is improved.

At the same time, certain conductive agents can serve as effective additives. For instance, AC's rich pore structure allows it to adsorb Zn^{2+} in aqueous solutions or create space for the deposition of zinc passivation during charge and discharge cycles [17–19]. Moreover, certain metals and their oxides or hydroxides can improve battery performance as additives such as ZnO and MgO. They can enhance the hydrogen evolution overpotential, regulate the battery's internal pH, prevent hydrogen evolution flatulence, and inhibit electrode corrosion. Gonzalez's research group developed layered bimetallic hydroxide ZnAl–LDH and Zn–Al–Cu–LDH solid additives, which effectively reduced the reaction overpotential of the zinc negative electrode, suppressed H_2 precipitation, and resulted in tighter deposition of zinc [20].

Binders such as PTFE, polyacrylic latex 133 (LA133), and polyvinylidene fluoride (PVDF) are frequently used in electrode preparation. PTFE and LA133 are commonly selected as aqueous binders, while PVDF is typically used as an oily binder. The aqueous binders use deionized or distilled water as a solvent, while *N*-methyl-2-pyrollidone (NMP) is usually used as the solvent for PVDF. It is crucial to use the appropriate amount of binder when preparing the electrode, as using too little can result in the electrode falling off while using too much can reduce the electrode's conductivity.

Preparation of zinc gel can use the grinding method or stirring method. Metal materials with good conductivity and lower electrochemical reaction activity than zinc should be selected for the current collector. Copper and zinc substrates are usually chosen. The methods of loading zinc paste include coating method and slurry method. The coating method is suitable for foil current collectors, and the process is consistent with the preparation of positive electrodes. The slurry process applies to three-dimensional or porous current collectors, such as copper mesh, foam copper, and three-dimensional zinc metal.

Table 7.1 shows the proportion of zinc powder, conductive agent, additive, and binder in the negative electrode of zinc paste. The ratio of the four shall be flexibly changed according to different application requirements.

(b) *Molding method*: The molding method involves evenly mixing electrode powder, which includes active substances, binders, and additives, according to the specified proportion. This mixture is then placed into a conductive framework, typically a conductive mesh, and pressed for molding. The pressure used for molding is typically around 10–15 MPa, depending on the desired thickness of the electrode.

(c) *Sintering method*: The active electrode material powder is first pressed into shape and then sintered at high temperatures to make electrodes. It can be sintered into an electrode substrate, impregnated with active material, and dried. Sintered electrode has high mechanical strength and large porosity, but the preparation cost is high, and the process is complex, which is not conducive to industrial production.

Table 7.1 Composition ratio of zinc paste cathode.

Composition	Zinc powder	Conductive agent	Additive	Binder
ω (%)	50–80%	10–30%	0–10%	~10%

Table 7.2 Composition ratio of coating.

Composition	Coating materials	Binder
Mass fraction (%)	70–90%	10–30%

7.3.2.2 Zinc Plate

The zinc plate has a long history of commercialization, which has resulted in its early application in AZIBs. The commercial high-purity zinc foil is polished with metallographic sandpaper or treated with dilute acid to remove impurities such as zinc oxide, zinc carbonate, and zinc hydroxide from the surface. After polishing, the zinc foil is cleaned with deionized water or absolute ethanol and dried directly in an oven before use.

Due to the susceptibility of high-purity zinc foil to dendrite formation, corrosion, and passivation during the process of charge and discharge, electrode failure or reduced cycle life may occur. As a solution, the construction of a zinc negative electrode interface control layer has become a commonly employed modification method. This layer can be achieved through artificial coating or in situ growth interface techniques. Once a protective layer is added, the zinc foil electrode can be utilized directly after being cut.

From the earliest nano $CaCO_3$ [21] to various organic or inorganic materials, the action mechanism of coating materials is different, and the coating process is constantly innovated. The zinc negative electrode coating mainly consists of coating material and binder. The ratio of them is given in Table 7.2. The binder can use oil binder PVDF or aqueous binder LA133. The oil-based binder has poor hydrophilicity, and too high proportion will reduce the wettability of the electrode and increase the contact angle between the electrode and the electrolyte, which is not conducive to the transmission and migration of zinc ions in the electrode reaction process, thus increasing the polarization of the electrode and reducing the electrochemical performance. If the ratio is too low, the coating adhesion is not firm and easily falls off from the negative electrode. The coating material particles are unevenly dispersed.

Moreover, the pores formed by stacking particles are disordered, and the distribution of zinc ion transmission channels is uneven, which makes it easier to induce dendrite formation. So, the modification effect is greatly reduced. After the coating material and binder are prepared proportionally, the slurry can be mixed by grinding or stirring, following the same preparation processes as those used for zinc paste. The slurry should not be too thick; otherwise, the coating will become thick and dense, making it difficult for the electrolyte to penetrate the electrode completely and increasing the internal resistance of the battery. However, if the slurry is too thin, the coating is difficult to form. The blocking effect will weaken when the coating thickness is thin, and the electrolyte absorption will be limited. Generally, the coating thickness of zinc foil controlled within 10–30 μm is better.

The preparation process of in situ growth interface layer includes chemical passivation and electrodeposition. Chemical passivation mainly involves the formation of a solid passivation film on the surface through the direct reaction of zinc metal with acids, bases, or strong oxides. This film plays a role in protecting the zinc negative electrode, regulating zinc deposition morphology, and homogenizing zinc ion transmission channel. The electrodeposition method uses zinc foil as the substrate and deposits a modified layer in situ on the zinc foil surface by applying a direct current. When preparing a passivated negative electrode, the passivation duration needs to be accurately controlled. If the time is too long, the passivation layer will be too thick, making it difficult for the electrolyte to contact the deep active zinc negative electrode. Thus, the migration rate of zinc ions will be slower, and the reaction polarization will be larger. If the passivation layer is too thin, it will lead to an inferior modification effect.

7.3.2.3 Galvanized Material

In recent years, galvanized materials have played a great role in the field of AZIBs. Active materials are deposited on specific electrode substrates (such as nickel foam) by electrochemical method in electrolytes, and then the required electrodes are made through subsequent processes (rolling and drying). It has the advantages of simple preparation process, short production cycle, strong controllability, and high utilization rate of active substances.

Electroplating technology offers a variety of options to produce different zinc deposits, which can be influenced by factors such as current density, electroplating time, and electrolyte concentration. Typically, two commonly used techniques in electroplating are constant current electroplating and pulse electroplating. Among them, pulse electroplating has gained popularity in the industry for its ability to enhance the quality of plated parts. This technique can significantly improve the coating structure and dispersion ability, reduce the porosity and internal stress of the coating, and enhance its corrosion resistance. Additionally, pulse electroplating offers the advantages of reducing raw material consumption and improving production efficiency.

7.3.2.4 Zinc Alloy

Inspired by the wide application of zinc alloys in anticorrosion materials, researchers often alloy zinc metal with other metals to improve the hydrogen evolution overpotential of negative electrodes and inhibit battery corrosion, in addition to the aforementioned modification methods of zinc negative electrodes. Generally speaking, there are two preparation methods for alloy zinc electrodes: (i) Eutectic: adding mercury substitute metal elements to high-purity zinc to form eutectic zinc alloy; (ii) Displacement electroplating method: metal elements (or compounds) containing mercury are added to the electrolyte and electroplated and deposited on the surface of zinc metal. The main preparation technologies are spray, low-temperature calcination, electroplating, etc. Although zinc alloy is widely used as an anticorrosion material, it is still rare as the negative electrode of AZIBs, and the research is still in its infancy and needs further development.

7.3.2.5 3D Zinc Anode

As our understanding of ZIBs deepens, we have discovered that using a collector with a three-dimensional structure has a significant advantage: it provides a high electrochemically active surface area, which helps to distribute the electric field uniformly and minimizes changes in zinc deposition morphology during charging and discharging. This, in turn, helps to prevent the formation of zinc dendrites. The three-dimensional zinc negative electrode is a type of electrode that is highly electrochemically active and reversible, and it is deposited on a three-dimensional current collector through electrodeposition. Studies have reported the existence of three-dimensional blocks of zinc sponge negative electrode [22], three-dimensional copper base zinc negative electrode [23], foam graphene composite zinc anode [24], CuO nanowire-modified zinc net negative pole [25], etc., further highlighting the potential of this technology.

While the three-dimensional zinc negative electrode can unify the distribution of electric field, regulate ion behavior, and mitigate issues related to zinc dendrite, self-corrosion, and passivation, it should be noted that the technology required to implement this approach is highly demanding.

7.3.3 Electrolyte

7.3.3.1 Aqueous Electrolyte

Using an aqueous solution as the electrolyte of a ZIB can achieve lower cost and higher safety performance. In addition, it has been widely studied because of the high ionic conductivity of aqueous solutions [26].

The commonly used zinc electrolyte salts of aqueous electrolytes include aqueous electrolytes such as $ZnSO_4$, $Zn(CH_3COO)_2$, $Zn(CF_3SO_3)_2$, $Zn(TFSI)_2$, and $ZnCl_2$. The preparation process for aqueous electrolytes is simple. According to the required concentration, a certain amount of electrolyte zinc salts is dissolved in distilled or deionized water.

(a) *Advantages of aqueous electrolytes*: Any electrolyte must meet its basic characteristics. The electrolyte composition strongly affects the electrochemical behavior and theoretical specific capacity of AZIBs. Therefore, it is necessary to comprehensively consider such issues as ionic conductivity, ionic migration number, electrochemical stability window, temperature adaptability, and electrolyte/electrode interface when designing electrolytes [27].

The electrolyte used in an aqueous solution system typically involves using an aqueous solvent and a zinc salt solute. Additional components or additives can also be incorporated to supplement or enhance the overall solution. By introducing these supplementary materials, the electrolyte can be customized to optimize its function and meet the unique needs of the application. The advantages of aqueous electrolytes are as follows.

The price of aqueous electrolytes is low due to the abundant resource of water on earth. Currently, the production cost of organic solvents such as acetonitrile, *N*,*N*-dimethylformamide (DMF), and trimethyl phosphate used in zinc batteries is relatively high. DMF costs about US$1100 per ton, while acetonitrile costs about US$3200 per ton.

The safety performance of aqueous electrolytes is higher than that of organic electrolytes. The safety performance of the water solvent system is shown in two aspects: one is low toxicity to the ecological environment (biological safety), and the other is the fire safety problem caused by electrolyte leakage or battery explosion (mechanical safety). Organic liquids are flammable and combustible, so most organic liquid battery safety accidents are often caused by the combustion of internal organic electrolytes. Organic liquid also has certain ecological toxicity. To reduce the damage of organic solution to the environment and human body, the battery with organic electrolytes generally needs to be assembled and prepared in the glove box, which greatly increases the production cost. The high safety of zinc salt solution can make the aqueous zinc-ion battery in a simple indoor environment. Although some aqueous electrolytes prepared at present contain organic additives, most of them contain less than 0.5 wt%, and their volatility in aqueous solution is far lower than that of purely organic substances, almost nonvolatile, with little impact on the human body and the environment.

The composition of by-products from aqueous electrolytes is comparatively simpler than those from organic systems. This simplicity facilitates the analysis of battery outcomes. In the electrochemical reaction process of AZIBs, the only significant side reaction is the decomposition of water, resulting in the by-product of OH^-. This by-product combines with zinc, making it easier to study and understand the overall reaction mechanism.

In addition, aqueous electrolytes have higher ionic conductivity. Conversely, the high viscosity of organic solution leads to low ionic conductivity (only 1/100 of that of aqueous electrolyte), which is the key factor determining the internal resistance and rate performance of the battery [28].

(b) *Limitations of aqueous electrolytes*: The aqueous zinc salt solution typically exhibits a limited electrochemical window due to its low concentration. The electrochemical window, which signifies the voltage range where the electrolyte remains stable, is narrow in such solutions. Unfortunately, low working voltage results in hydrogen evolution reactions on metal-negative electrodes.

One commonly used strategy to widen the potential window is through the use of a high-concentration salt electrolyte, commonly known as "water in salt." In this type of electrolyte, the large amount of solute present means that there is not enough solvent available for effective cationic solvation. Consequently, there are progressively fewer free solvent molecules, which maximizes the anion–cation interaction in the solution. As a result, the activity of water is significantly reduced when compared to conventional aqueous solutions.

This reduced activity, in turn, weakens the decomposition trend at high potential, thereby effectively widening the electrochemical window. For example, when the concentration of $ZnCl_2$ electrolyte solution increases from 1 to 30 M, the absolute working voltage with the vanadium-based positive electrode increases by 0.4 V [29].

(c) *Matching principle of aqueous electrolytes*: In the preparation of electrolytes, attention should be paid to the matching principle in the electrolyte system, that is, the selection of appropriate electrolyte type, appropriate concentration, and additives with positive effects. The matching principle can be summarized as follows:

  (i) *Salt selection*. For example, when carrying out the strategy of water in salt, zinc salt with high solubility in water should be selected as the main part. The solubility of commonly used zinc sulfate in aqueous solution is low; that is, it cannot be the main role of the salt in this strategy. To inhibit the side reaction of hydrogen evolution in water, zinc salts with low solvation effects, such as $Zn(CF_3SO_3)_2$ and $Zn(TFSI)_2$, can be used to inhibit the active water from participating in the electrode reaction during the de-solvation process. To effectively promote the commercial development of aqueous AZIBs, it is necessary to avoid using high-cost salts such as $Zn(CF_3SO_3)_2$ and $Zn(TFSI)_2$ in research.

  (ii) *Salt coordination effect*. The difference in association/coordination behavior of different zinc salts in an aqueous solution should be considered.

  (iii) *Concentration effect*. The electrolyte concentration will affect the electrode performance, and a high concentration of electrolyte can prolong the cycle life of the battery. However, the increase in zinc salt concentration will also reduce the ionic conductivity of the electrolyte.

  (iv) *Additive species*. Such as inorganic metal cationic additives, organic anionic/cationic additives, nonionic additives, and bi-ionic additives.

7.3.3.2 Gel Electrolyte

Although liquid electrolyte has strong availability, the corresponding battery is susceptible to electrolyte leakage. Also, the need for advanced, flexible electronics necessitates electrolytes with a high degree of flexibility and integrity [2]. Gel electrolytes with excellent electrochemical properties are prepared by introducing polymers into aqueous solutions. Gel electrolytes typically fall into one of three categories: those without cross-linking, those with physical cross-linking, and those with chemical cross-linking networks [30].

To create a gel electrolyte without cross-linking, a water-soluble polymer is dispersed in water, forming hydrogen bonds between the polymer segments and water molecules. This increases the overall viscosity and reduces fluidity, resulting in a gel electrolyte without cross-linking. However, these gel electrolytes often have limited mechanical properties, such as poor resistance to pressure, and may require additional diaphragms for added protection.

Through simple heat treatment, hydrogen bonds can be formed between the segments of polymer molecules, while the polymer backbone can have stronger mechanical properties. The obtained physically cross-linked hydrogel electrolyte can maintain a more stable form. At the same time, the physically cross-linked hydrogel electrolyte has higher ionic conductivity because it maintains sufficient water. For example, physical cross-linking PVA/$Zn(CF_3SO_3)_2$ hydrogel electrolytes can self-heal by hydrogen bonding without external stimulation [31].

Through polymerization, more stable and stronger covalent bonds can be formed between different chains of molecules. The mechanical properties of hydrogel electrolytes can be greatly improved. The corresponding chemically crosslinked hydrogel electrolyte has great potential for practical application. For example, chemical cross-linking polyacrylamide (PAM) electrolyte shows good tensile strength (273 kPa) and high tensile capacity with a tensile strain of 3000% [32]. However, due to the existence of water molecules in polymer networks, the traditional hydrogel electrolyte will inevitably freeze and lose its elasticity below zero degrees, which will affect its electrochemical performance and its practical application in an extremely cold environment.

7.3.4 Assembly Process of Full Battery

7.3.4.1 Coin Cell

The button battery is easy to assemble. Significantly, AZIBs can be assembled indoors, and there is no special requirement for the water oxygen value.

The assembly steps of a coin cell are to assemble the negative shell (spring, spacer, or others), negative electrode, separator, electrolyte (or electrolyte membrane), positive electrode, and positive shell in order, and finally, press them with a sealing machine.

7.3.4.2 Soft-Packed Cell

This section describes how to prepare a simple soft-packed battery in the laboratory. The second-half process of soft-packed battery preparation (when the active material is prepared after a series of the previous process of mixing, coating, rolling, and drying) is described as the object. The steps include electrode forming, cell preparation, tab connection, drying, aluminum–plastic film forming, prepackaging, injection, formation, hot lam, and packing (Figure 7.5).

(a) *Electrode forming*: The pole piece must be cut to the appropriate specifications based on the size of the pocket bag and the chosen method of cell preparation, whether laminated or wound.

    For laminated cores, the pole piece must be cut into rectangular sheets of the specified size. On the other hand, if the winding method is used, both the pole piece and membrane must be cut into long strips of the specified length.

    During the cutting process, it is possible to reserve the tab for later use. However, it is also possible to weld or paste the tab with insulating glue at the time of cutting for added convenience.

(b) *Cell preparation*: In laminated batteries, the electrode and separator layers are stacked and laminated together in a flat configuration. This creates a thin and flexible battery that can be easily integrated into devices with limited space. Laminated batteries are typically used in low-power applications such as mobile phones, laptops, and portable electronic devices.

In wound batteries, the electrode and separator layers are wound together in a spiral configuration. This creates a higher-capacity battery with better thermal management compared to laminated batteries. Wound batteries are typically used in high-power applications such as electric vehicles, power tools, and energy-storage systems.

(c) *Tab connection*: There are generally three ways to connect the pole tab and pole piece (the pole piece with reserved pole tab still needs to be connected with a longer pole tab): riveting, ultrasonic spot welding, and laser spot welding.

Riveting connects the pole ear and the fluid collector during acupuncture. The effect is lower than that of ultrasonic spot welding and laser spot welding, and it is easy to cause the deformation of the pole piece at the connection, but the price of the machine is cheaper.

Ultrasonic spot welding and laser spot welding are two kinds of connection methods to melt and bond the joint. They have a good effect and are commonly used in industrial production, but the machine price is slightly higher. No matter what welding method is selected, ensure the pole lug is fastened. After the tab welding is complete, check whether the cell is short-circuited.

(d) *Drying*: Put the cell into the vacuum drying oven for drying.

(e) *Aluminum–plastic film forming*: The aluminum–plastic film undergoes a punching process to form a pocket for the cell. Using a punching die, the aluminum–plastic film is skillfully deformed, creating a pocket ideal for accommodating the cell. Then, the aluminum–plastic film is washed and shaped, which gives it its distinct pocket-like appearance. In the case of thin cells, a single pit is flushed, while for thick cells, double pits are chosen. It is important to note that excessive deformation of the aluminum–plastic film can lead to breaking beyond its deformation limit, resulting in fractures. As such, care must be taken when forming the pocket to ensure that the film is not subjected to too much stress.

(f) *Prepackaging*: Pre-encapsulation refers to top and side sealing, leaving a seal for electrolyte injection. After putting the cell into the pit, the whole aluminum–plastic film can be put into the fixture for top and side sealing in the top and side sealing machine. When packaged, the two heads have a certain temperature (~180 °C). When closing, they are pressed on the aluminum–plastic film, and the polypropylene (PP) layer of the aluminum–plastic film is melted and bonded together. The polar tab is mainly sealed by polar tab glue. Short-circuit test and moisture test shall be carried out after prepackaging, and injection shall be carried out after passing the test.

(g) *Injection*: Using a vacuum prepackaging machine, seal the reserved compartment after injecting the electrolyte. If it is a zinc-ion battery using organic electrolyte, this step needs to be conducted in the glove box.

(h) *Standing, formation, and baking*: After one seal is completed, the cell needs to be put to rest first. According to different processes, it can be divided into high-temperature standing and room-temperature standing. The purpose of standing is to make the injected electrolyte fully infiltrate the electrode.

In addition to the special chemical cabinet, there is the option to use a clamp to secure the cell. This can be done by placing the cell between two glass plates and securing it with a clamp. This method provides a simpler alternative for cell formation.

After formation, some cells, especially thick cells, may be deformed due to large internal stress. Therefore, a fixture-shaping process can be set after formation, which is also called fixture baking. Baking can improve the flatness of a soft-packed battery, make the thickness of the battery cell meet the requirements, and have high consistency. Reduce the diffusion distance between the positive and negative electrodes of the battery by pressing the zinc core closely with the diaphragm to reduce the diffusion distance between the positive and negative electrodes.

(i) *Vacuum-packed*: As gas will be generated during the formation process, it is necessary to puncture the airbag of the battery and vacuumize it simultaneously to extract any excess gas and electrolyte before the secondary sealing stage. This ensures that the airbag is free from any gas or liquid that could compromise the quality of the final product. The battery is then immediately packaged in the second sealing area to ensure it is airtight. Once the second sealing is complete, the soft-packed battery is essentially formed and ready for use.

(j) *Subsequent process*: After being packed, the edge shall be cut and folded, that is, the first seal and the second seal shall be cut to the appropriate width and then folded to ensure that the width of the cell does not exceed the standard. The folded cell can be subjected to a capacity division test (capacity test) to detect whether the capacity of the cell reaches the specified minimum value or other performance tests.

7.4 Aqueous Zinc-Ion Battery Manufacturers

The energy-storage/conversion market has recently seen a surge in the production of AZIBs, with a number of notable manufacturers leading the way. While this overview will touch on some of the most prominent players in the field, it is important to keep in mind that there are numerous other manufacturers and start-ups making significant strides with innovative chemistries. Nonetheless, our intention is to provide a comprehensive yet concise introduction to the leading manufacturers in this rapidly evolving sector.

Hanwei Technology is a company dedicated to the research and production of aqueous batteries that aim to provide low-cost, high-power, and safe energy-storage solutions for industrial and commercial fields. The company utilizes high-density, high-specific-energy zinc metal anodes in its battery cells, which can reach an energy density of 200–300 Wh L^{-1}. This enables the batteries to be adapted to various applications, such as fast charging and grid power compensation. Hanwei Technology uses a microfluidic battery configuration, which solves the problem of zinc dendrites

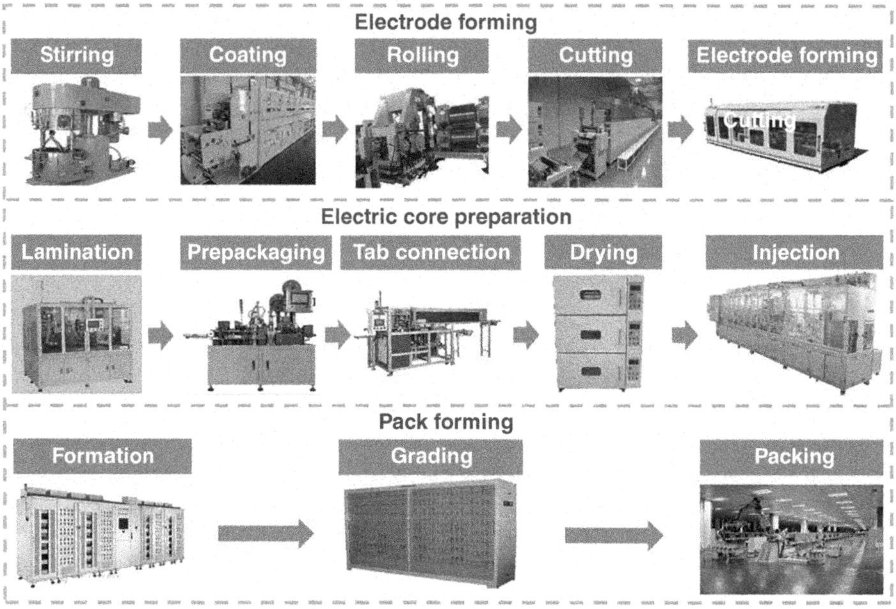

Figure 7.5 Assembly process of soft-packed cell. Source: Huang et al. [31]/John Wiley & Sons.

and hydrogen evolution and extends the device's life without requiring a membrane or large storage tank, resulting in lower costs and a wider range of usage scenarios. Hanwei Technology has the advantage of China's leading zinc mine reserves, enabling them to control the cost of their AZIB cell at 350–400 RMB per kWh, which is about 50% lower than lithium-ion batteries. The company has completed the construction of its AZIB production line and can offer scientific research and production capabilities from laboratory to industrialization. The company has pioneered the use of some lithium-ion and lead–acid equipment transformation, breaking through the scale effect constraints of the battery cell industry represented by lithium-ion batteries and achieving systematic cost reduction and efficiency improvement. Compared to traditional lithium-ion battery production lines, Hanwei Technology reduces fixed costs by about 38% and operating costs by about 41% at the production line level. The company's first-phase production line is expected to reach an annual capacity of 30–60 MWh, and the second-phase production line is under construction, which will increase the total capacity of the production line to about 80–150 MWh per year. Hanwei Technology focuses on the three major subdivided markets of industrial and commercial energy storage, which are photovoltaic supporting energy storage, enterprise peak-valley arbitrage, and new energy vehicle charging and battery-swapping scenes.

***HanShu Technology*'s** AZIB electrolyte utilizes a neutral aqueous solution sourced from the abundant and inexpensive metal zinc, resulting in a more straightforward structure and environmentally friendly composition. With inherent high safety features, including resistance to impact, puncture, and damage, it eliminates the risk of fire or explosion, making it the perfect choice for large-scale energy-storage applications. The company has successfully raised funds in two previous rounds, in 2014 and 2018, and recently, on 30 January 2023, completed another round of funding. This latest round of funding is aimed at upgrading the assembly line for AZIB devices and the trial production line for layered oxide and Prussian blue-positive electrode materials. It will also cover the company's operating expenses and working capital to accelerate the industrialization process in the core materials of zinc-ion and sodium-ion batteries. HanShu Technology holds exclusive patents and key technologies for AZIB, focusing on energy storage. They offer circular and square zinc-ion batteries (groups). Besides, they have successfully developed rechargeable flexible batteries by printing environmentally friendly battery materials on thin film substrates. These cutting-edge batteries boast numerous advantages, including flexibility, low cost, lightweight, customization, safety, and environmental friendliness, which makes them suitable for a wide range of applications, such as wearable devices, electronic tags, and heart monitoring. Notably, in 2021, HanShu Technology successfully developed zinc-based single-flow batteries and battery management systems (BMS). They collaborated with listed company KeLue Electronics to demonstrate kWh-level energy storage using zinc-ion batteries. The 2 kWh energy-storage battery module assembled by HanShu Technology has been purchased by Yichun KeLue Energy Storage Technology Co., Ltd.

Ruihai Lake (Qingdao) Energy Technology Co., Ltd. is a pioneering Sino-foreign joint venture high-tech enterprise. It focuses on the development of cutting-edge technology in the field of new energy, catering to the demand for high-performance, high-security, and pollution-free secondary battery materials, battery technology, and system integration in the future green energy market. The company boasts a team of highly professional technical experts who have spent years researching the field of aqueous batteries, both domestically and internationally. With a state-of-the-art research and development laboratory and a pilot production line, the company has successfully overcome several significant hurdles in battery materials, battery technology, battery production process, and production equipment, providing unwavering support for the rapid expansion of the enterprise. The company is dedicated to the development of Rechargeable Hybrid Aqueous Battery (ReHAB) technology, which is a novel electrochemical system that utilizes "water" as the electrolyte solvent and comprises two distinct half-cell mechanisms: the positive electrode uses the "insertion–extraction" principle, and the negative electrode uses the "dissolution–deposition" principle. The aqueous battery has undergone two generations of improvements, with a nominal voltage of up to 12 V, a nominal capacity of up to 100 Ah, a cycle life of up to 4000 cycles, a capacity retention rate of 95%, and a temperature range of −20 to 50 °C. It also features several functions, including overtemperature, overvoltage, undervoltage, overcurrent protection functions, balanced charging and discharging, and high safety standards. Besides, the bidirectional inverters allow for the conversion and control of AC/DC electrical energy and energy flow direction, with multiple operating modes such as grid connection and off-grid operation.

Moreover, the friendly human–machine interaction interface is customizable and equipped with remote monitoring real-time upload of state of charge (SOC) and state of health (SOH). Currently, the company is working on the third generation of aqueous batteries, which improves the electrolyte by using semisolid or solid-state electrolytes to replace the flowing electrolyte, expands the electrochemical window, and improves cycle performance. Leveraging the aforementioned technology, the company has developed a unique energy-storage system consisting of battery cells, battery modules, battery clusters, BMS, inverters, high-voltage cabinets, air conditioning systems, main control systems, and back-end monitoring. The system boasts an installation capacity of 50 kWh and can be customized to meet the specific requirements of the customers. It caters to the application needs of the emerging market of power storage and is used for grid-connected energy storage of city solar street lights, wind power stations, off-grid energy storage in isolated and central-western regions, and provides energy-storage devices for various nodes of power transmission, distribution, and consumption. The energy-storage system can also be utilized for backup power and small- and medium-sized off-grid and grid-connected energy-storage systems composed of wind and solar energy to improve the controllability of new energy-grid connection, achieve smart frequency and peak regulation of the power grid, improve power grid stability, and reduce user electricity costs.

Salient Energy, a Canadian start-up, has developed a groundbreaking zinc-ion battery for stationary energy storage in residential applications. This innovative battery technology offers several advantages over traditional lithium-ion batteries, including increased safety, lower cost, and greater environmental sustainability. ZIBs use zinc ions as the active species in the battery's chemistry, making them a safe and sustainable alternative to lithium-ion batteries that have a higher risk of fire. Their zinc-ion battery has a rated capacity of 60 Ah, a nominal voltage of 1.3 V, and a volumetric energy density of 100 Wh L^{-1}. The battery measures 26 cm × 24 cm × 1.2 cm and weighs 1.3 kg. Its shelf life for 2 weeks is 100%, and that for 6 months is 95%. Currently, Salient Energy is manufacturing these zinc-ion batteries at its facility in Dartmouth, Nova Scotia, Canada, with a capacity of 100 batteries per month. The company is also in the process of ramping up production to support pilot projects in the residential space. Salient Energy is developing its first gigafactory plant design, which will be located in the United States or Canada. According to the manufacturer, the ZIB has similar power capabilities to the existing lithium-ion storage systems, with the important differences between the two are cost, safety, and supply chain abundance as opposed to performance. It can respond almost instantaneously to store or deliver clean power as needed. It is well-suited for short-duration storage, a term for energy-storage systems that can completely charge or discharge in just a few hours. Recognizing the importance of safety in the residential market, Salient Energy believes that its zero fire-risk systems will significantly simplify installation and provide homeowners with peace of mind. As a result, their initial focus is on the residential market, beginning with targeting homebuilders who are developing net-zero homes and communities. The company plans to expand from residential into industrial energy-storage and utility-scale systems built for the grid in the near future.

Eos Energy Storage is a leading provider of stationary energy-storage solutions that have been designed to overcome the limitations of traditional battery technologies. Their Znyth battery modules have been rigorously tested and validated for over a decade, setting new industry standards in terms of simplicity, safety, durability, flexibility, and availability. Each Znyth battery module is a self-contained unit with a closed-system design, making it easy to maintain and manufacture. They can last at least 6000 cycles, almost twice the operational life of most conventional battery chemistries. Additionally, they are inherently safe to use, with a water-based electrolyte and flame-retardant polymer framing that eliminates the risk of thermal runaway. Besides, Znyth battery modules are capable of withstanding a variety of harsh conditions, and they can recover fully from 90 °C+ abuse cycles with just a short rest period. Their overall degradation curve is well below conventional standards, with less than 9% degradation over the first 3 years and zero degradation thereafter. The Znyth chemistry is highly tolerant of significant variation in operational requirements, with no impact on degradation when a module's storage duration ranges from 3 to 12 hours. The maximum DoD can be reduced for applications demanding round-trip efficiency in the mid-1980s. Beyond that, Eos Znyth technology requires just five low-cost, widely used, earth-abundant commodities with no geopolitical issues connected to their extraction. It enables local sourcing that minimizes the risk of supply chain disruptions and related price swings. Znyth battery modules are the building blocks of all Eos energy-storage systems, and their single battery module specifications are impressive. With a nominal voltage of 64 VDC and a maximum rated energy of 3.2 kWh, they have a nominal discharge capacity of 51.8 Ah and a maximum current of 18 ADC. Their round-trip DC efficiency is approximately 75%, and they can operate in temperatures ranging from −20 to 55/65 °C (charge/discharge). The battery module dimensions are 386H × 428W × 585D mm, weighing 98 kg.

7.5 Summary and Outlook

AZIBs have the advantages of low cost, intrinsic safety, and environmental friendliness and show great potential in large-scale energy-storage systems. However, in its development process, it still faces many challenges. According to the previous achievements in the field of ZIB, this chapter mainly introduces and discusses the ZIB devices, battery performance parameters, component preparation process, and full battery assembly process. Additionally, considering the difficulties and future development of AZIBs, this chapter proposes possible prospects for battery preparation process and devices.

One important point to consider is that the current preparation methods for positive and negative electrodes for AZIBs are not suitable for large-scale production. Therefore, research needs to be done on appropriate active substances, current collectors, electrode formulas, and coating processes. For instance, developing a current collector that can realize an anode-free battery system can improve the utilization rate of metal zinc anodes.

Another important point is that flexible and safe AZIBs equipped with hydrogel electrolytes and various flexible collectors are becoming strong competitors to lithium batteries in the application field of biocompatible electronic devices. More perfect designs of flexible AZIBs will enable aqueous zinc electricity to occupy a place in the field of electrochemical energy storage.

Additionally, preparing functional devices and diversifying the application fields of zinc batteries can also enhance their competitiveness. Examples include compressible zinc batteries and gel batteries with self-healing functions. Ideally, zinc batteries integrating energy storage and application can simplify energy-storage networks and reduce energy loss. Thermal-sensitive zinc batteries can also be prepared and applied to energy-storage systems sensitive to temperature environments.

Furthermore, the development of large-scale energy-storage power plants with lower requirements on mass-specific capacity but higher requirements for safety can also be sought. In this case, the performance parameters of AZIBs, such as coulomb efficiency, self-discharge performance, capacity, and charging and discharging platform, are particularly important and should be the research focus.

References

1 Yamamoto, T. and Shoji, T. (1986). *Inorganica Chimica Acta* 117: L27–L28.

2 Yu, P., Zeng, Y., Zhang, H. et al. (2019). *Small* 15: 1804760.

3 Gao, Q. and Liu, S. (1984). *China Nonferrous Metallurgy* 59–67. +38.

4 Shi, B. (2001). *China Packaging* 21: 2.

5 Xu, C., Li, B., Du, H., and Kang, F. (2012). *Angewandte Chemie International Edition* 51: 933–935.

6 Ma, L., Chen, S., Li, N. et al. (2020). *Advanced Materials* 32: 1908121.

7 Luo, H., Wang, B., Wu, F. et al. (2021). *Nano Energy* 81: 105601.

8 Oldham, K., Myland, J., and Bond, A. (2011). *Electrochemical Science and Technology: Fundamentals and Applications*, vol. 10, 193–212. Wiley.

9 Deng, C., Xie, X., Han, J. et al. (2020). *Advanced Functional Materials* 30: 2000599.

10 Xie, C., Liu, S., Zhang, W. et al. (2023). *Angewandte Chemie International Edition* 135 (28): e202304259. https://doi.org/10.1002/ange.202304259.

11 Fu, X., Zhang, W., Lan, B. et al. (2020). *ACS Applied Energy Materials* 3: 12360–12367.

12 Huang, Z., Song, Y., Feng, D. et al. (2018). *ACS Nano* 12: 3557–3567.

13 Jiang, H., Zhang, Y., Liu, Y. et al. (2020). *Journal of Materials Chemistry A* 8: 15130–15139.

14 Cheng, Y., Luo, L., Zhong, L. et al. (2016). *ACS Applied Materials & Interfaces* 8: 13673–13677.

15 Chae, M., Heo, J., Lim, S., and Hong, S. (2016). *Inorganic Chemistry* 55: 3294–3301.

16 Li, W., Wang, K., Cheng, S., and Jiang, K. (2019). *Advanced Energy Materials* 9: 1900993.

17 Bani, A., Kasiri, G., Glenneberg, J. et al. (2018). *ChemElectroChem* 5: 2073–2079.

18 Yanagisawa, H., Matsumoto, Y., and Machida, M. (2010). *Applied Surface Science* 256: 1619–1623.

19 González, P. and Cuervo, Y. (2014). *Chemical Engineering Research and Design* 92: 2715–2724.

20 Mohan, D. and Singh, K.P. (2002). *Water Research.* 36: 2304–2318.

21 Kang, L., Cui, M., Jiang, F. et al. (2018). *Advanced Energy Materials* 8: 1801090.

22 Parker, J., Chervin, C., Nelson, E. et al. (2014). *Energy & Environmental Science* 7: 1117–1124.

23 Zhang, Q., Luan, J., Fu, L. et al. (2019). *Angewandte Chemie International Edition* 58: 15841–15847.

24 Chao, D., Zhu, C.R., Song, M. et al. (2018). *Advanced Materials* 30: e1803181.

25 Zhang, Q., Luan, J., Fu, L. et al. (2020). *Small* 16: 2000929.

26 Song, M., Tan, H., Chao, D., and Fan, H.J. (2018). *Advanced Functional Materials* 28: 1802564.

27 Zhang, T., Tang, Y., Guo, S. et al. (2020). *Energy & Environmental Science* 13: 4625–4665.

28 Kundu, D., Adams, B.D., Duffort, V. et al. (2016). *Nature Energy* 1: 16119.

29 Zhang, L., Rodríguez-Pérez, I.A., Jiang, H. et al. (2019). *Advanced Functional Materials* 29: 1902653.

30 Huang, S., Zhu, J., Tian, J., and Niu, Z. (2019). *Chemistry Europe* 25: 14480–14494.

31 Huang, S., Wan, F., Bi, S. et al. (2019). *Angew Chem Int Ed* 58: 4313–4317.

32 Li, H., Liu, Z., Liang, G. et al. (2018). *ACS Nano* 12: 3140–3148.

8

Advanced Characterization Tools and Theoretical Research Methods

With the growing concerns about the environmental problems caused by fossil fuels such as coal, oil, and natural gas, it is urgent to apply sustainable and renewable energy sources. However, the efficient application of most sustainable energy sources, such as solar and wind energies, depends heavily on efficient energy storage systems [1–3]. Among the many alternative technologies, rechargeable batteries have attracted more attention than other energy storage technologies due to their economical maintenance, flexible scale, and energy storage efficiency [4]. As one of the most widely used rechargeable battery systems, lithium-ion batteries (LIBs) with high energy density, excellent charge/discharge stability, and life span have been rapidly developed and commercialized in portable electronic devices and electric vehicles [1, 5, 6]. However, the scarcity of lithium resources on Earth has limited the widespread adoption of lithium batteries in energy storage systems and electric vehicles. Therefore, there is an urgent need to develop next-generation rechargeable battery systems with safety, long cycle life, and low cost.

Aqueous zinc-ion batteries (AZIBs) have attracted considerable attention due to the abundance of Zn in nature (Zn is the ninth most abundant element on Earth's crust), environmental benignity, safety, and low cost. However, the current energy density and cycle stability of AZIBs are inadequate for electric vehicles and energy storage systems. During zinc insertion or extraction, various changes, including volume expansion/contraction, phase transition, morphological evolution, and surface reconstruction, occur at the electrode and electrode/electrolyte interface. Monitoring and understanding these physical and chemical changes in electrodes and interfaces are essential to guide the improvement of the electrochemical performance of AZIBs. The main concerns in AZIBs investigation are as follows: (i) the relationship of crystal structure, local structures, and zinc storage mechanism; (ii) the evolution of surface characteristics and solid–electrolyte interphase; (iii) the relationship between performance and reaction anisotropy; (iv) thermodynamics and kinetics of the electrode materials. Therefore, morphological and structural characterization tools play a key role in understanding the main concerns.

In recent years, great efforts have been made to develop advanced techniques to characterize AZIBs in electrochemical processes from different perspectives, dimensions, and scales in order to optimize ZIB systems. Among many characterization methods, direct observation of the structural and morphological distribution

Aqueous Zinc Ion Batteries: Fundamentals, Materials, and Design, First Edition.
Haiyan Wang, Qi Zhang, Yixin Li, and Yougen Tang.

and evolution of the electrode materials for AZIBs is very important to understand their electrochemical behaviors. Advanced imaging techniques, such as scanning electron microscopy (SEM), transmission electron microscopy (TEM), and scanning transmission electron microscopy (STEM), provide direct visualization of structural/morphological changes caused by electrochemical processes, as well as the Zn–ion distribution and diffusion in a wide range of atomic level to the particle level, even the whole electrode level. Some important electrochemical reaction mechanisms can also be obtained from electron energy loss spectroscopy (EELS). X-ray diffraction (XRD) can be used to study the structure and structural changes of electrode materials before and after and during the electrochemical processes, such as structure evolution, lattice parameters, and atomic positions of elements. However, XRD is not good at studying amorphous electrodes.

In contrast, X-ray and neutron pair distribution function (PDF) analysis is a powerful tool to provide information on the local structures of electrode materials, even for amorphous or aperiodic disordered systems. The capacity of the electrode materials is closely related to the participating electrons from the redox couples of cation or anion. Therefore, it is of great significance to determine the valence state of each element in the original electrode and its electronic structure evolution during the electrochemical processes for studying the charge storage mechanism of electrode materials and designing new electrode materials. Special attention should be paid to the multi-element oxide systems and materials with multi-electron transfer from single-element atoms. Soft and hard X-ray absorption spectroscopy (sXAS/hXAS) can monitor the valence of various elements in electrode materials and their changes during the charging and discharging cycle. Meanwhile, the local structural information (such as coordination number and bond length) can be obtained by extending the fine structure fitting and data analysis of X-ray absorption. Moreover, X-ray photoelectron spectroscopy (XPS), infrared (IR) spectrometers, and Raman spectroscopy are also used to determine the electronic structure, oxidation state, and local environment information of electrode materials.

In addition, many in situ techniques have been designed and applied to battery research using specially designed batteries or systems. These in-situ techniques are important because they can provide real-time information about changes in materials under operating conditions.

In this chapter, we conclude recent advances in advanced characterization techniques, including ex situ and in situ techniques, through some selected examples. Due to the size and scope of this chapter, we cannot cover all the important results in this field but we wish to provide a good reference as a starting point. Through comparing various advanced characterizations, a complete understanding of the electrode and electrode/electrolyte interface is essential to explore new materials and optimize systems for AZIBs. This review is divided into the following sections. First, the basic characterization of the various techniques for AZIBs is detailed in Section 8.1. The applications of advanced in situ techniques for AZIBs are concluded in Section 8.2. We can conclude that an in-depth understanding of the electrodes and electrode/electrolyte interfaces is essential to optimize current systems and explore new materials for AZIBs through comparing various basic and in situ characterization techniques.

8.1 Characterization Techniques

The macroscopic structure of rechargeable batteries, such as electrode topography, particle size and shape, and the nature of the solid electrolyte interface (SEI), has a significant impact on the performance of the batteries. The instrumental analysis method, which can be used to characterize the electrode material, has become an indispensable experimental means in designing electrode materials and electrolytes for AZIBs. In order to better understand the role of these characterization techniques in electrode materials, we mainly introduce the composition, morphology, and structure of the materials.

8.1.1 Apparent and Morphological Observations

Many important electrochemical properties of electrode materials are determined by their morphological characteristics. For example, the properties of granular electrode materials are very different from those of nanowires and tubes. Therefore, we also need to use some techniques, such as electron microscope (EM), atomic force microscope (AFM), and scanning tunnel microscope (STM) to characterize the morphology of electrode materials.

8.1.1.1 Electron Microscope (EM)

EM, an indispensable tool in modern science and technology, is widely used in microscopic research in the fields of life science, physics, chemistry, justice, earth science, materials science, and industrial production. It can be divided into SEM, TEM, and STEM. TEM is often used to observe the microscopic structures of substances that cannot be distinguished by ordinary microscopes. SEM is mainly used to observe the morphology of the solid surfaces. It can also be combined with an electron energy spectrometer to form an electron microprobe for substance composition analysis. Meanwhile, EM, the most widely used measurement technique for AZIBs, is mainly used to detect the surface morphology of the electrode and the separator.

SEM SEM is an effective tool to characterize particle size and morphology of materials, and the spatial resolution of field emission scanning electron microscopy (FESEM) can be up to 10 nm. The information provided by SEM mainly includes the geometric morphology of the material, the dispersion state of the powder, the size and distribution of the nanoparticles, and the elemental composition and phase structure of the specific morphology region. SEM has relatively low requirements for samples, and the morphology of either powder or bulk samples can be directly observed. Therefore, it has been widely used in AZIBs to characterize the surface morphology of electrodes (including anode [7] and cathode [8]) and separators [9] after the charge and discharge processes. For example, Kang et al. used SEM to study the bare and the nano-$CaCO_3$-coated Zn anodes before and after 100 cycles of Zn-stripping/plating (Figure 8.1A) [9]. As shown in Figure 8.1A, compared with bare Zn foil, the nano-$CaCO_3$-coated Zn foil displayed a smooth, dense, and flat surface after 100 striping/plating cycles. This result reflected the smooth and stable Zn stripping/plating reactions enabled by the nano-$CaCO_3$-coating. In addition, to understand Zn stripping/plating behaviors, they also used SEM to observe the morphology of a fresh separator (filter paper), which was found very inhomogeneous microscopically (Figure 8.1B). There are tens of micron pores in some parts of the filter paper, while the rest parts are much denser. Interestingly, the size and distribution of the pores in filter paper are very similar to those of the large protrusions on the cycled Zn foils, suggesting a potential relationship between them (Figure 8.1B(a,b) vs. (c,d)) [9]. Their results indicated that the influences of pores on Zn protrusion/dendrite evolution should be size-dependent, i.e. small pores promote the formation of small protrusions/dendrites. The analysis of the abovementioned anodes and separators during Zn stripping/plating processes demonstrated that SEM is a powerful tool for studying the morphology in AZIBs. Meanwhile, SEM was used by Shi et al. to investigate the morphologies of the Zn@metal collectors (Zn@Cu foam, Zn@Cu foil, and Zn@Ni foam) after the plating/stripping processes, as shown in Figure 8.1C [10]. The surface of the metal collectors can be found neat and smooth in their states (Figure 8.1C(a–c)). Among these metal collectors, Cu and Ni foam (Figure 8.1C(a,c)) possessed a characteristic three-dimensional (3D) structure, while Cu foil (Figure 8.1C(b)) displayed an ordinary 2D structure. After the electrodeposition process, sheet-like zinc arrays grow on the surface of the metal collector, as shown in Figure 8.1C(d–f). The thickness and uniformity of deposited zinc arrays for different metal collectors were diverse due to the different zinc nucleation overpotentials (ZNO). Compared with the other collectors (Cu foil and Ni foam), Cu foam presented possessed the lowest ZNO, resulting in the most uniform zinc array morphology. This further verified that lower ZNO effectively contributes to the more uniform zinc deposition behavior. Figure 8.1C(g–i) displayed the morphologies of Cu foam, Cu foil, and Ni foam during the stripping process to compare the reversibility of sheet-like zinc arrays on different metal collectors. The results showed that the zinc arrays on the surface of Cu foam and Cu foil almost totally exfoliated, suggesting that the plating/stripping behavior was highly reversible. Beyond that, an energy dispersive spectrometer (EDS) as an important accessory to the SEM also can be used to study the energy storage mechanism for Zn^{2+}-intercalated cathode materials during charging/discharging processes in AZIBs [13]. For instance, the EDS mapping was used by Rubel et al. to investigate the elemental and morphological changes of electrolytic manganese dioxide (EMD) electrodes after five cycles in 1 M $ZnSO_4$ electrolytes with different pH values, as shown in Figure 8.1D [11]. During the discharge process, the electrode surface for pH 4 (Figure 8.1D(a)) was covered with a zinc hydroxide sulfate (ZHS) sheet, and a new layer of Zn–Mn oxide, consisting of hetaerolite and Zn-intercalated Mn oxide. The ZHS disappeared during the charge process while the new Zn–Mn–oxide layer remained. Additionally, as another

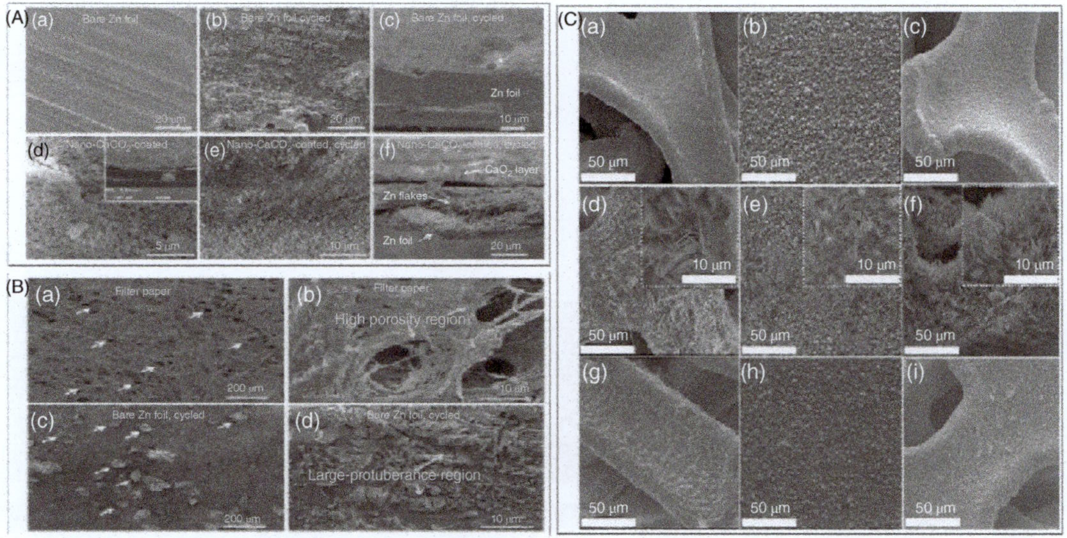

Figure 8.1 (A) SEM images of (a–c) bare and (d–f) nano–$CaCO_3$–coated Zn foils (a, d) before and (b, c) and (e, f) after 100 stripping/plating cycles. The arrow in panel (c) points to a Zn protrusion detaching from the Zn foil. The inset of panel (d) manifests a typical cross-sectional SEM image of the nano–$CaCO_3$–coated Zn foil [9]; Copyright 2018, Wiley-VCH. (B) SEM images of (a, b) a fresh filter paper and (c, d) a cycled bare Zn foil (0.25 mA cm^2, 0.05 mAh cm^{-2}, 100 stripping/plating cycles with a filter paper separator). Arrows in panels (a, c) and dotted lines in panels (b, d) highlight some typical pores in the filter paper and protrusions on the cycled Zn foil; Copyright 2018, Wiley-VCH. (C) SEM images of (a) Cu foam, (b) Cu foil, and (c) Ni foam in the pristine state. SEM images of (d) Zn@Cu foam, (e) Zn@Cu foil, and (f) Zn@Ni foam after the plating process. SEM images of (g) Zn@Cu foam, (h) Zn@Cu foil, and (i) Zn@Ni foam after the stripping process; Copyright 2019, American Chemical Society. (D) SEM SE images and corresponding EDX maps of EMD electrodes after five cycles in electrolytes with different pH values: (a) pH 4 and (b) pH 2.5. Fully discharged and fully charged states are denoted as FD and FC, respectively; Copyright 2019, American Chemical Society. (E) EBSD inverse pole figure (IPF) maps (a) and pole figures (b) of pristine Zn electrode, (c) EBSD IPF maps and (d) pole figures of (002)-textured Zn electrode, revealing the (002) dominated crystal plane of the as-prepared (002)-Zn electrode; Copyright 2022, Elsevier. Source: (a, b) Li et al. [9], © 2018 / John Wiley & Sons. (c) Shi et al. [10], © 2020 / American Chemical Society. (d) Rubel et al. [11], © 2019, American Chemical Society. (e) Wang et al. [12]/With permission of Elsevier.

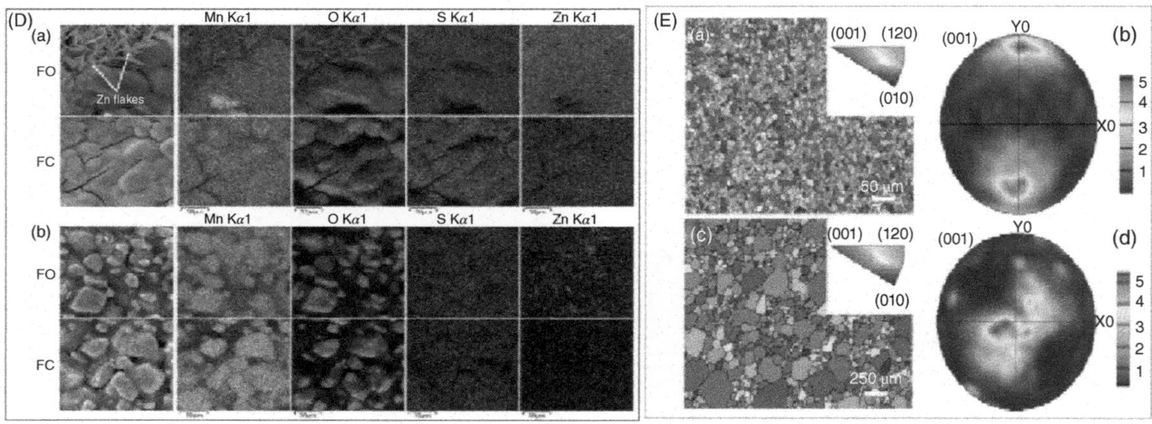

Figure 8.1 (*Continued*)

important adjunct to SEM, electron backscattered diffraction (EBSD) is currently diffractive with submicron spatial resolution while retaining the conventional features of SEM, which can be successfully applied to further analyze the texture information of electrodes for AZIBs. Wang et al. used EBSD to in-dept study the texture information of (002)-Zn foil due to its highly reversible performance in AZIBs [12]. The EBSD results (Figure 8.1E(a,c)) displayed that the fabricated Zn foil through thermal treatment possessed more (002)–dominated surface in comparison to pristine Zn foil. The (001) polar figures further demonstrated the texture change from the pristine Zn foil to (002)-dominated Zn foil (Figure 8.1E(b,d)). Based on the above analysis, SEM and its accessories play an important role in the changes in the morphology and element distribution of the electrode material and separator of AZIBs during the plating/stripping processes.

TEM TEM, a kind of high resolution and high magnification EM, is one of the important tools to observe and analyze the morphology, element, organization, and structure of materials. TEM projects a beam of accelerated and concentrated electrons onto a very thin sample. The electrons collide with the atoms in the sample and change direction, resulting in solid Angle scattering. The size of the scattering Angle is related to the density and thickness of the sample, so the light and dark images can be formed. In general, TEM possesses a resolution of 0.1–0.2 nm and a magnification of tens of thousands to millions of times and is used to observe ultrastructures, that is, structures smaller than 0.2 μm. First, TEM was mainly applied to investigate the structural, morphological, and elemental evolution of the materials during the charging/discharging processes and further to elucidate the reaction mechanism of cathode materials for AZIBs. For example, TEM and its accessory (EDS Mapping) were used by Hou et al. to investigate the morphological and elemental changes of the $2ZnCO_3 \cdot 3Zn(OH)_2$@graphene (ZZG) composite electrode during cycles in 2 M K_2CO_3 electrolyte [14]. As shown in Figure 8.2A(a–d), the reduced Zn particles were independently scattered on the graphene sheet substrate, exhibiting a hexagonal shape at the fully discharged state. After charging, $2ZnCO_3 \cdot 3Zn(OH)_2$, which were oxidized by Zn particles with homogeneous rice-like particles, were uniformly distributed on the graphene sheet, which was very similar to the pristine ZZG. Meanwhile, the structural evolution of electrode materials during the charge/discharge process can be directly observed by TEM, which can establish a relationship between the structure of the electrode material and battery performance. Li et al. used the TEM test to study the elemental and structural changes of sheet-like $Zn_3V_2O_7(OH)_2 \cdot H_2O$ cathode material during charge/discharge processes (Figure 8.2B) [15]. It can be seen from the figure that the sheet structure is layered and interwoven to a certain extent, which can enhance the connection between the atoms, thereby increasing the conductivity of the material during the charging/discharging process in AZIBs. The boundary of $Zn_3V_2O_7(OH)_2 \cdot H_2O$ material is clear, and there is no other phase substance. The elemental mapping reveals uniform distributions of Zn, V, and O within the material. TEM was used by Soundharrajan et al. to study the effect of Mn^{2+} additives in electrolytes on the performance of spinel $ZnMn_2O_4$ (ZMO) cathode [16].

The TEM results further indicated that the exceptional performance was facilitated by the electrochemical equilibrium between Zn^{2+} and Mn^{2+} ions via a stable Zn^{2+} insertion/desertion in bulk, a reversible electro–deposition/dissolution of MnO_x from the Mn^{2+} additive in the electrolyte onto (from) the surface of the cathode, and the reversible Zn-insertion into the undissolved surface MnO_x layer. To further investigate the high initial contribution of the $ZnMn_2O_4$ (ZMO) cathode in the $MnSO_4$ (MS) based on MnO_x formation, Soundharrajan et al. conducted TEM analysis of a ZMO sample after the initial charge/discharge processes. It was evident that the porous microbar structure was retained after the first charging cycle of the ZMO cathode, surrounded by an abundant porous matrix (Figure 8.2C(a)). Figure 8.2C(b) displayed the high-resolution TEM (HRTEM) image that revealed the ZMO matrix surrounded by many new thin layers of nanosheets. As shown in Figure 8.2C(I), the formation of nanosheets on the ZMO microrods displayed possible electrolytic Mn deposition, confirming the uniform distribution of Mn and O determined by EDS-mapping. However, the nanoparticles remained alongside the parent porous ZMO microrods (Figure 8.2C(c,d)). The corresponding EDS mapping analysis, as shown in Figure 8.2C(II), revealed that the discharging products were composed of Zn, Mn, O, and S, suggesting that the presence of sulfur indicated the formation of hydroxyl zinc sulfate (ZHS) during the zinc insertion process. The disappearance of the nanosheets during the charging process showed the deposition and dissolution of Zn on the cathode materials during the electrochemical cycle. In addition, TEM was also used to investigate the interfacial region of the self-supported anode materials for AZIBs. As shown in Figure 8.2D, the Zn_xCu_y/Zn interfacial region was measured through HRTEM testing, revealing the seamless integration of Zn_xCu_y alloy shell on Zn core [17]. According to the characteristic fast Fourier transform (FFT) pattern recognition of the selected region in Figure 8.2D(b,c), two regions with different crystal structures can be observed, corresponding to the cubic Zn_xCu_y and the hcp Zn, respectively. Based on the aforementioned analysis, TEM technique can provide high-resolution images of electrode materials in AZIBs for investigating the structural, morphological, and elemental evolution of these materials during charging/discharging processes at a nanometer scale.

STEM STEM, a microscope with both TEM and SEM, which possesses two modes (high-angle annular-dark-filed (HAADF) and annular-bright-field (ABF)) for capturing high-resolution images, has been considered a superior technique for providing structural, chemical, and electronic information on an atomic scale. HAADF STEM has a strong ability to image heavy atoms due to the correlation between image contrast and atomic number Z (close to $Z1.7$); meanwhile, ABF-STEM provides image contrast proportional to $Z^{1/3}$ and is sensitive to light elements such as Li and Na, making both light and heavy elements visible [18]. STEM techniques have proven to be highly successful tools for determining precise atomic positions and structural defects on the atomic scale [19]. STEM can be considered a very successful technique for determining structural defects and precise atomic positions on the atomic scale. STEM can be used to monitor the states, compositions, and microstructures of cathode materials for AZIBs. Chao et al. found that the formation

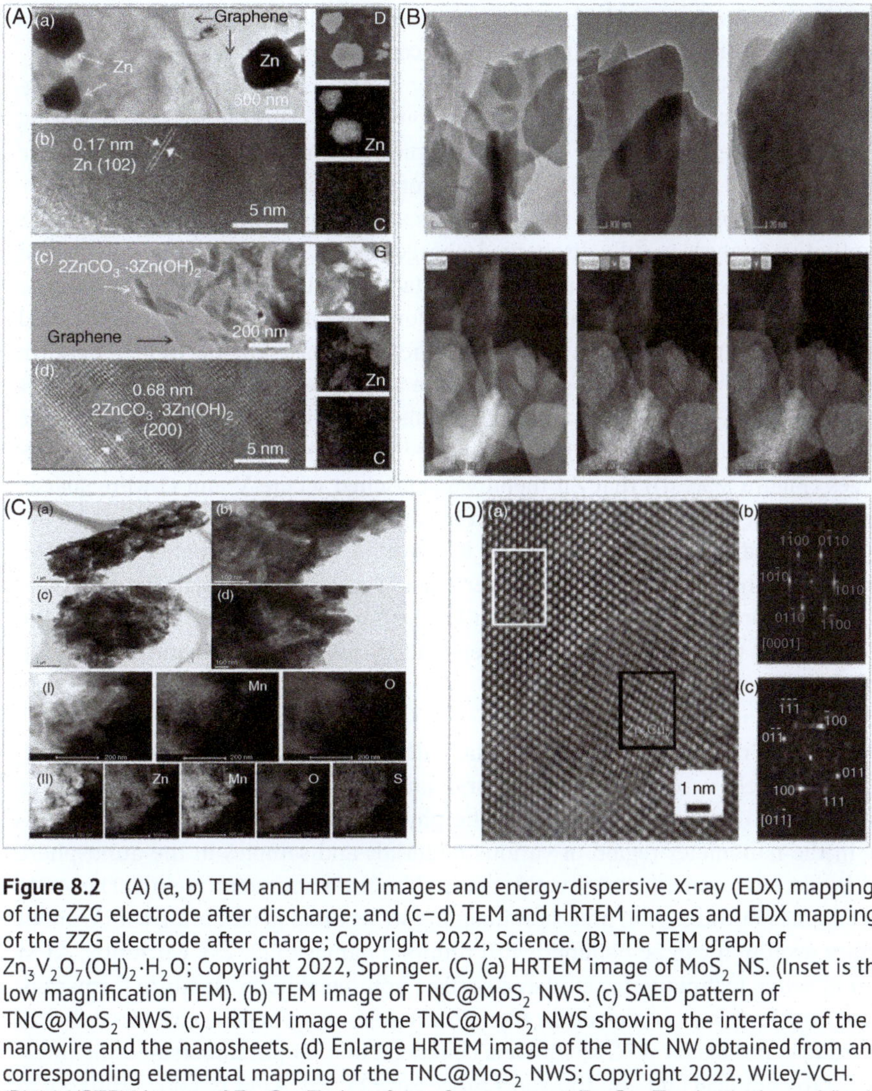

Figure 8.2 (A) (a, b) TEM and HRTEM images and energy-dispersive X-ray (EDX) mapping of the ZZG electrode after discharge; and (c–d) TEM and HRTEM images and EDX mapping of the ZZG electrode after charge; Copyright 2022, Science. (B) The TEM graph of $Zn_3V_2O_7(OH)_2{\cdot}H_2O$; Copyright 2022, Springer. (C) (a) HRTEM image of MoS_2 NS. (Inset is the low magnification TEM). (b) TEM image of TNC@MoS_2 NWS. (c) SAED pattern of TNC@MoS_2 NWS. (c) HRTEM image of the TNC@MoS_2 NWS showing the interface of the nanowire and the nanosheets. (d) Enlarge HRTEM image of the TNC NW obtained from and corresponding elemental mapping of the TNC@MoS_2 NWS; Copyright 2022, Wiley-VCH. (D) (a) HRTEM image of Zn_xCu_y/Zn interface of nanoporous Zn_xCu_y/Zn electrode; and (b, c) FFT patterns of HCP Zn (b) and cubic Zn_xCu_y phases (c) corresponding to write and black squares in (a); Copyright 2022, Springer. Source: (a) Hao et al. [14], © 2020/American Association for the Advancement of Science – AAAS. (b) Li et al. [15], © 2022 / Springer Nature. (c) Soundharrajan et al. [16], © 2020/John Wiley and Sons, Inc. (d) Meng et al. [17], © 2022/Springer Nature/CC BY 4.0.

of the oxide vacancies in MnO_2 in the acidic environment was believed to be associated with different zinc electrodeposition mechanisms for AZIBs [20]. They used the EELS, an important accessory to the STEM, to monitor Mn 3s spin-energy splitting of the deposited MnO_2 in the AZIBs with 0.1 M H_2SO_4 and without H_2SO_4. The EELS results indicated the existence of mixed oxidation states of Mn^{4+}/Mn^{3+} in the MnO_2 fluff in the battery with 0.1 M H_2SO_4, suggesting that the Mn vacancies were dependent on electrolyte pH. They further verified the micropores and Mn vacancies in the MnO_2 fluff within the exposed (101) facets using HADDF-STEM.

In addition, STEM was also used by Wang et al. to investigate the morphological and structural evolution of layered porous $Zn_{0.3}V_2O_5$ (ZVO) cathode during charge/discharge to elucidate the storage mechanism of Zn/ZVO battery. The fully discharged ZVO cathode is composed of nanoflowers and flaky precipitate. The intercalation reaction of zinc in ZVO cathode was verified by STEM–EDS quantitative analysis in fully discharged/charged states, suggesting that the layered porous structure can provide abundant contact with the electrolyte, shorten the ion diffusion path, and provide a buffer for alleviating the stream generated during the electrochemical, which is beneficial for fast kinetics and long-term stability [12]. By performing FFT toward the atomic-level HAADF-STEM images (Figure 8.3(a,b)), Zhu et al. can observe Zn metal distribution under the existence of $Zn_x(OTf)_y(OH)_{2x-y}\cdot nH_2O$ [21]. It can be seen from the EDS-mapping and EELS images (Figure 8.3(c,d)) that Zn metal was enriched where V and O were sparse, indicating that the deposition sites of Zn in cathodic materials were not dependent on VO_2. Based on the above analysis, it can be seen that compared with TEM, the STEM technique can provide higher resolution of cathode material at atomic scale. Therefore, STEM technology can be further applied to AZIBs to study the mechanism of zinc intercalation/de-intercalation during charging and discharging processes.

AFM AFM, a novel and useful technique with high resolution at the atomic level, has been widely used in semiconductors, nano-functional materials, and various nano-related disciplines. It can detect the physical properties, including morphology in the nanometer region of various materials and samples in the atmosphere and liquid environment, or directly manipulate the nanometer region. The AFM tip remains neutral during precision imaging, so accurate in situ information about surface evolution can be obtained with minimal damage [22, 23]. Currently, AFM is an effective technique for studying surface volume change, dendrite formation, phase transition, ion and electron transport, interfacial reaction, and instability of electrode materials for AZIBs, which is beneficial to in-depth understand the internal electrochemical and ion–transport processes [24, 25]. AFM is a powerful tool to characterize the changes in surface morphology of anode for AZIBs during charging/discharging processes. Li et al. used AFM technique to investigate the nucleation and growth mechanism of the Zn deposition on the Cu–Zn alloy layer on Zn foil (Cu–Zn@Zn) electrode and bare Zn electrode at 0.2 and 1 mAh cm^{-2} [26]. The AFM images displayed that many small and dense metallic Zn crystal nuclei formed on the Cu–Zn@Zn electrode at 0.2 mAh cm^{-2} (Figure 8.4A(a)).

In contrast, the accumulation of metallic Zn particles was observed on the bare Zn electrode (Figure 8.4A(b)). Surprisingly, when the electroplating capacity was increased to 1 mAh cm^{-2}, the Zn particles could still be uniformly and densely deposited on Cu–Zn@Zn electrode. The AFM results confirmed the homogeneous Zn particles without the growth of Zn dendrite can be deposited on the Cu–Zn layer with sufficient Zn crystal nucleus sites. As we know, the coating polymers

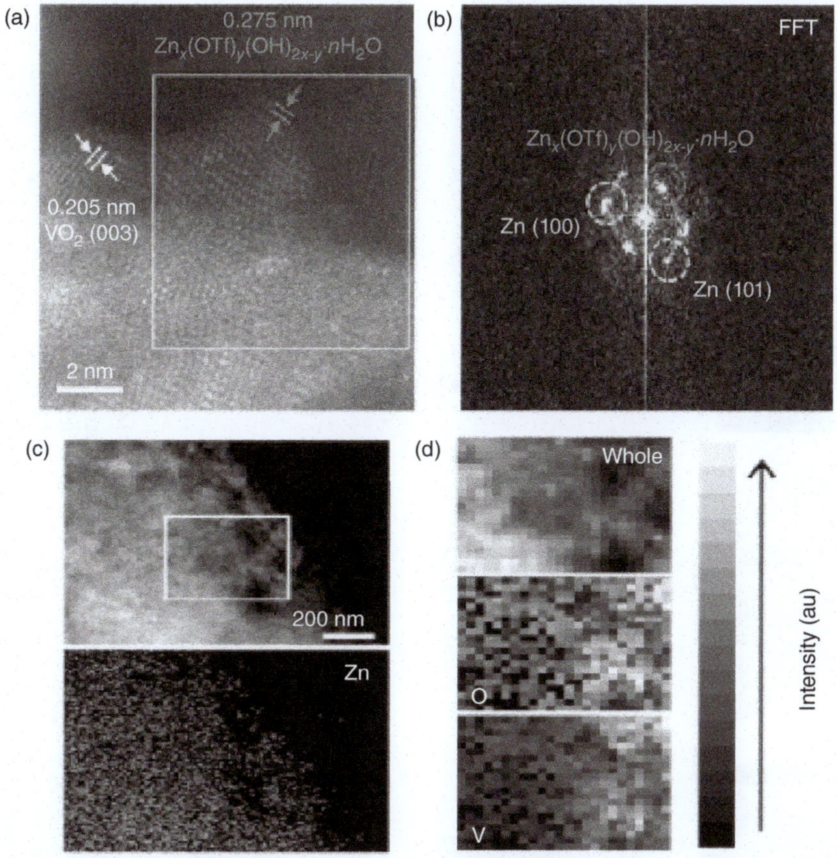

Figure 8.3 Investigation of the form of cathodic Zn deposition. (a, b) HAADF-STEM images and the corresponding Fourier transform at 0.3 V. (c) HAADF-STEM image and the corresponding EDS mapping of Zn element. (d) EELS mapping collected from the white box in (c). Source: Zhu et al. [21], © 2022/with permission from Elsevier.

of microporous polymers (PMI) can effectively isolate the interface between zinc anode and electrolyte to avoid subsequent side reactions, and its hard skeleton helps to protect the integrity of the coating over a long period. Su et al. [27] used AFM technique to investigate the morphology of Zn anode coated by PIM-1, PIM-2, and PIM-13 layer (PIM-1@Zn, PIM-2@Zn, and PIM-13@Zn for short) before cycling and after 150 charging/discharging processes (Figure 8.4B). As shown in Figure 8.4B(a), the AFM images of PIM-1@Zn, PIM-2@Zn, and PIM-13@Zn displayed relatively homogeneous coatings on the Zn surfaces. After 150 charging/discharging cycles, the PIM-1 and PIM-13 protection coatings were still homogeneous and tightly bound to the Zn foil from the AFM images (Figure 8.4B(b)).

In contrast, the PIM-2 coating is cracked due to brittleness caused by the harder decafluorobiphenyl units in the polymer chain (Figure 8.4B(b)). After 50 cycles, the surface of bare zinc was rough and uneven, which is related to the irreversible side reactions during the cycle. The AFM results demonstrated that the PIM coating can

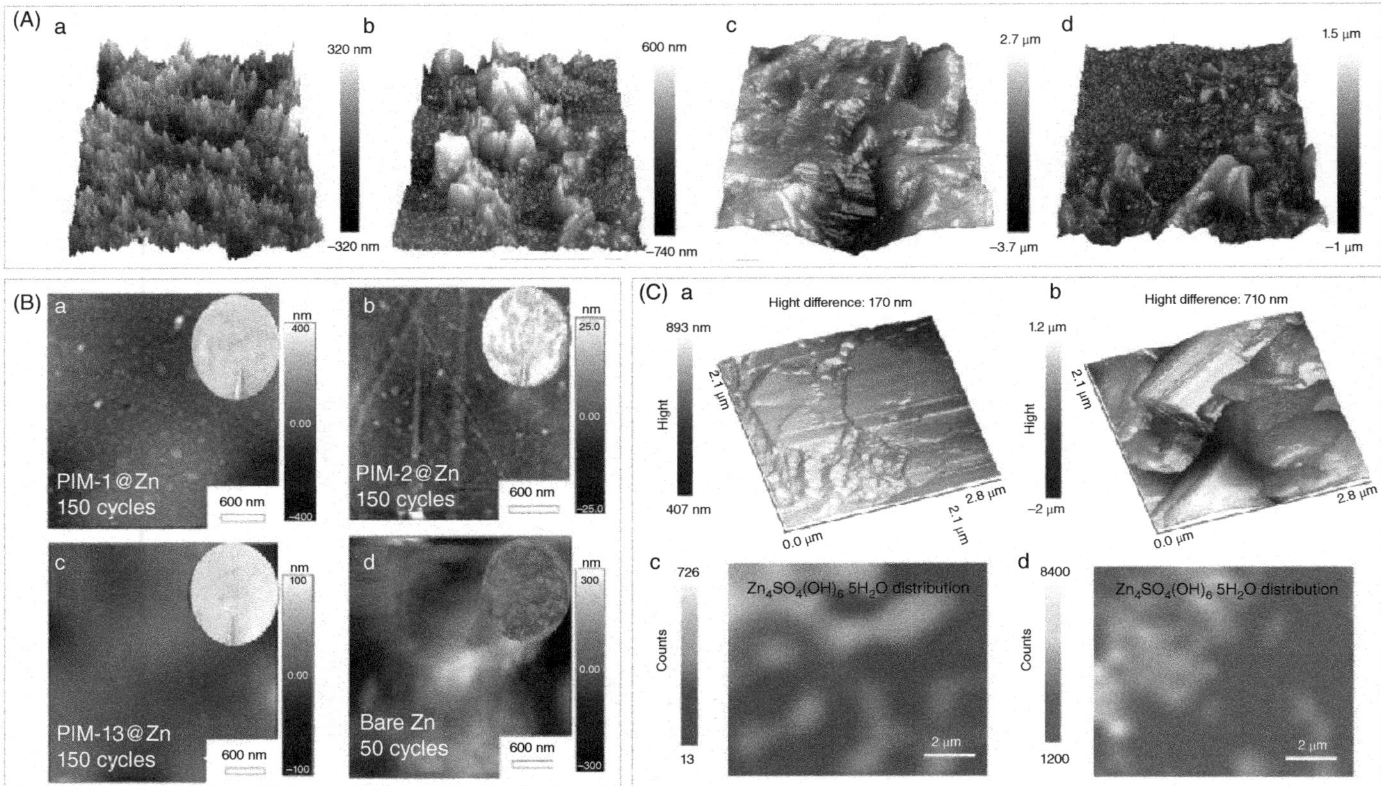

Figure 8.4 (A) Morphology evolution and theoretical simulation of Cu-Zn@Zn and bare Zn electrodes; (a) AFM image of Cu-Zn@Zn electrode and (b) bare Zn electrode with 0.2 mAh cm^{-2} Zn deposited; (c) AFM image of Cu-Zn@Zn electrode and (d) bare Zn electrode with 1 mAh cm^{-2} Zn deposited [26]; Copyright 2022, Wiley-VCH. (B) AFM images and optical photos of PIM-1@Zn (a), PIM-2@Zn (b), PIM-13@Zn (c) and bare Zn (d) after charging and discharging cycles; Copyright 2022, Elsevier. (C) AFM 3D height imaging on the surfaces of Zn deposits after 200 cycles plating/stripping (1 mAh cm^{-2}, 5 mA cm^{-2}), (a), (c) on FCOF film-coated Ti and (b), (d) bare Ti; Copyright 2021, Nature. Source: (B) Su et al. [27], © 2022/with permission from Elsevier. (C) Zhao et al. [28], © 2021/Springer Nature/CC BY 4.0.

effectively reduce the Zn corrosion and thus inhibit the formation of the “dead” Zn. In addition, information about homogeneity and height difference can also be obtained using height images and 3D AFM images. Therefore, the AFM image combined with height image can provide vital insight into the morphology, homogeneity, and height difference of anode materials at the nanoscale. Using AFM technique, Zhao et al. [28] revealed the planar Zn deposition morphology, fast Zn^{2+} transport, and corrosion resistance properties of the ultrathin, fluorinated 2D porous covalent organic framework (FCOF) film, which were expected to significantly improve the electrochemical performance of Zn anodes. As evidenced by the AFM height and phase imaging (Figure 8.4C(a)), the horizontally arranged platelet morphology of the deposited Zn underneath FCOF remained homogeneous and flat after 100 cycles during charging/discharging processes in AZIBs. However, after 100 cycles, the deposited Zn on bare Ti foil displayed fluctuating and rough patterns with a much higher height of 710 nm (Figure 8.4C(b)).

Similarly, Wang et al. used AFM and height images to investigate the persistent impact of $Zn_3(BO_3)_2$-coated Zn (Zn@ZBO) and bare Zn anode on the Zn deposition with 100 charging/discharging cycles. The AFM images demonstrate the significant dendrite suppression by ZBO. The Zn deposition on bare Zn resulted in a rough surface with a height difference of about 4.7 μm. In contrast, the Zn deposition on Zn@ZBO displayed a homogeneous surface with about 0.6 μm, suggesting ZBO coating possessed excellent dendrite-suppressing ability. The abovementioned analysis suggests that AFM can monitor battery health and provide fault information, which is critical for optimizing and developing better electrolytes and electrode materials.

8.1.1.2 Laser Scanning Confocal Microscope (LSCM)

LSCM, a new technique developed in the mid-1980s and widely used, is an advanced analytical instrument for cell and molecular biology produced by the penetration of modern high-tech means, including laser, electronic imaging, and computer image processing, and combined with the traditional optical microscope [29, 30]. Currently, some researchers have attempted to use LSCM to provide high-definition color motion images for the surface or cross-section observation of electrodes during charging/discharging processes [31]. Therefore, LSCM was widely applied in the observation and measurement of the electrochemical reactions in AZIBs, which can be used to observe the distribution of electrochemical reactions on the electrode or the whole depth of the AZIBs and further analyze the dendrite formation mechanism on the anode surface. Qin et al. used LSCM to observe the morphology and roughness of Zn deposition on an anode surface at current density of 0.5 mA cm^{-2} in 3 M $ZnSO_4$ electrolytes with different ethylene glycol (EG) volume concentrations, including 0 vol% EG (0E), 10 vol% EG (10E) and 68 vol% EG (68E) [32]. As is visible in Figure 8.5A, the higher the concentration of EG in the electrolyte, the more glabrous the Zn deposition, which supports the claim that EG can be conducive to dendrite growth during charging/discharging processes. The surface of Zn anode in 0 EG electrolyte was rough after plating at 0.5 mA cm^{-2} for 1 hour, and isolated dendrites with a height of about 32.6 μm were formed.

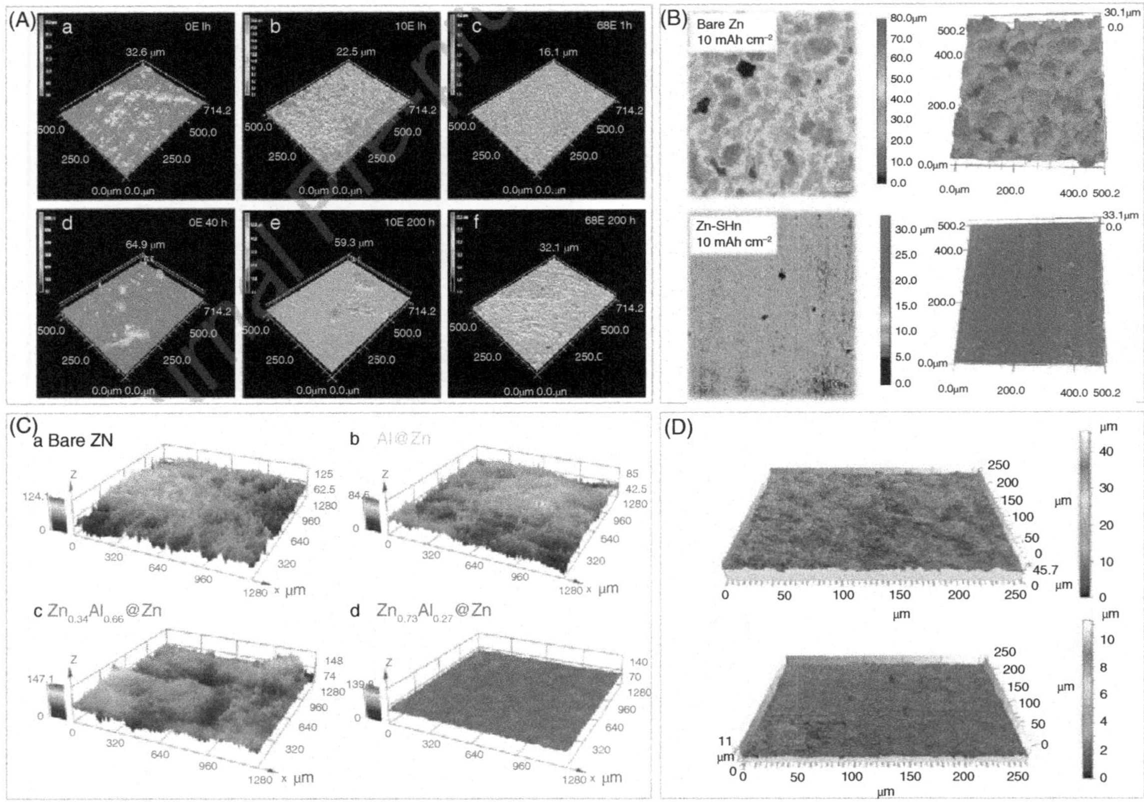

Figure 8.5 (A) LCSM images of Zn plate at different cycling times in 3 M $ZnSO_4/H_2O$/EG electrolytes. LCSM images of Zn plate after plating at 0.5 mA cm^{-2} for 1 h in (a) 0E, (b) 10E, and (c) 68E electrolytes; (d) LCSM images of Zn plate after cycling at 0.5 mA cm^{-2} for 40 h in 0E electrolytes; and LCSM images of Zn plate after cycling at 0.5 mA cm^{-2} for 200 h in (e) 10E and (f) 68E electrolytes [32]; Copyright 2020, Elsevier. (B) LCSM images of bare Zn and Zn−SHn electrodes after Zn deposition [33]; Copyright 2022, Wiley-VCH. (C) The 3D CLSM images of (a) bare Zn, (b) Al@Zn, (c) $Zn_{0.34}Al_{0.66}$@Zn, and (d) $Zn_{0.73}Al_{0.27}$@Zn after 24 h cycling; Copyright 2022, American Chemical Society. (D) 3D CLSM of Zn plating on bare Zn and Al_2O_3@Zn anodes; Copyright 2022, Springer. Source: (C) Zheng et al. [34], (D) Wang et al. [35]/Springer Nature.

In contrast, Zn deposits in the 10 EG and 68 EG electrolytes were more uniform and denser. Similar to AFM technique, LSCM also provides 3D images to further investigate the structural evolution of anode surface in AZIBs at the micron scale. Through LSCM technique, Yang et al. elucidated the compact and uniform Zn plating on the Zn anode with the hydrogel layer [33]. As shown in Figure 8.5B, the surface of exposed Zn formed agglomeration dendrites with high roughness due to the limited nucleation sites. Under the protection of hydrogel layer, the existence of a large number of zincophilic nucleation sites makes zinc nucleation uniform.

Similarly, Zheng et al. conducted 3D LSCM characterization of the bare Zn, Al@Zn, $Zn_{0.34}Al_{0.66}$@Zn, and $Zn_{0.73}Al_{0.27}$@Zn symmetric cells after cycling for 24 hours at 1 mAh cm^{-2} to further investigate the impact of the Zn–Al alloy films on the structural evolution of the Zn anodes [34]. Figure 8.5C displayed that the $Zn_{0.73}Al_{0.27}$@Zn anode surface with a large area of 1280 μm × 1280 μm was still flat after cycling, while the other three anodes were rougher with obvious Zn dendrites, suggesting that the $Zn_{0.73}Al_{0.27}$ film can effectively suppress the Zn dendrites' growth. Therefore, dense Zn deposits can be obtained under the insulated hydrogel without dendrite formation, and the surface of the hydrogel layer is maintained flat and smooth according to 3D LSCM images. Furthermore, 3D LSCM was measured by Wang and his co-workers to study the surface topography of Zn foil with Al_2O_3 coating (Al_2O_3@Zn) and bare Zn symmetric cells after 50 cycles at 1 mA cm^{-2} and 1 mAh cm^{-2} (Figure 8.5D) [35]. The result revealed that the Zn deposition on the bare Zn formed distinct dendrites. Conversely, the Zn deposits onto the Al_2O_3@Zn surface were more uniform than those onto the bare Zn surface, indicating that the Al_2O_3 coatings can effectively inhibit the growth of Zn dendrites. All of these application examples display that LSCM is an excellent technique to study the morphological changes of anode surface during charging/discharging processes for AZIBs.

8.1.1.3 Other Apparent and Morphological Techniques

In addition to the abovementioned morphological and apparent characteristic techniques, researchers tried to use other techniques, such as scanning Kelvin probe microscope (SKPM) and micro-computerized tomography (micro-CT), to achieve morphology and apparent characterization of electrode materials. SKPM was used by Yan et al. to investigate the passivation evolution of the Volta potential of the bare Zn foil with/without montmorillonite (MMT) coating immersed in the electrolyte for 7 days, as shown in Figure 8.6A. The bare Zn foil as anode displayed narrow ranges of Volta potential with an average value of about −1284.7 mV, which was very close to the theoretical Volta potential of Zn metal (about −1320 mV) [36]. As depicted in Figure 8.6A(a, b), the Volta potential of MMT-Zn and bare Zn foil exhibited a significant positive shift to −609.9 and 1732 mV, respectively, owing to the passivation film's ability to enhance the escape work of Zn atoms on the surface. [38]. Furthermore, the Zn foil protected by the MMT coating possessed a narrow range of −700 and −500 mV, indicating an evenly thin passivation film. In contrast, the Volta potential of bare Zn fluctuated significantly, with an average value of 1732 mV

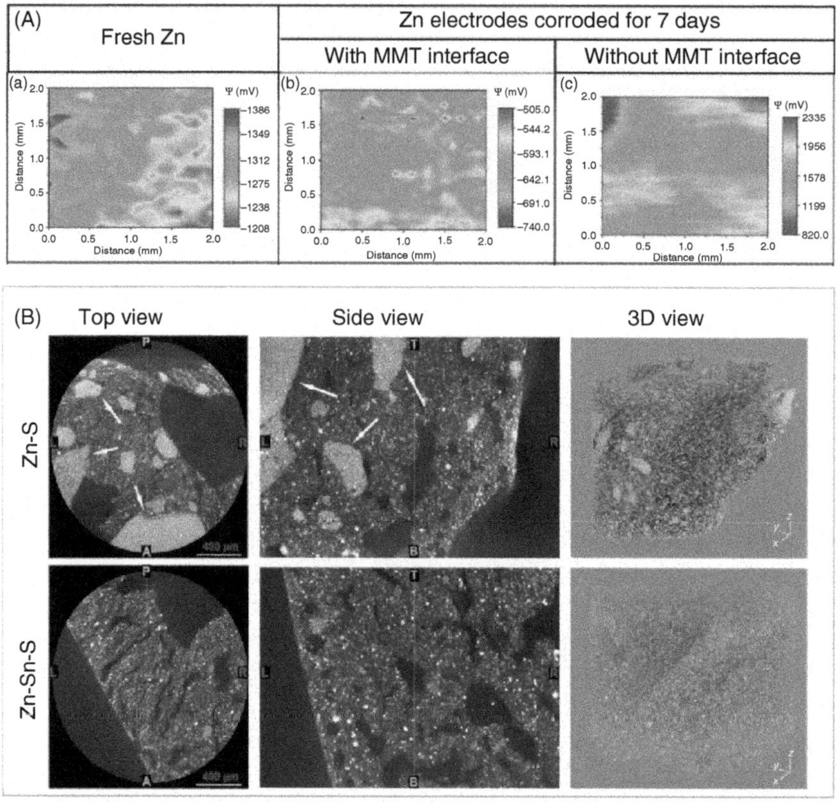

Figure 8.6 (A) Volta potential maps via SKP of (a) polished Zn foil, (b) MMT-Zn, and (c) bare Zn electrodes immersed into 2 M $ZnSO_4$ for a week; Copyright 2021, Wiley-VCH. (B) X-ray macro-CT images of Zn–Sn–S slurry anode in a half cell after cycling at 0.5 mA cm^{-2} for 20 cycles. Write arrows refer to agglomerative zinc deposition. Source: (a) Yan et al. [36]/John Wiley & Sons. (b) Yang et al. [37], © 2023/John Wiley & Sons, Inc.

(Figure 8.6A(c)), indicating the nonuniform formation of a thick passivation film, which markedly increased the escape work of the Zn substrate. The SKPM results indicated that Zn-based MMT coating isolated the corrosion of aqueous electrolyte, maintained the activity of Zn anode, and reduced the passivation film on the electrode. X-ray micro-CT was applied to study AZIBs for in-depth observation of the metallic Zn electrodeposition distribution in 3D space on 3D printing graphene array scaffolds (3DGs) with a tube structure (3DGT) and pillar structure (3DGP) [39]. As shown in Figure 8.6B, our group also used 3D X-ray micro-CT test to characterize the uniform dissolution/deposition behavior of zinc in the semi-solid slurry anode before (Zn–S) and after adding tin powers (Zn–Sn–S) for AZIBs. As seen in Figure 8.6B, the Zn–Sn–S slurry anode showed uniform distribution of light spots corresponding to zinc deposition on the transparent 3D images, while the Zn–S slurry anodes showed accumulated gray spots (large zinc deposits) and large chunks on the 3D image [37].

8.1.2 Structural and Spectroscopic Techniques

In recent years, in order to induce the optimization of AZIB systems, researchers have developed structural characterization techniques (such as XRD, Raman, IR, EPR, XPS, EXAFS, and sXAS/hXAS) for structurally characterizing ZIBs from different angles, dimensions, and scales during electrochemical processes.

8.1.2.1 X-Ray Diffraction

Although neutron diffraction, electron diffraction, IR spectroscopy, and other methods can be used to analyze the structure of materials, XRD, the most effective and widely used method, has been widely used in physics, chemistry, material science, and various engineering and technical sciences. It has become an important experimental and structural analysis technique, which has the advantage of nondestructive sample. XRD can qualitatively analyze the phase composition of the sample by comparing the XRD pattern of the testing sample and the standard substance. Meanwhile, by analyzing and calculating the diffraction intensity data of the sample, the phase composition can be quantitatively analyzed. Besides phase analysis, crystallinity, lattice constant, and dislocation density can also be measured by XRD. Based on the abovementioned characteristics, XRD is a widely applied technique to study the crystal structure, cation mixing, and phase transformation of electron materials for Zn ion batteries according to the relative strength and position of the peak. Specifically, XRD is used to study the structure, evolution, and phase transformation of various anode (such as Zn with different microstructures and exposed facets) and cathode (such as manganese oxides, vanadium oxides, polyanionic compounds, Prussian blue compounds, and transition metal disulfides) materials and membrane surfaces during cycling. For example, Ma et al. investigated the intercalation/delamination process of Zn^{2+}, and the concomitant phase evolution of $CoFe(CN)_6$ cathodes has been studied by ex situ XRD. As shown in Figure 8.7a, their results revealed that the structural transition could be detected from the changes in XRD pattern (011) and (200) reflections with peaks at 17.4° and 18.4°, respectively. When the battery is fully discharged, two apparent peaks appear at 17.4° and 18.4°, and the peak located at 15.3° disappears, revealing the phase transition from cubic phase to monoclinic phase. During the charging process, these peaks reversibly return to their initial state, indicating a reversible change in the structure. Meanwhile, our group performed XRD to understand the deposition behavior of Zn on F–TiO_2-modified cathode [7]. As shown in Figure 8.7b, the broader (004) peak in F–TiO_2 demonstrates the larger exposure of (001) facet, and the narrower (200) peak in F–TiO_2 corresponds to the larger crystal size parallel to (001) facet. It also proved that the relatively higher exposed area of (001) facet of F–TiO_2 on the anode surface plays an important role in adjusting Zn deposition behavior. The TiO_2-protective layer on the anode surface with highly exposed (001) facet (ZF@F-TiO_2) can completely confine the Zn deposition in the restricted space and improve cycling stability of the cells. Specifically, ZF@F–TiO_2 exhibits a longer cycling life than ZF@C-TiO_2, indicating its better Zn reversibility achieved by the specific exposed facets. Meanwhile, XRD is also used to study the effect of different

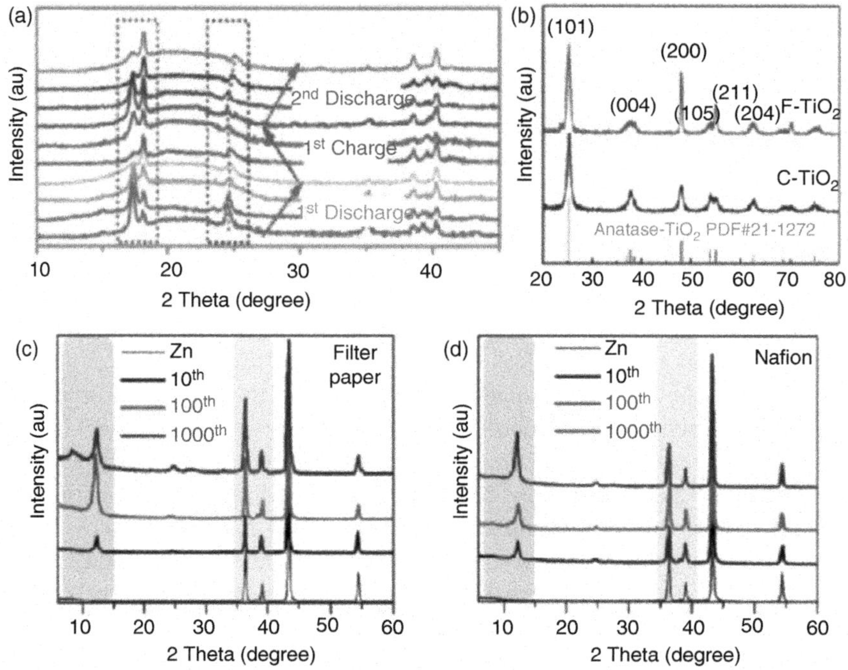

Figure 8.7 (a) The galvanostatic charge/discharge profile at 0.3 A g^{-1} and the different color spheres indicate the potentials where ex-XRD patterns are collected for structural analysis of $CoFe(CN)_6$ cathode; Copyright 2015, American Chemical Society. (b) XRD patterns of $F-TiO_2$ and $C-TiO_2$; Copyright 2019, Wiley-VCH. (c, d) Ex situ grazing-incidence XRD patterns for Zn metal at different stripping/plating cycles with filter paper and (d) Nafion at 0.2 mA cm^{-2}; Copyright 2021, Wiley-VCH. Source: (b) Adapted from Naveed et al. [7], (c, d) Adapted from Zhao et al. [40].

separators on the deposition of Zn during charging/discharging processes in AZIBs system. For instance, Du et al. investigated the effects of separators (filter paper and Nafion-based membranes) on the growth of an effective SEI and the deposition of Zn through ex situ XRD (Figure 8.7c,d), indicating that Nafion-based membranes can enable the formation of a planar SEI, which protects the metal surface and prevents short-circuiting. In addition, Zhao et al. used XRD technique to study the effect of bare Zn anode and Zn anode decorated by an anti-corrosion elastic constraint (AEC) coating (i.e. TiO_2 nanoparticles with polyvinylidene fluoride [PVDF] matrix modification) on the by-products during charging/discharging processes in AZIB system [40]. According to the abovementioned analysis, XRD is an indispensable method to study the surface phase transition, by-products, and structural changes of electrode materials in AZIBs. It provides important information for studying the mechanism during charging/discharging processes in AZIB system.

8.1.2.2 Raman Spectroscopy

Raman spectroscopy, a scattering spectrum, is based on the Raman scattering effect discovered by scientist C.V. Raman from Indian. The advantages of Raman spectroscopy, such as its simple system operation, short measurement period, and higher sensitivity, make it widely applied in chemistry, physics, biology, and medicine. The application of confocal Raman microscopy in battery research is the most important step, opening the way for more complex new developments. At present, Raman spectroscopy is mainly a combination of surface-enhanced Raman scattering, tip-enhanced Raman spectroscopy, and spatial offset Raman spectroscopy, in which the Raman spectrometer is integrated with a nonoptical microscope to enable the simultaneous analysis of two or more techniques within the same area of the sample. It has become increasingly obvious that the development of Raman techniques is promoting the development of spectroelectrochemical research on AZIB components. Raman spectroscopy is widely used to study the solvation structure of electrolytes with different solutions in AZIBs [41]. For example, Yang et al. proposed a novel idea of developing a saturated electrolyte front surface, which can be achieved by a metal–organic frame (MOF) coating, resulting in uniform Zn deposition [42]. A porous nonconductive layer placed on the surface of the Zn regulates the solvation structure before accepting charge in the aqueous electrolyte. They first checked the solvation structure of different concentrations of $ZnSO_4$ solutions in MOF pores through Raman spectroscopy. As shown in Figure 8.8A, the Raman peaks were measured in a series of solutions. The concentration gradient between 0.4 and 3.3 M (saturated) was studied to illuminate the Raman peaks' variation. It appeared that the ν-SO_4^{2-} band center was located in the 983–985 cm^{-1} area with a sharp shoulder shift to higher frequency with increasing electrolyte concentration. Higher concentration complexes were characterized by excessive coordination between Zn^{2+} and anion, and the solvation structure of the typically dilute $[Zn(H_2O)^{6+}]$ ion complex was remarkably suppressed.

Meanwhile, the marked shift of the ν-SO_4^{2-} band to greater frequencies in MOF films indicates a distinctive solvation structure characterized by the high dependence of Zn^{2+} on anions, forming a much closer ion connection $[Zn^{2+}SO_4^{2-}]$. The Raman band (260 cm^{-1}) was assigned to a $[Zn^{2+}{\cdot}OSO_3^{2-}]$ ligand mode in the intron-pair structure of the internal sphere complexes ion–pair as the concentrations were increased. In the study of Zn–O mode vibration in the MOF pore, the amplification peak of the $[Zn^{2+}{\cdot}OSO_3^{2-}]$ ligand mode confirmed a more condensed complex ion, which seemed impossible in $ZnSO_4$ solutions with limited solubility. In addition, they further found that the intensity of O–H stretch vibration was significantly inhibited, confirming a small minority of water in MOF pores. The abovementioned results indicated that by removing water molecules as effectively as possible inside the MOF passages, a supersaturated electrolyte was formed, which was beneficial to the uniform Zn deposition on anode.

Similarly, Raman spectroscopy was performed by Zhang group to study the evolution of the H-bonds and the formation of the solvation configurations in the electrolyte with increasing $ZnCl_2$ concentration [43]. In Figure 8.8B, O–H strain oscillations of water molecules (3000–3700 cm^{-1}) were observed with an obvious wide peak, which tended to convolve with three components. With $ZnCl_2$, the Raman peak to frequency was narrowed to the high frequency, showing the reduced intensity of H-bond proportion in the solution. With the increase of $ZnCl_2$

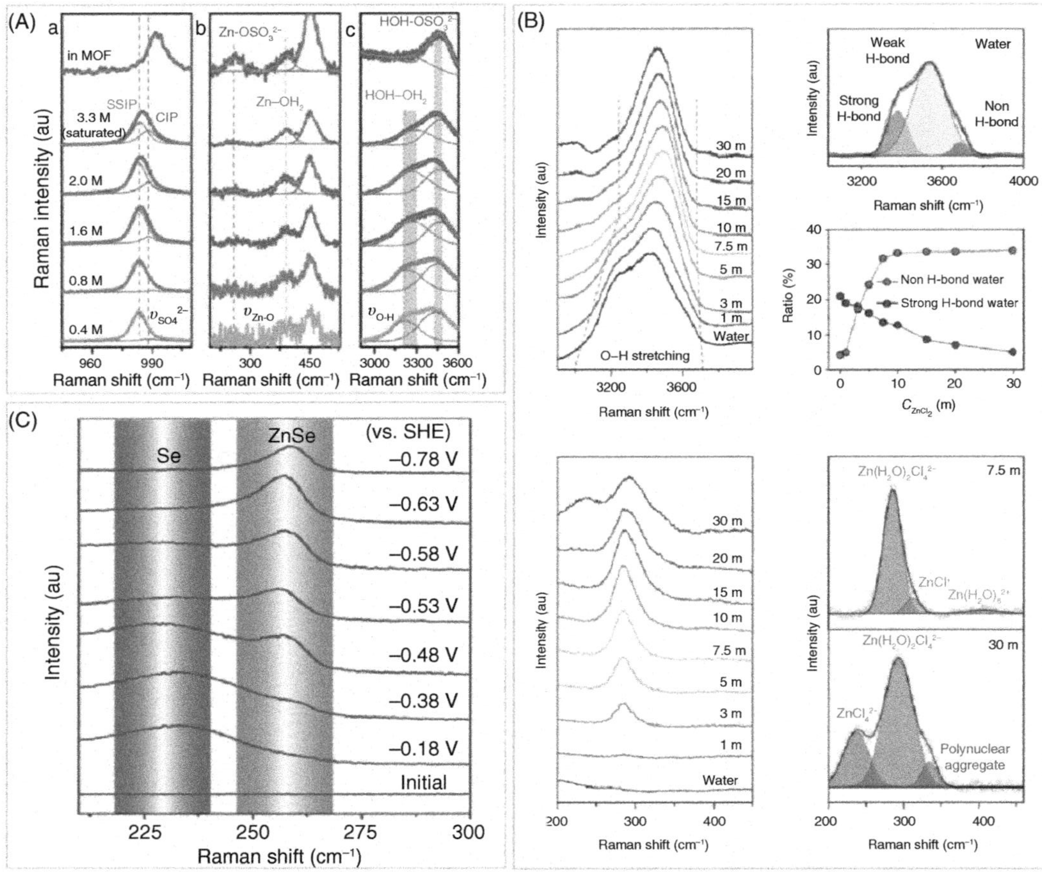

Figure 8.8 (A) Raman spectroscopy of concentration gradient $ZnSO_4$ aqueous solutions and in MOF channels: (a) ν-$SO_4{}^{2-}$ band; (b) ZnO stretch; (c) OH stretch vibration; Copyright 2020, Wiley-VCH. (B) The electrolyte structures including H–bonds and ion interactions; Copyright 2020, Nature. (C) Raman signals of products deposited on copper foil at different potentials. Source: (A) Yang et al. [42]/John Wiley & Sons. (B) Zhang et al. [43]/Springer Nature/Licensed under CC BY 4.0. (C) [44]; Copyright 2022, Wiley-VCH.

concentration, the strong H-bond water decreased, and the non-H-bond water increased. This indicates that the H-bond structure was gradually destroyed with the increase in salt concentration.

Meanwhile, they also found that the solvation configuration of Zn^{2+} in the various $ZnCl_2$ concentration electrolytes was unveiled by Raman spectroscopy in the region of 200–450 cm^{-1}. After introducing $ZnCl_2$, a new peak appeared at 284 cm^{-1}, and the solvated Zn^{2+} configuration was formed with the increase in the $ZnCl_2$ concentration. When the concentration further increased to 30 m (mol kg^{-1}), the initial peaks became two peaks, proving the new coordination mode of Zn^{2+}. Therefore, the in-depth study on the solvation mechanism of electrolytes is helpful to optimize the electrolyte and improve the circulation stability of AZIB.

In addition to the study about the solvation of electrolytes in AZIB, Raman spectroscopy can also be used to verify the deposition behavior of Zn or Zn alloys on the electrode surface. A recent study found that ZnSe coating can effectively inhibit the side reaction, promote efficient dissolution, and accelerate the deposition kinetics of Zn^{2+} on the Zn anode [44]. Zhe et al. used Raman spectroscopy to precisely monitor the electrodeposition of ZnSe on copper foil surface at varied potentiostatic levels ranging from −0.18 to −0.78 V, which can obtain the key voltage parameters of steady deposition of ZnSe on Zn foil surface [45]. As shown in Figure 8.8C, as the voltage decreased from −0.18 to −0.78 V, the Raman peak of Se at 230 cm^{-1} gradually changed to that of ZnSe at 256 cm^{-1}. They further found that ZnSe depositions could be triggered simultaneously at the voltage of −0.78 V, and the characteristic peak intensity of ZnSe in the Raman spectra was slightly lower than that at −0.63 V, which indicated that ZnSe can be codeposited at voltages lower than −0.63 V.

8.1.2.3 Infrared (IR) Spectroscopy

IR analysis refers to the analysis and identification of substance molecules by means of IR. A beam of IR rays of different wavelengths is irradiated onto the molecule of a substance, and some specific wavelengths of IR rays are absorbed to form the IR absorption spectrum of this molecule. Each molecule has a unique IR absorption spectrum determined by its composition and structure, from which the molecular structure can be analyzed and identified. IR absorption spectrum is produced by the continuous vibration and rotational motion of molecules. Molecular vibration refers to the relative motion of each atom in the molecule near the equilibrium position. Multi-atomic molecules can form a variety of vibration patterns. The IR spectrometer is divided into prism and grating spectrometer, and Fourier transform infrared spectrometer (FTIR). Therein, FTIR spectrometer, based on the principle of the Fourier transform of IR interference spectrum, can be collected in four modes: transmittance, reflectance, attenuated total reflectance (ATR), and diffuse reflectance. FTIR can be used to study the structure and chemical bonds of molecules and can also be used as an experimental technique to characterize and identify chemical species; meanwhile, the IR spectrum is highly characteristic and can be analyzed and identified by comparing the IR spectrum with that of standard compounds; in addition, the characteristic wave number of chemical bond is used to identify the type of compound and can be used for quantitative determination. Due to these characteristics, FTIR is widely used in the research of configuration, conformation, and mechanical properties of polymers, as well as in the fields of physics, astronomy, meteorology, remote sensing, biology, and medicine [46]. Currently, in the field of AZIBs, FTIR technique is mainly applied to study the effect of additives on the solvation structure of electrolytes so as to provide effective information for electrolyte optimization [47]. Sun et al. found that the addition of glucose has the function of weakening the solvation interaction of Zn^{2+} with water through FTIR [48]. As shown in Figure 8.9a, three peaks at near 1078, 1031.7, and 989.3 cm^{-1} were ascribed to C—O stretching from glucose molecules [50]. The introduction of glucose with varying concentrations resulted in a significant blue shift in the stretching vibration of SO_4^{2-} at 1085.7 cm^{-1} in pure $ZnSO_4$ electrolyte, consistent with the peak position observed for $ZnSO_4$ powder. These results indicated that the restriction in SO_4^{2-} region had been relieved, which was due to an impairment in electrostatic coupling between Zn^{2+} and SO_4^{2-} by introduction of glucose additive, thus further validating the regulated Zn^{2+} solvation shell [49], which was beneficial to inhibit the occurrence of side-reaction and guide the uniform deposition of zinc during charging/discharging process in AZIBs.

Similarly, FTIR was conducted by Ming et al. to probe the solvation structure evolution of the hybrid electrolytes (i.e. zinc trifluoromethanesulfonate ($Zn(OTf)_2$) + propylene carbonate (PC)/water mixtures). In Figure 8.9b, the broad peak range of 1040 to 1070 cm^{-1} was attributable to the partial and complete coordination of OTf^- anions, which demonstrated that there was no peak in this region in the pure aqueous solution. However, with the increase in PC content, the broad peak gradually enhanced and redshifted, which may result from the interaction between OTf^- anions and PC or Zn^{2+} ions. They also observed a similar tendency in the vibrations of $—CF_3$ and $—SO_3^-$. These results indicated that the presence of PC can effectively regulate the coordination environment of the OTf^- anions, which can provide an effective salination-induced mixed electrolyte to induce uniform Zn deposition. The abovementioned two examples display that FTIR technique plays an important role in the solvation research of electrolytes for AZIBs. The abovementioned analysis demonstrated the FTIR as a powerful technique that can be used to study the solvation structure of electrolytes in AZIBs, thus providing reliable information for the optimization of electrolytes.

8.1.2.4 X-Ray Photoelectron Spectroscopy (XPS)

XPS is a quantitative energy spectroscopy technique used to determine the composition of elements in materials, as well as the chemical and electronic states of the elements contained therein. This technique uses X-rays to illuminate the material to be analyzed while measuring the kinetic energy and the amount of electrons escaping from the material's surface from 1 to 10 nm below, resulting in XPS. As an advanced technique, XPS possesses the following characteristics: (i) Sensitivity to all elements of the same order of magnitude; (ii) easy to observe chemical shift information; (iii) Qualitative or semi-quantitative identification of each element; (iv) the sample damage is small; (v) the surface is sensitive, and the surface

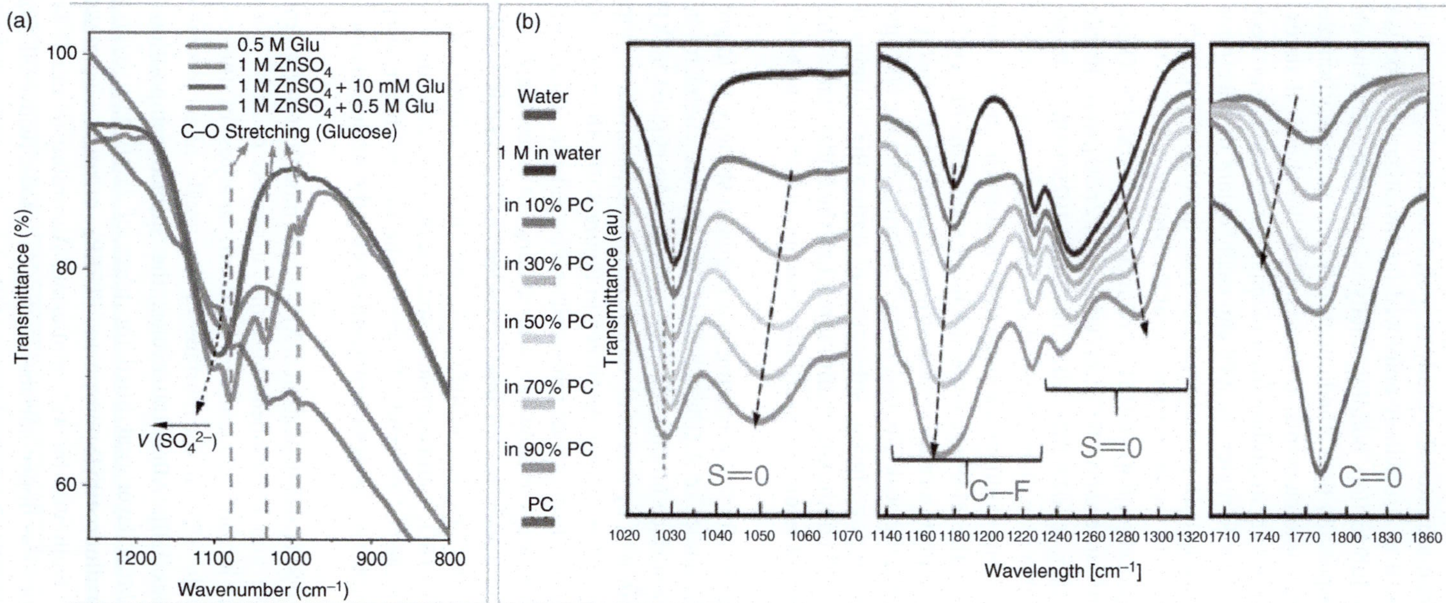

Figure 8.9 (a) FTIR spectra for 1 M $ZnSO_4$, 1 M $ZnSO_4$ with 10 mM glucose, 1 M $ZnSO_4$ with 0.5 M glucose and 0.5 M glucose; Copyright 2021, Wiley-VCH. (b) FTIR spectra of pure water, PC, and various electrolytes with 1 M concentration. Source: (a) Sun et al. [48]/John Wiley & Sons. (b) Hao et al. [49]; Copyright 2022, American Chemical Society.

analysis depth is about 2–5 nm. This technique finds extensive applications in the fields of metals, inorganic materials, catalysts, polymers, coating materials, ores, and other substances. It is also widely employed for investigating processes such as corrosion, friction, lubrication, bonding, catalysis, and oxidation. The electrode materials of AZIBs, especially the cathode materials, are commonly complex compounds containing multiple elements. Understanding the chemical states of elements in raw materials is very important for understanding the reaction mechanism and designing new electrode materials. XPS can easily obtain the electronic states of positive and negative ions in the original electrode material. First, XPS is used to study the role of Zn^{2+} in the energy storage of Zn/NaV_3O_8 (Zn/NVO) system in the selected charging/discharging state [51]. In the Zn 2p spectra of NVO electrodes (Figure 8.10a), the Zn^{2+} inserted in NVO, $Zn(OH)_2$, and $ZnSO_4$ were distributed in three pairs of Zn^{2+} peaks at 1023/1046, 1024/1047, and 1025/1048 eV, respectively. The results displayed that a small quantity of the inserted Zn^{2+} in NVO was detected at the initial state of second cycling (discharge state: line a to h; charge state: line h to a). The XPS spectra in the selected charge–discharge state of the first cycle further confirmed that the partially inserted Zn^{2+} was still not extracted from the NVO. However, in the second cycle, the peak intensity of Zn^{2+} inserted into the NVO gradually increased during discharging process; subsequently, during the charging process, the peak value decreased continuously until it reached the initial state, which suggested the continuous and reversible insertion/extraction of Zn^{2+} in the NVO during cycling.

Meanwhile, surface modification/coating is a common strategy to improve the electrochemical performance of AZIBs. The surface interface chemistry between the electrode and the surface modification layer can also be studied by XPS [52]. For instance, XPS was used by Jiao et al. to investigate the passivation property of Zn anode with polymer glue coating. As shown in Figure 8.10B, the composition of the polymer glue–protected zinc anode remained unchanged after immersing in the aqueous electrolyte for 7 days; meanwhile, after cycling six times, the Zn composition of the polymer glue–protected Zn anode remained as Zn^0, and it was the same for O and S composition. In contrast, the cyclic bare Zn anode showed Zn^{II} peak, with O corresponding to SO_4^{2-} and complex S species, indicating the anode surface passivation of the unprotected Zn anode. Therefore, the XPS technique can provide crucial reference information for optimizing and treating the surface of zinc anodes in AZIBs.

In addition to the large-scale application of XPS for the modification/optimization of zinc anode surface and the embedding/removal process of Zn^{2+} in cathode materials in AIZBs, it can also conduct in-depth analysis of the compositions of self-repairable SEI on the surface of Zn anode after several charging/discharging cycles [30]. After 100 cycles, the surface composition of Zn anode with a thin MOF layer that filled the pores of MOF with hydrophobic $Zn(TFSI)_2$-tris(2,2,2-trifluoroethyl)-phosphate (TFEP) organic electrolyte was analyzed using XPS facilitated by Ar^+ sputtering. According to the F 1s spectrum in Figure 8.10c, the upper surface of the Zn anode recovered from the $Zn(TFSI)_2$/TFEP@MOF/H_2O electrolyte contained organic CF_3 (94.7%) and

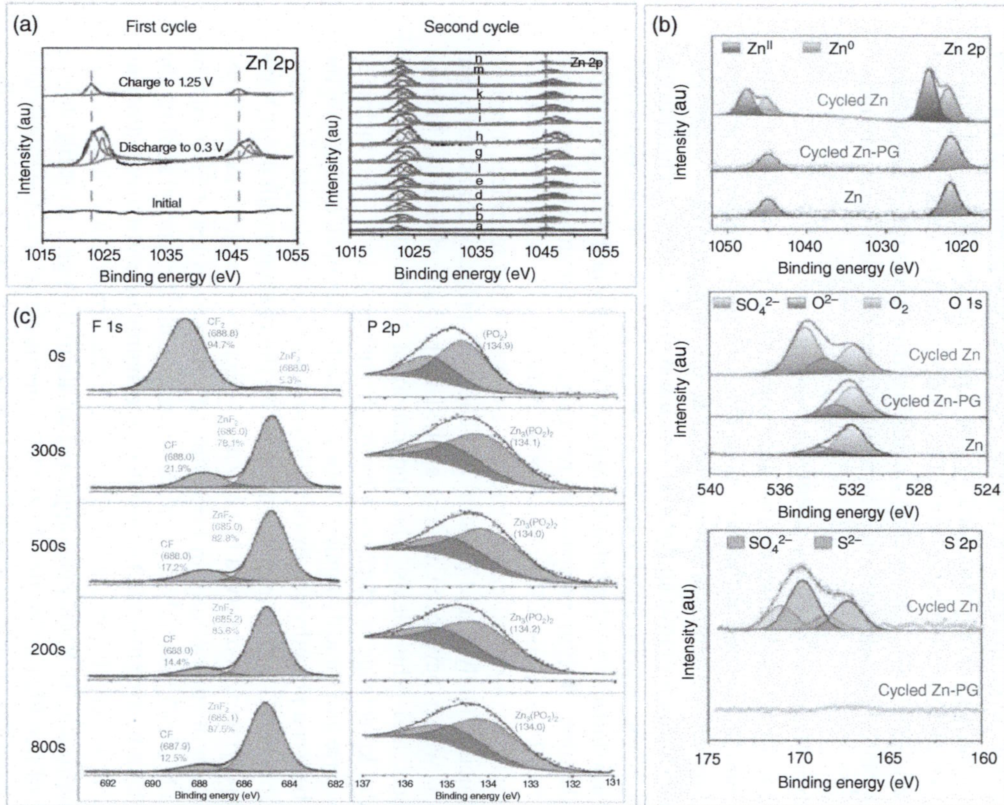

Figure 8.10 (a) Zn 2p and V 2p at selected states; Copyright 2018, Nature. (b) Zn 2p, O 1s, and S 2p X-ray photoelectron spectroscopy curves of pristine Zn foil and cycled Zn anodes, respectively. (c) XPS features of the SEI on Zn after 100 charge/discharge cycles in $Zn(TFSI)_2$-TFEP@MOF/H_2O electrolyte. The F 1s and P 2p spectra are displayed in rows, with corresponding Ar^+ sputtering depth profiling results in columns. Source: (a) Wan et al. [51]/Springer Nature/Licensed under CC BY 4.0. (b) Wang et al. [52]; Copyright 2021, Wiley-VCH. (c) Qiu et al. [30]; Copyright 2020, Wiley-VCH.

inorganic ZnF_2 (5.3%). After 300 seconds of sputtering on the surface of the recovered $Zn(TFSI)_2$/TFEP@MOF/H_2O electrolyte, the peak of CF_3 disappeared; however, a new peak appeared at 688.0 eV, corresponding to the CF produced by the reduction of $Zn(TFSI)_2$-TFEP. At further sputtering to 1800 seconds, the inorganic ZnF_2 content increased from 78.1% at 300 s to 87.5%. In addition, for P 2p spectrum, the upper surface of $Zn(TFSI)_2$-TFEP@MOF/H_2O electrolyte recovered mainly contains organic PO_3, which may be the incomplete reduction product of TFEP. After 300 seconds of sputtering, the organic PO_3 peak disappeared, and a new inorganic $Zn_3(PO_4)_2$ peak appeared at 134.1 eV. The peak $Zn_3(PO_4)_2$ still existed when further sputtering occurred in the 1800s. The XPS results indicated that the dense ZnF_2–$Zn_3(PO_4)_2$ SEI can separate Zn surface from trace water dissolved in the $Zn(TFSI)_2$–TFEP organic electrolyte layer but allows Zn^{2+} to diffuse through.

8.1.2.5 Nuclear Magnetic Resonance (NMR) Spectroscopy

NMR spectroscopy is used to study the absorption of nuclear nuclei to radio-frequency radiation. It is one of the most powerful tools for qualitative analysis of the composition and structure of various organic and inorganic substances, and sometimes also for quantitative analysis. In organic synthesis, NMR technique can not only analyze the structure and determine the configuration of the reactants or products but also has a wide range of applications in the study of the charge distribution and localization effect in the synthesis reaction and exploration of the reaction mechanism. NMR techniquc, which can provide molecular chemical structure and molecular dynamics information, has become a conventional technical means of molecular structure analysis and physical and chemical property characterization of substances, which have been widely applied in physics, chemistry, biology, medicine, food, and other fields. The application of NMR spectroscopy vastly accelerates the knowledge of electrode/electrolyte materials in AZIBs, including the local structure and ionic dynamics. NMR was used by Jin et al. to elucidate the interactions between polyethylene oxide (PEO) and aqueous electrolyte species on the molecular level [53]. As shown in Figure 8.11a, the chemical shift of ^{1}H NMR peak of PEO polymer was observed in 1 M $ZnSO_4$ electrolyte, which was attributed to the interaction of ether oxygen in PEO with Zn^{2+} cations. A lower PEO content (0.1 wt%) in 1 M $ZnSO_4$ electrolytes produced a larger upfield chemical shift for the ^{1}H NMR peak, indicating that PEO molecules were mainly involved in coordinating Zn^{2+} cations at lower PEO content. When PEO concentration was increased to 1 wt%, an additional ^{1}H NMR peak appeared at about 3.7 ppm, probably due to the presence of free PEO molecules. In $ZnSO_4$ aqueous electrolyte, the interaction between PEO and Zn^{2+} ions can effectively slow down the reaction kinetics of Zn^{2+} ions, inhibit the electrolyte flux, and make Zn^{2+} more evenly distributed at the Zn anode-electrolyte interface, resulting in stable interface reaction and stable Zn deposition.

Similarly, liquid-state NMR spectroscopy was employed by Lv et al. to investigate the interaction among Zn^{2+}, 3-(1-methylimidazole) propanesulfonate (ImS), and water in the ImS/2 M $ZnSO_4$ electrolyte (ImS/ZSO) [54]. As shown in Figure 8.11B, in the electrolyte, the proton signals of all ImS showed gradually increased field

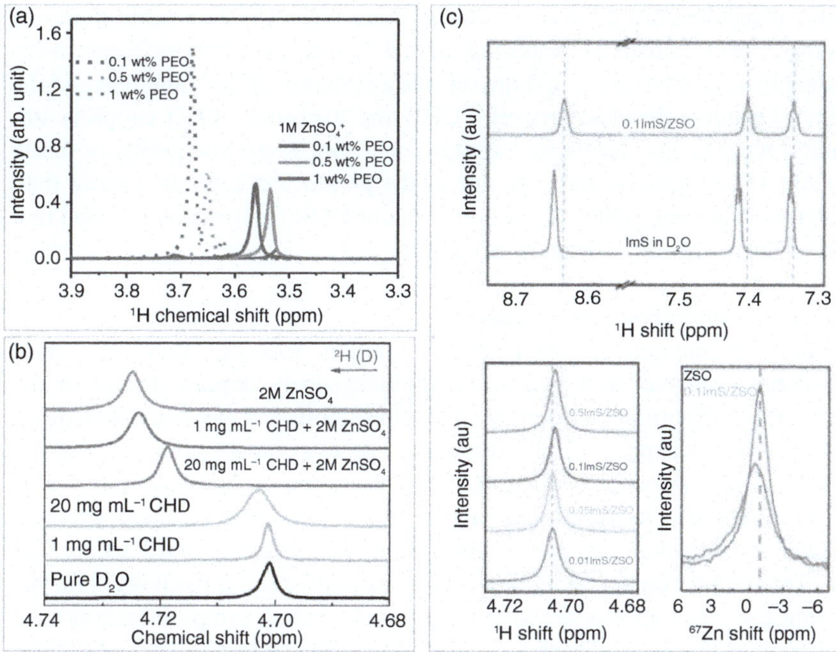

Figure 8.11 (a) Characterizations of 1 m $ZnSO_4$ + PEO aqueous electrolytes; ^{1}H NMR spectra of ether groups (O—CH_2) of PEO molecules in H_2O and 1 M $ZnSO_4$ electrolytes with different PEO contents. PEO molecules show interactions with H_2O molecules and Zn^{2+} ions; Copyright 2020, Wiley-VCH. (b) H-NMR spectra of H_2O from pure D_2O, 1 mg mL^{-1} CHD and 20 mg mL^{-1} CHD, 2 M $ZnSO_4$, 1 mg mL^{-1} CHD+2 M $ZnSO_4$, and 10 mg mL^{-1} CHD+2 M $ZnSO_4$ in D_2O. (c) Liquid-state NMR spectra of ^{1}H and 67Zn in different electrolytes. Source: (a) Jin et al. [53]/John Wiley & Sons. (b) Lv et al. [54]; Copyright 2022, Springer. (c) Zhou et al. [55]; Copyright 2022, Royal Society of Chemistry.

shift relative to the internal standard, which is due to the electron absorption effect caused by the interaction between —SO_3^- group and Zn^{2+}. With the increase of ImS concentration, the ^{1}H chemical shift field in H_2O becomes larger, revealing that the change of water molecular chemical environment leads to the decrease of electron cloud density; meanwhile, the 65 Zn chemical shift in ZSO and ImS/ZSO solution displayed a downfield shift, indicating an intensification of the shielding effect on Zn^{2+} via possible formation of —SO^3–Zn^{2+} pairing, which signified the coupling between —SO^3 and Zn^{2+}. These NMR results indicated that the strong interaction of ImS with Zn^{2+} enabled it to alter the distribution and deposition path way of Zn^{2+} in the electric double layer.

NMR spectroscopy, as a significant spectroscopic technique, can also be utilized to investigate the operational mechanisms of organic additive-assisted electrolyte solvation structures for AZIBs. [55]. Specifically, Wu et al. used NMR spectroscopy to interpret the working mechanism of cyclohexanedodecol (CHD)-assisted electrolyte structure (i.e. $[Zn(H_2O)_5(CHD)]^{2+}$) and monitor the change of coordinated H_2O molecules. As shown in Figure 8.11c, the ^{2}H peak shifted from 4.7009 ppm (pure D_2O) to 4.7259 ppm (2 M $ZnSO_4$) due to the reduced electron density around the water molecules during solvation. Furthermore, when CHD concentrations of 1 and 20 mg L^{-1} were added, the peaks moved to 4.7238 and 4.7188 ppm, respectively, suggesting that the CHD regulated the solvation structure of Zn^{2+} and released the coordinated water molecules. 1 mg mL^{-1} CHD and 20 mg mL^{-1} CHD samples without $ZnSO_4$ were further tested for better comparison, where the peaks moved to 4.7012 and 4.7029 ppm, respectively, due to the presence of more free water than electrolytes containing $ZnSO_4$. The observation of NMR results demonstrated that CHD can change the Zn^{2+} solvation structure.

8.1.2.6 X-Ray Absorption Spectroscopy

With the construction of synchrotron radiation sources, XAS has made unprecedented progress and is playing an increasingly important role in the characterization of substance structure (including atomic and electronic structures) and physicochemical property interpretation (such as single-atom catalyst site study and in situ/operando testing). XAS has been one of the most widely used synchrotron radiation techniques among various characterization methods based on synchrotron radiation. Because the transmitted light intensity of this technology is related to atomic number and atomic mass, qualitative and quantitative analysis of relevant test elements can be carried out for all kinds of samples, such as solid (crystal or amorphous), liquid, and gas. Techniques related to XAS are called X-ray absorption fine structure (XAFS) or extended XAFS (EXAFS). Meanwhile, the X-ray absorption spectrum segment has a steep rise near the electron excitation of the shell of the target atom, which is often called X-ray absorption near-edge structures (XANES). In the cathode materials for AZIBs, XAS is usually used to determine the valence states and coordination of various elements and their changes during the electrochemical reaction processes [56]. Zhang et al. used XAS to investigate the evolution of $ZnMn_2O_4$ spinel cathode during Zn insertion/extraction process [57]. Their results (Figure 8.12A) revealed that Mn L-edge XAS gave a direct probe of 3d unoccupied states of Mn through dipole–allowed 2p-to-3d transitions. In the L_3 region, peaks at 640.5 and 642.9 eV are considered the fingerprint of Mn^{4+}, and peaks at 639.7 and 641.7 eV correspond to Mn^{3+}'s oxidation state [59]. The gradual appearance and disappearance of these peaks clearly illustrate the electron transfer in the Mn^{3+}/Mn^{4+} redox coupling during charge and discharge.

Similarly, XAS technique was used by Kumar and his group to study the coordination environments of Fe and V centers of the amorphous-$FeVO_4$ in AZIB [58]. Figure 8.12B displays the XANES spectra of Fe and V K-edges, which offer insights into their local coordination environments and electronic structures. On discharge, peak C shifts to a lower energy value and is recovered on charge. This 0.8 eV energy transfer indicates the reduction of Fe^{3+}. It is important to note that part of this reduction may occur overall, as full reduction of Fe^{3+} to Fe^{2+} is expected to transfer 4 eV downward. The peak B intensity decreases irreversibly upon discharge, indicating that there is no occupied nearest-neighbor 3D state of the iron atom. For V–K-edge, the XANES spectrum of the original sample showed a pre-edge feature, indicating the presence of a tetrahedral environment. The results show that the front peak intensity decreased during discharge, and the rising edge

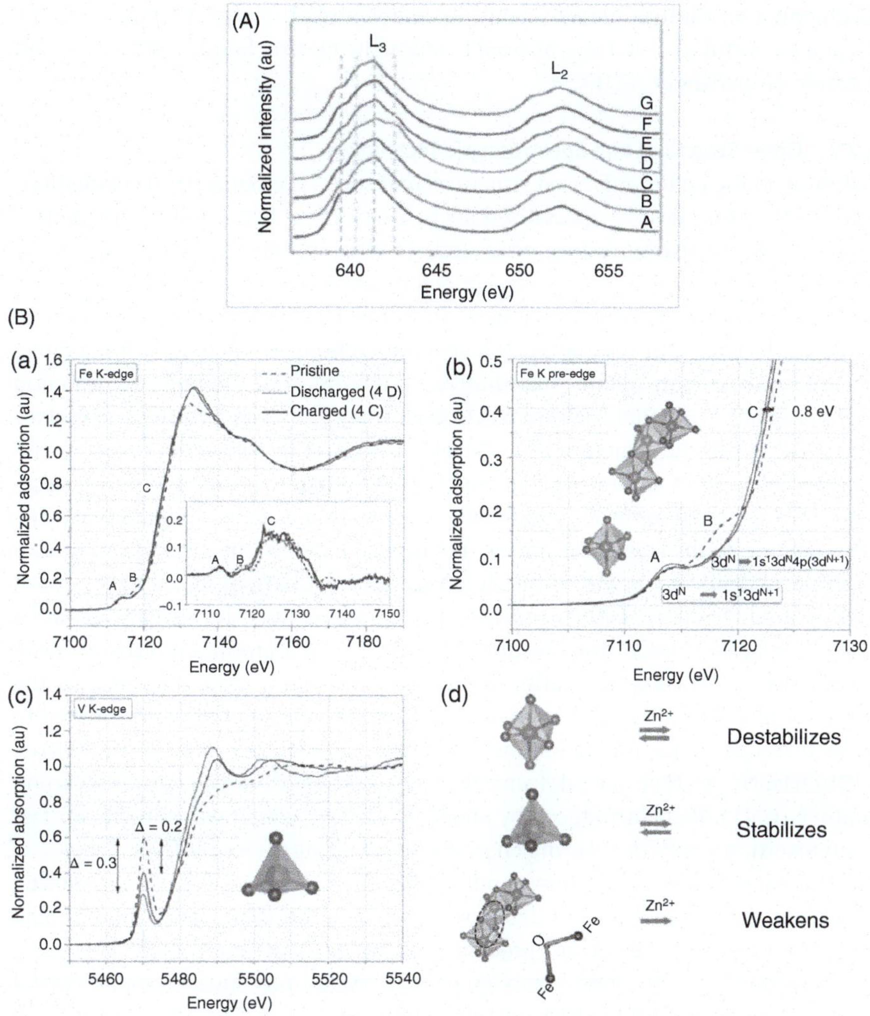

Figure 8.12 (A) SXAS of Mn L-edge spectra. (B) Local structural study of the pristine, discharged, and charged electrodes: (a) Background subtracted and normalized XANES spectra for the Fe K-edge (b) Zoomed in XANES spectra for the Fe K-edge showing the change in the pre-edge intensities and the shift in rising edge upon cycling, (c) background subtracted and normalized XANES spectra for the VK-edge showing the change in the pre-edge intensity and the shift in rising edge upon cycling, and (d) summary of local structural changes upon Zn. Source: (A) Zhang et al. [57]; Copyright 2016, American Chemical Society. (C) Kumar et al. [58]/John Wiley & Sons. (D) Kumar et al. [58]; Copyright 2020, Wiley-VCH.

reversibly transferred to the lower energy, which indicated that the oxidation state of V decreased during the discharge process. The local structure of Fe becomes unstable due to the loss of inversion symmetry when zinc ions enter; the V becomes more symmetrical and tends to be stable, which suggests that the $FeVO_4$ is modified upon discharging and recovered upon subsequent charging. The analysis of the abovementioned cathode materials during charge and discharge demonstrated the XAS as a powerful tool that can be used to study the electrochemical structural and electronic properties of AZIBs.

8.1.2.7 Other Structural and Spectroscopic Techniques

In addition to the commonly used structural and spectral characterization methods mentioned above, several specialized techniques have been applied to AZIBs, including electron paramagnetic resonance (EPR) [30, 60], 2D wide-angle X-ray scattering (2D-WAXS) [61, 62], electrospray ionization mass spectroscopy (ESI-MS) [63, 64], are widely utilized in LIBs and sodium-ion batteries. For instance, EPR, also known as electron spin resonance, is a spectral technique employed for studying. EPR spectroscopy can obtain quantitative, structural, and dynamic information about materials. It is widely used in electrochemistry due to its sensitivity to surface defects and electronic structure of materials, as well as its selectivity for paramagnetic species. Guo et al. developed a highly self-adaptable poly(dimethylsiloxane) (PDMS)/TiO_{2-x} coating on Zn anode, which possessed excellent conductivity due to the coating with oxygen vacancies, resulting in the volume stabilization and inhibition of dendrite growth of the Zn anode during charging/discharging cycles in AZIBs [65]. They used EPR spectroscopy to characterize the oxygen vacancies in TiO_{2-x}. As can be seen from Figure 8.13a, there was a prominent vacancy peak with a g value of 2.003, indicating a large number of oxygen vacancies in the coating. Similarly, EPR spectroscopy in Figure 8.13b by Pan et al. was recorded to deliver direct evidence for the abundant oxygen deficiencies (g-factor of 1.98) of the DVOH@NC@CNTF (i.e., defect-rich $V_2O_{5-x}\cdot nH_2O$@N-doped carbon@carbon nanotube (CNT) fiber) introduced by anodic oxidation, which was suitable for fast Zn-ion insertion/extraction during charging/discharging cycles [66].

In XRD experiment, the XRD method is called 2D-WAXS, in which the scattering of θ greater than 5° and the 2D detector are used at the same time, and the data recorded by the 2D detector are processed, analyzed, and interpreted. This technique is widely used in phase analysis, crystal crystallinity measurement, crystal orientation analysis, film thickness measurement, and so on. For AZIBs, Zhao et al. applied 2D-WAXS to investigate the crystal structure of deposited Zn on bare Zn foil surface and Zn surface modified by FCOF film during charging process (i.e. 2D porous covalent organic framework) [28]. The 2D-WAXS patterns of deposited Zn underneath FCOF film displayed some strong, discrete diffraction spots in the ring plane, while for bare Zn deposits, the results were continuous diffraction rings (Figure 8.13c). These results showed that the bare zinc deposit was polycrystalline and nondirectional, and the zinc grain size affected by FCOF film was large and directional [67].

Electrospray ion source (ESI) is the most commonly used interface for liquid mass spectrometry (MS), which can be used to study thermally unstable and polar compounds. In particular, ESI-MS was used to characterize the solvation species in the electrolyte solutions [68]. In Figure 8.13D, the ESI-MS spectra of $Zn(TFSI)_2$/EMC-TMP were nearly indistinguishable from those of $Zn(TFSI)_2$/TMP, exhibiting prominent peaks corresponding to $[ZnTFSI(TMP)_2]^+$, $Zn(TFSI)_2(TMP)$,

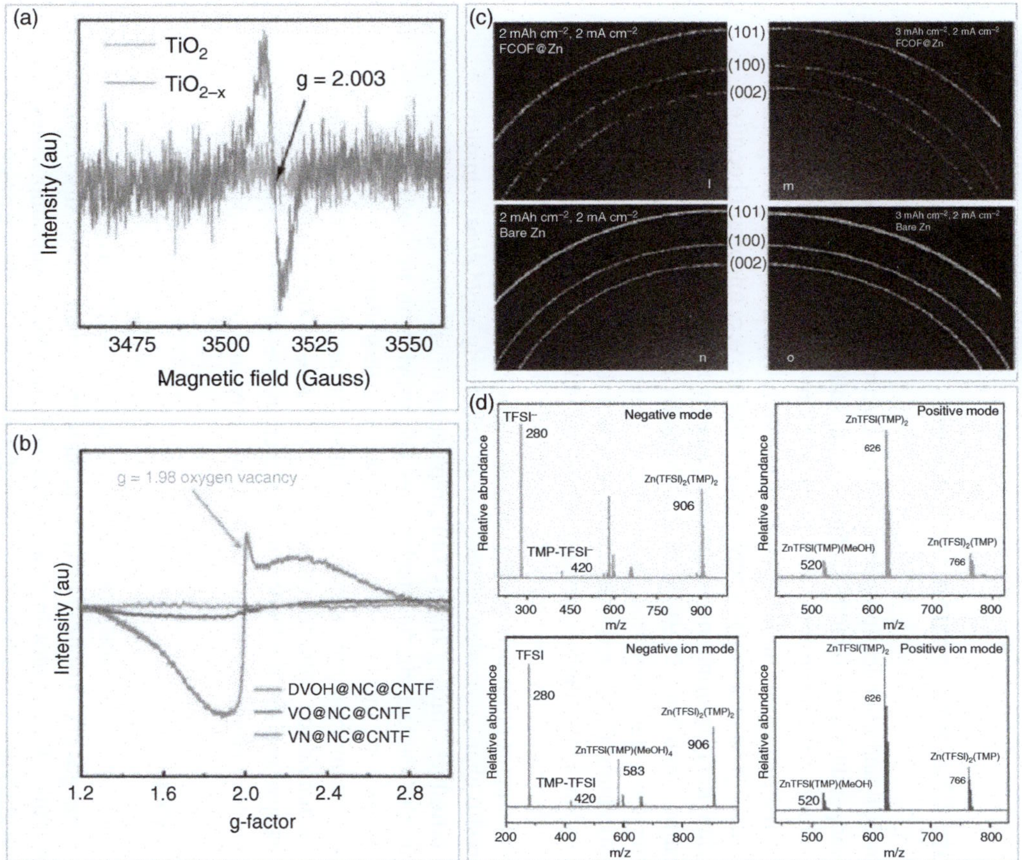

Figure 8.13 (a) EPR to measure oxygen vacancies. (b) EPR spectra of VN@NC@CNTF, DVOH@NC@CNTF, VO@NC@CNTF. (c) The WAXS results of Zn deposits underneath FCOF film and on bare Ti. (d) Electrospray ionization mass spectroscopy of 1.5 M $Zn(TFSI)_2$/EMC−TMP (1:3 by volume) electrolyte solution in negative ion mode and positive ion mode; Electrospray ionization mass spectroscopy of 1.5 M $Zn(TFSI)_2$/TMP electrolyte solution in negative ion mode and positive ion mode. Source: (a) Guo et al. [65]/John Wiley & Sons. (b) [66]; Copyright 2022, Royal Society of Chemistry. (c) Zheng et al. [67]; Copyright 2021, Nature. (d) [68]; Copyright 2020, Wiley-VCH.

and $Zn(TFSI)_2(TMP)_2$ over a wide mass-to-charge range. These results suggest that significant amounts of ion pairs and aggregates involving TMP are present. In addition, they further found that no peaks of EMC solvation EMC-$TFSi^-$ or Zn^{2+} were detected, indicating that EMC molecules were less likely to participate in $TFSI^-$ or Zn^{2+} solvation in the presence of TMP. It should be noted that the strength of TMP–$TFSi^-$ is much lower than that of Zn^{2+} solvated with TMP, possibly due to the weak interaction between $TFSI^-$ and TMP.

8.2 In Situ Characterization Techniques

Real-time physical and chemical properties of the electrodes and interface, especially their changes during the operation of the Zn-ion batteries, are quite important for the full picture of the electrochemical process. Many in situ tools and experiments have been designed and applied to battery research, especially using designed cells and systems. These experiments performed in situ are very important because they can provide timely information on the changes in material and composition at or near operating conditions. Recently, in situ characterization techniques during battery cycling have developed rapidly, enabling them to successfully capture real-time information. Compared with ex situ characterization, in situ characterization can display real battery-operating conditions and provide more reliable and accurate information by continuously monitoring the evolution of electrode structures and complex physicochemical processes throughout the electrochemical cycles. The in situ characterization technique can effectively avoid contamination that may be introduced during the ex situ preparation of samples. Due to these advantages, in situ characterization techniques can provide an in-depth understanding of the working mechanisms, the performance degradation process, and so on. Recently, many researchers have developed various in situ characterization techniques for Zn-ion batteries (such as in situ XRD, in situ Raman, in situ optical microscopy, and in situ FTIR). In this section, we present these in situ characterization techniques, including typical application and operation principles.

8.2.1 In Situ FTIR

As FTIR spectroscopy is a surface-sensitive technology, in situ FTIR can be used to detect surface changes in electrode or electrolyte materials during charge/discharge processes. To study the changes in a separator with functional supramolecules (GF@SM)- and glass microfiber (GF)-based cells during charge/discharge processes (Figure 8.14), in situ FTIR was conducted by Liu et al. [69]. The FTIR spectrum of GF@SM-based cell maintains a similar trend as Zn deposited on the Cu grid (Figure 8.14a), indicating that the internal state of Zn metal was stable and undisturbed, which may be due to a large number of zincophilic sites in GF@SM that promoted the uniformity of Zn-ion flux, avoided the change in electrolyte concentration, and ensured the stability of Zn deposition. However, for GF-based cells (Figure 8.14b), the peak strength located at 1080 cm^{-2} (SO_4^{2-} bond) and

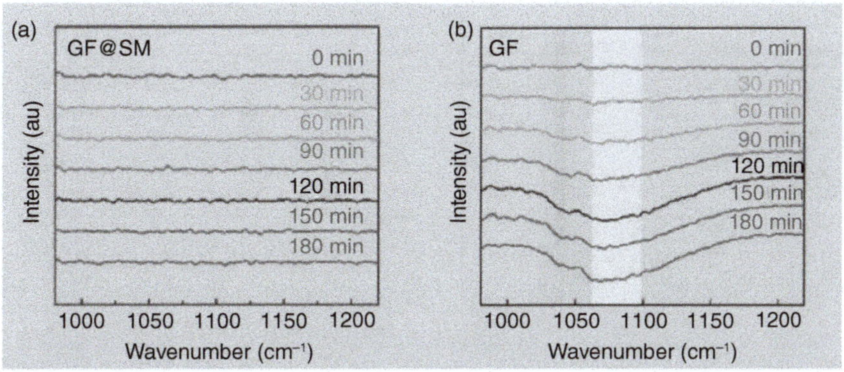

Figure 8.14 FTIR spectra at various plating stages of (a) GF@SM-based and (b) GF-based cells. Source: Liu et al. [69]/with permission of Elsevier.

1030 cm^{-2} (Si—O bond) increased gradually with Zn-plating time, which may be due to the side effect of the hydrogen evolution reaction, leading to the increase in SO_4^{2-} concentration in the electrolyte. The comparison between the in-situ FTIR spectra of GF@SM- and GF-based cells is helpful in verifying the dominant role of GF@SM in achieving dendrite-free deposition and inhibiting side reactions. Therefore, in-situ FTIR is a promising surface technique to research the changes in electrode or electrolyte materials in Zn-ion batteries during charge/discharge processes.

8.2.2 In Situ XRD

XRD is a very convenient and powerful method to characterize the structure and phase composition of crystal materials. In situ XRD technology can be used to detect the structural evolution and phase transformation of Zn-ion battery materials during charge and discharge processes. Compared with other in-situ characterization techniques, cells for in situ XRD are relatively simple to design and operate. In general, in situ XRD cells contain an X-ray window that allows X-rays to reach the electrode material and detector. Depending on experimental conditions, Beryllium foils and Kapton bands are commonly used as X-ray windows.

An excellent example can be observed in the research by Jing et al., where the in situ XRD patterns provide real-time information on structural evolution upon cycling [70]. As shown in Figure 8.15a, the XRD patterns of a self-supporting V_2O_5 nanosheet cathode were collected continuously during complete discharge/charge cycles. The remaining diffraction peak is thought to be the evolution of the crystal structure change of V_2O_5 during the discharge/charging process. They found that water molecules in the electrolyte were embedded into the V_2O_5 nanosheets under electrochemical action, and the V_2O_5 nanosheets underwent irreversible phase transformation to finally form the amorphous $V_2O_5{\cdot}nH_2O$ structure. The typical study can further validate the advantages of in situ XRD in providing more in-depth structural evolution and Zn-ion diffusion dynamics.

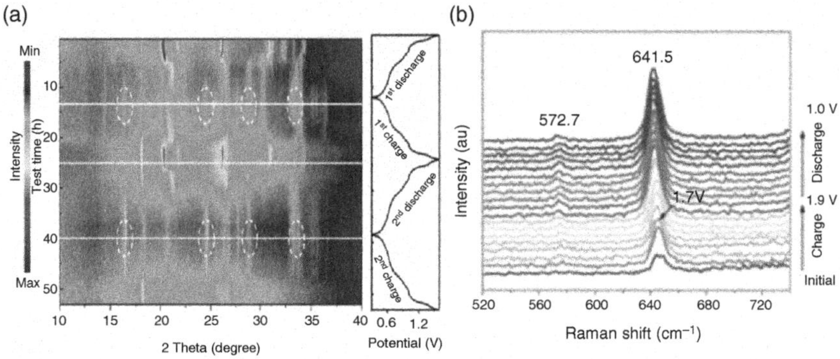

Figure 8.15 (a) 2D contour plots of the in-situ XRD patterns during two complete discharge/charge cycles. The marked series of diffraction peaks can be assigned to the $Zn_xV_2O_5 \cdot nH_2O$ phase. (b) In situ Raman spectra of the MnO−NC cathode during charge/discharge cycles. Source: (a) Jing et al. [70]/With permission of Elsevier. (b) Yang et al. [71]/With permission of Elsevier.

8.2.3 In Situ Raman

In situ Raman spectroscopy can be used to study the changes in structure and electronic properties during the battery cycle by detecting low-frequency transition patterns, such as laser-induced vibration and rotation in electrode materials. A special in situ electrochemical cell has been designed and used for in situ Raman measurements to study real-time information about electrode materials during the cell cycle [72]. Yang et al. explored the energy storage mechanism of Zn/MnO-nanoparticle system during charge/discharge processes through the device of the in situ Raman (results are shown in Figure 8.15b) [71]. They found that the MnO-nanoparticle cathode was partially oxidized to form layered MnO_2 during the first charging process, accelerating the insertion and extraction of Zn ions. A dissolution-redeposition reaction mechanism of the MnO_2 cathode is directly visualized in Zn-ion batteries via in situ Raman spectroscopy by Wu et al. [73]. In situ, Raman spectra of MnO_2 obtained at different voltages during charge and discharge provided valuable chemical information about MnO_2. These results displayed that MnO_2 was reduced to Mn^{3+} during the discharge process, followed by a disproportionation reaction to form Mn^{2+} and Mn^{4+}; meanwhile, the redeposition of Mn^{2+} formed a species with high Zn-content on the surface of the MnO_2 cathode in the high-potential window during the charging process. These two examples indicate that in situ Raman spectroscopy is a powerful tool to probe the subtle changes in electrode structures during the charge/discharge processes in Zn–ion battery.

8.2.4 In Situ AFM

Although in situ spectroscopy techniques can be used to study the structural changes of electrode materials and electrolytes, they are unsuitable for studying the electrode-electrolyte interface behavior. In this regard, the in situ AFM technique is a powerful surface-sensitive method based on a sharp cantilever indentation that enables tracking of surface evolution by scanning the interface between the electrode and the liquid electrolyte. In situ AFM characterization can obtain direct information about the interface reaction process and morphology evolution. This is a realistic device that does not destroy the electrode materials. In particular, the in situ AFM method can be used to obtain the relationship between the evolution of SEI layer on the electrode surface and electrolyte composition/additives, electrode potential, etc. In addition, the in–situ AFM characterization technique based on nanoindentation method can also measure the mechanical properties of electrode materials in situ during the battery cycle. In situ, AFM experimental results demonstrated that polished Zn foils with a flattened initial surface induced a uniformly plated/stripped Zn structure, lowering the occurrence of dendrites and short-circuit failure in batteries [74]. Figure 8.16A displayed the in situ AFM images at different stages during the Zn-plating procedure for three polished Zn foils that were polished with P1200, P2500, and P4000 grind papers. After 1 minute of electroplating on the P1200 polished Zn foil, a large number of particles were formed throughout the surface, with individual particles and clusters ranging in size from 0.5 to 5 μm. As the flux to scratches increased, the larger particles grew faster, eventually causing the AFM image to capture jumpers and fail (after 10 minutes of plating). However, the nucleation uniformity of P2500- and P4000-polished zinc foils differed after 1 minute of electroplating. Fine zinc particles were evenly distributed throughout the surface of the P4000 polished zinc foil, with no obvious aggregation, but some particles were found to preferentially nucleate at the scratched edges. During subsequent plating steps, the particles grow almost uniformly. The in situ AFM results displayed that the smoother the Zn surface was, the lower the probability of Zn roughness and dendrite formation, suggesting electrochemical evidence.

Meanwhile, in situ AFM was employed by Wang et al. to directly observe the morphologies of Zn in the early plating stage, which can better present the growth process of Zn [75]. In Figure 8.16B, the morphologies of Zn deposited on substrate (i.e. highly oriented pyrolytic graphite [HOPG]) in both 0.5 and 0.1 M $ZnBr_2$ electrolyte displayed a hexagonal shape parallel to the substrate. According to the AFM images obtained in 0.5 M electrolyte after plating Zn for 150 seconds, Zn crystals exhibited a smoother hexagonal surface and a pyramidal appearance with sharp edges. However, the upper surface of galvanized coating was rough and undulating, and the edge of hexagonal coating was fuzzy after plating Zn for 150 seconds in 0.1 M electrolyte, indicating that there should be more defects on the surface. Compared with that in 0.5 M electrolyte, the middle crystal did not grow dramatically, which may be related to the competitive relationship between nucleation and growth. These results also indicated that the nucleation process should continue when there are new active nucleation sites. Because of the low nucleation and growth energy of lattice matching, zinc crystals grew layer by layer on the surface of the existing zinc crystal.

Meanwhile, it can be seen that the deposition height of Zn in the 0.5 M electrolyte is much higher than that of 0.1 M electrolyte. The height curve of the upper surface of the Zn crystal in the 0.1 M electrolyte has obvious drastic fluctuations,

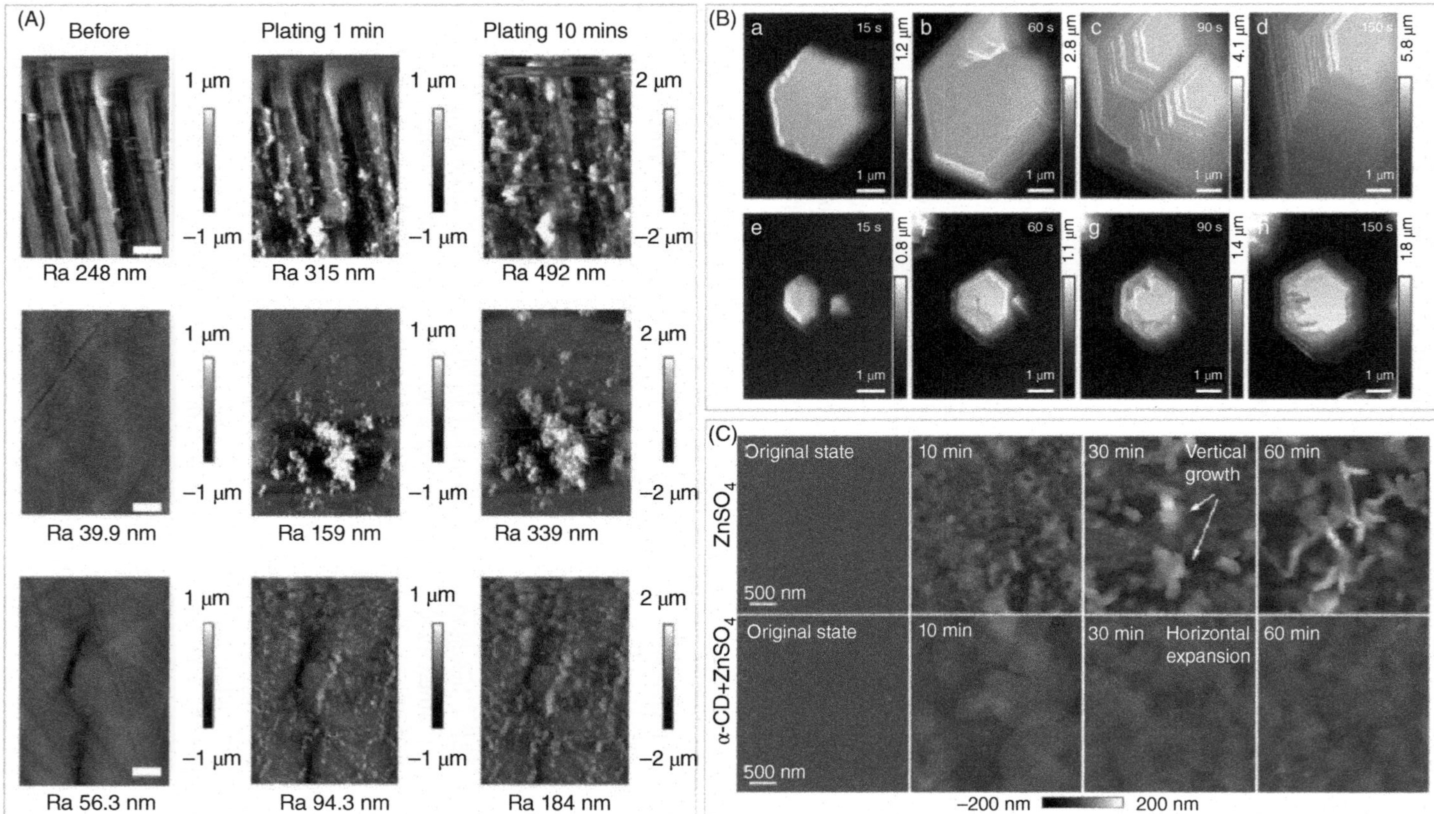

Figure 8.16 (A) In situ EC-AFM images of polished Zn foil electrodes before and after 1 min and 10 minutes of plating. The Zn foils were polished with P1200, P2500, and P4000 grind papers. Scale bars are 10 mm. (B) In situ AFM images of Zn electrodeposits on HOPG with a current of 10 mA cm^{-2} in (a–d) 0.5 and (e–h) 0.1 M Zn^{2+} electrolyte; Copyright 2021, Royal Society of Chemistry. (C) In-situ AFM images of the deposited Zn using $ZnSO_4$ electrolytes with/without α-CD at 40 μA cm^{-2}. Source: (A) Zhang et al. [74], © 2021/Royal Society of Chemistry/CC BY 3.0. (B) Wang et al. [75], © 2021/Royal Society of Chemistry. (C) Zhao et al. [76], © 2022/American Chemical Society.

further proving the phenomenon of Zn growth accompanied by nucleation in the low-concentration electrolyte. These results showed that the growth of 0.5 M electrolyte is much faster than that of 0.1 M in both parallel and perpendicular directions to the substrate. Similarly, the evolution of Zn deposition in $ZnSO_4$ electrolyte with/without cyclodextrin (α-CD) additive through in situ AFM technique by Zhao et al., as shown in Figure 8.16C [76]. In pure $ZnSO_4$ electrolyte, the initial Zn nucleation was heterogeneous and displayed a spotty distribution at a current density of 40 μA cm^{-2}. However, in the electrolyte containing α-CD, the plane–distributed Zn nuclei extended and grew horizontally, resulting in planar Zn deposits. These results suggested that the deposits with the electrolyte containing α-CD additive possessed lower roughness, which indicated the uniform morphology of plated Zn during charging process. These results demonstrate the unique ability of in situ AFM characterization techniques to study the morphology of electrode materials in AZIBs.

8.2.5 In Situ Optical Microscopy (OM)

OM, an optical instrument that uses the principle of optics to enlarge images of tiny objects, can extract information about fine structures. Although OMs are not as powerful as EMs, their advantages of being simple, inexpensive, and able to be observed in the environment make it easy to make in situ observations. Therefore, in zinc-ion batteries, the change in anode surface morphology can be observed through in situ OM technique, and the deposition mechanism of Zn during charge/discharge processes can be further studied. For instance, the Zn deposition behavior on the Zn surface with (CN–Zn) and without copper nitride (B–Zn) protective layer at different plating times was investigated through in–situ OM to verify the positive effect of the limited 2D diffusion mechanism on Zn nucleation and growth by our group [77]. As shown in Figure 8.17a, due to the simple 2D diffusion inducing Zn atoms to select the best deposition site with the lowest binding energy, there was local uneven nucleation of B-Zn (gray and black areas refer to zinc deposition area) in the initial electroplating stage. With the extension of deposition time, Zn dendrites can be observed in part region and another region without Zn deposition. However, the uniform Zn nucleation and growth induced by ultrathin CN layer can be observed with deposition time. In addition, the other research by our group also used in situ OM to prove the uniform nucleation and growth behavior of the Zn after electrochemical interface reconstruction (R–Zn) (Figure 8.17b) [78]. Tiny protuberance can be observed on the initial Zn surface after 10 minutes of plating, and the formation of dendrites can be observed during the continuous plating, which may be due to the preferential Zn deposition on the protuberance. However, the smooth morphology of the R–Zn surface after Zn deposition can be observed, which is due to the homogeneous surface-benefiting uniform Zn deposition. These two typical examples can further verify the advantages of in situ OM in providing a more in-depth understanding of the deposition mechanism of Zn during charge/discharge processes.

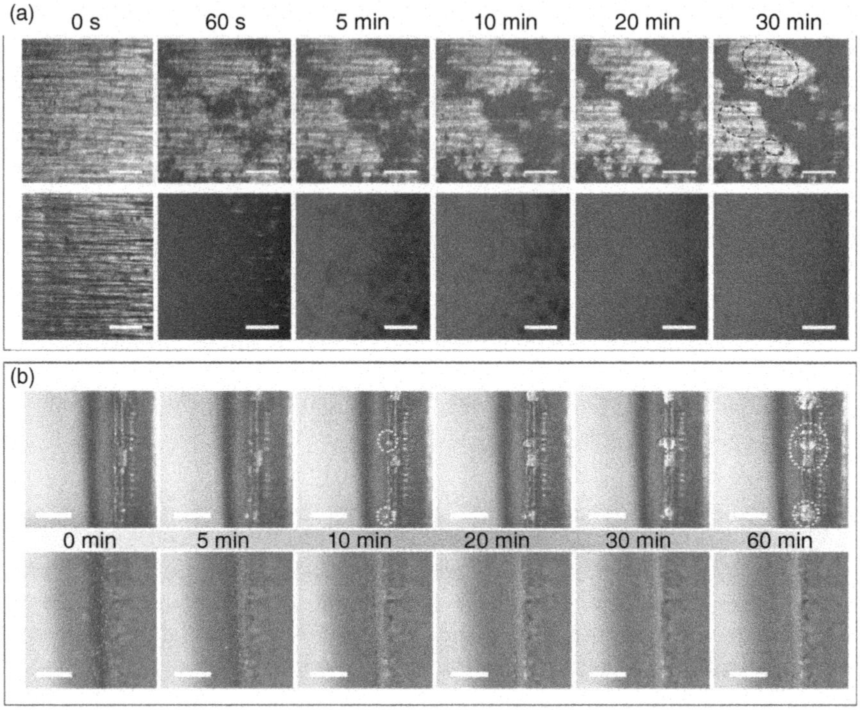

Figure 8.17 In situ optical microscope observations of zinc plating on the surface of (a) B-Zn and CN-Zn at 1 mA cm^{-2}, the regions marked by black circles denote as inert areas. Scale bars: 100 μm. In situ optical microscopic observations of zinc deposition on (b) P-Zn and R-Zn anodes, the write circles refer to the dendrites. Scale bar: 200 μm. Source: (a, b) Yang et al. [77], ©2022/John Wiley & Sons, Inc. (c, d) Yang et al. [78] Copyright 2022, Elsevier.

8.3 Theoretical Research Methods

8.3.1 Simulations in AZIBs

Simulation is reliable tool for simulating a variety of physical and chemical processes through mathematical modeling and solving, including those in the fields of mechanics, hydrology, electromagnetism, optics, acoustics, electrochemistry, and semiconductors. Therein, the finite element method (FEM) is a kind of calculation method for solving partial differential equations, which has been widely used in simulation. Several researchers have simulated the effects of electric field and ion concentration field distribution on dendrite growth in AZIBs, providing a solid theoretical basis for better coping strategies.

8.3.1.1 Simulations of Electric Field Distribution

Lu et al. constructed a three-dimensional CNT framework as a Zn host to realize a dendrite-free Zn anode and investigated the reason why CNT framework was beneficial to smooth galvanizing by simulating the electric field distribution of

Zn/carbon cloth (CC) and Zn/CNT anode after Zn nucleation [79]. As can be seen from Figure 8.18a, electric field distribution on the surface of bare CC was uneven. The charge region near the nucleus site of Zn in vitro was larger than other regions. This significant intensity gradient can promote the adsorption of more Zn^{2+} to the nucleation center, resulting in uneven initial deposition of Zn. Zn^{2+} was deposited on the prominences continuously during the subsequent cycle, resulting in the growth of Zn dendrites.

In contrast, the electric field distribution becomes more uniform after introducing the CNT framework, which can be attributed to the increase in specific surface area and the extremely small Zn nuclei. The evenly distributed electric field effectively avoided the tip effect on the CC substrate, thus ensuring that more Zn^{2+} was evenly deposited throughout the anode surface. Therefore, these results indicated that CNT can guarantee the homogeneous deposition of Zn at an early stage and that the growth of Zn dendrites will not be triggered even if the cycle time is further prolonged.

A vertical graphene (VG) carpet on the one side of a commercial fiberglass as Janus separator was developed by Li et al. to enhance the electrochemical performance of AZIBs [9]. In this work, the electric field distribution and local current density at anode–electrolyte interface were simulated by FEM (Figure 8.18b). They found that the 3D VG scaffold with an increased surface area can effectively reduce the current density from 1 to 0.24 mA cm^{-2}, better than that of 2D planar graphene (0.38 mA cm^{-2}). Furthermore, the 3D VG scaffold can induce a uniform distribution of the electric field at an early stage and maintain it in subsequent cycles. On the other hand, the 2D planar graphene was easy to form a single nucleation point and further resulted in an uneven electric field, which led to the subsequent growth of Zn dendrites.

8.3.1.2 Simulations of Zn^{2+} Concentration Field Distribution

Zhi et al. designed hydrogen-replaced graphene with sub-Angstrom ion tunnels as an artificial interface layer (AIL) to solve the problems in AZIBs [80]. By simulating the electric field and ion concentration field in this study, the inhibition mechanism of the protective layer on dendrites was revealed. As shown in Figure 8.18c, the local electric field inhomogeneity of bare Zn symmetric batteries at the surface protrusion was obvious, and the electric field intensity increased with the distortion direction. Although the AIL layer covered the rough Zn surface well, there was little change in the electric field after introducing semiconductor properties. The distributions of the Zn^{2+} concentration field were further simulated based on the simulated electric fields. The Zn@AIL symmetric battery exhibited even concentrations (0.6–0.7) throughout the Zn surface, including tips and flat regions.

In contrast, for the bare Zn symmetric battery, after only 5 seconds of diffusion driven by electric field, Zn^{2+} rapidly accumulated at the surface tips, and the maximum local normalized concentration could reach 1, while the concentration in adjacent regions was at most 0.2. The reason for the large difference is the homogenization of Zn^{2+} flux during the migration. Due to limited ionic flow, Zn^{2+} diffused from the electrolyte and concentrated on the outer surface of AIL (0.7–0.9 μm

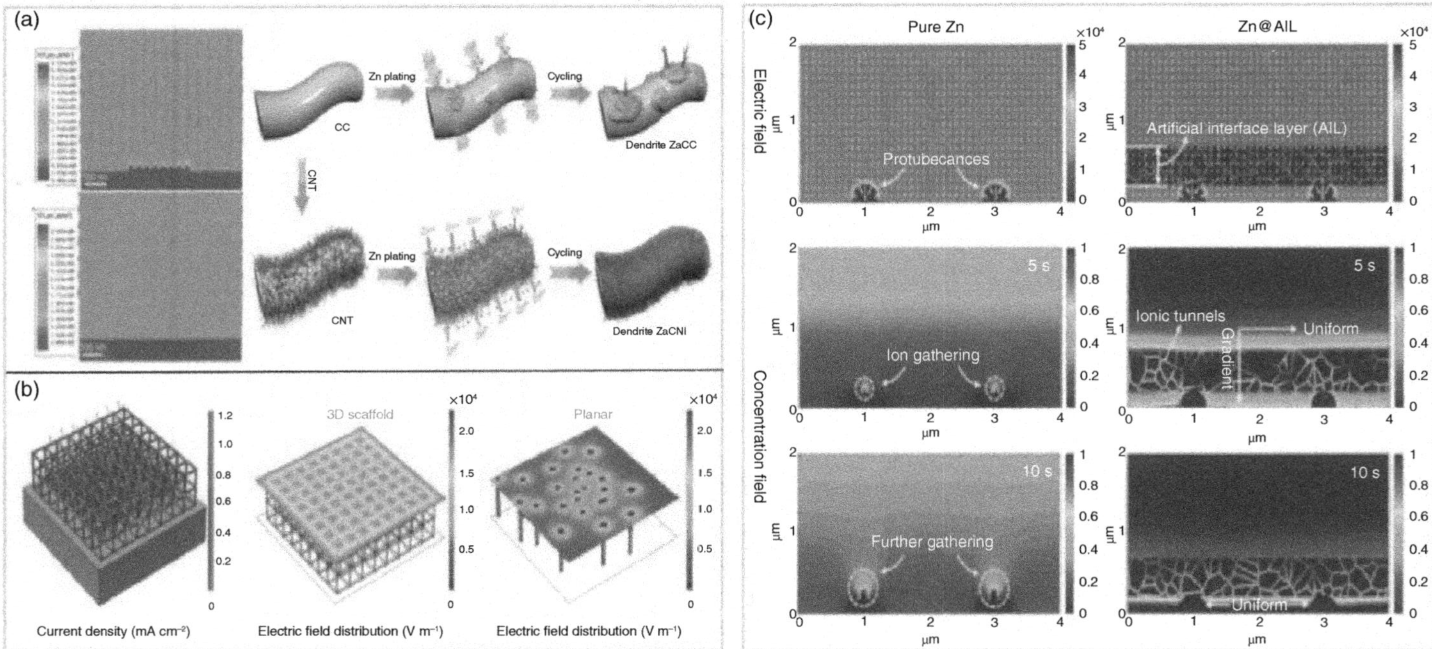

Figure 8.18 (a) Models of the electric field distributions for the Zn/CC and Zn/CNT anodes. (b) The current distribution in the 3D VG scaffold. Electric field distributions of 3D VG scaffold and 2D planar structure. (c) Electric field simulations on the bare Zn and Zn@AIL anodes; Zn^{2+} concentration field simulations after 5 s of diffusion for the bare Zn and Zn@AIL symmetric cells. Source: (a) Zeng et al. [79]/John Wiley & Sons. (b) Li et al. [9]/John Wiley & Sons. (c) Yang et al. [80]/John Wiley & Sons.

in height). The subsequent ion diffusion in the tunnel was evenly distributed in the transverse direction and possessed an obvious concentration gradient in the longitudinal direction. Therefore, Zn^{2+} was uniformly distributed along the surface of Zn after passing through AIL. The redistribution effect of the AIL was further verified with simulated diffusion time extended to 10 seconds. The results showed that the Zn^{2+} enriched area was formed on the convex surface, while the Zn^{2+} was still sparsely distributed in the adjacent area. On the contrary, the concentration of Zn^{2+} in Zn@AIL symmetric element was again evenly distributed along the entire surface of Zn, increasing to 0.8–1.0. Therefore, the ion-tunneling AIL redistributed the Zn^{2+} to deviate from the direction of the electric field, forming a relatively uniform concentration field at the AIL-Zn interface.

8.3.2 Theoretical Calculation in AZIBs

Theoretical calculations are mainly based on density functional theory (DFT), which is a quantum mechanical approach to study the electronic structure of multi-electron systems. Many studies on AZIBs have successfully applied the DFT to calculate the binding energy, migration energy, charge density distribution, state density, etc., and realized the real analysis at the atomic level, which promotes the further development of AZIBs.

8.3.2.1 Calculations of Adsorption Energy for Evaluating Zincophilicity

It is well known that zincophilicity is the key parameter for investigating different methods to inhibit dendrite growth. For example, a highly conductive protective layer must have a poor zinc affinity; on the contrary, Zn will be deposited above the protective layer, resulting in the failure of the protective layer, and for the host with high conductivity, a better Zn affinity will reduce the formation of nuclear barrier, which can effectively inhibit dendrite growth [81, 82]. Adsorption energy calculation, as a suitable tool, can evaluate the zincophilicity, which not only verifies the experimental phenomenon but also theoretically induces the selection of materials, providing strong support for improving protection methods.

The Zn affinity of different materials was compared by calculating the adsorption energy by our group [83]. In Figure 8.19, the interaction models of Zn with Cu, $CuZn_5$, Zn, and polyacrylamide (PAM) were established. The adsorption energy calculation results show that there was a strong interaction between Zn and Cu, and the adsorption energy was −1.58 eV, while the Cu—Zn solid solution ($CuZn_5$) can increase the adsorption energy to −1.94 eV. Interestingly, the Cu—Zn composition of copper and brass had a small effect on the binding kinetics with Zn since the Cu—Zn content of brass is relatively low (−1.37 eV). The polar acyl of PAM gives it strong adsorption energy for Zn (−1.25 eV) and can transfer Zn^{2+} uniformly from the electrolyte to the anode surface.

In addition, we also calculated the adsorption energy, which further revealed the effect of crystal orientation on the stability of zinc anodes [84]. The interactions between Zn and TiO_2 on different crystal planes were studied for the first time through DFT calculation. The calculated results showed that the adsorption energy

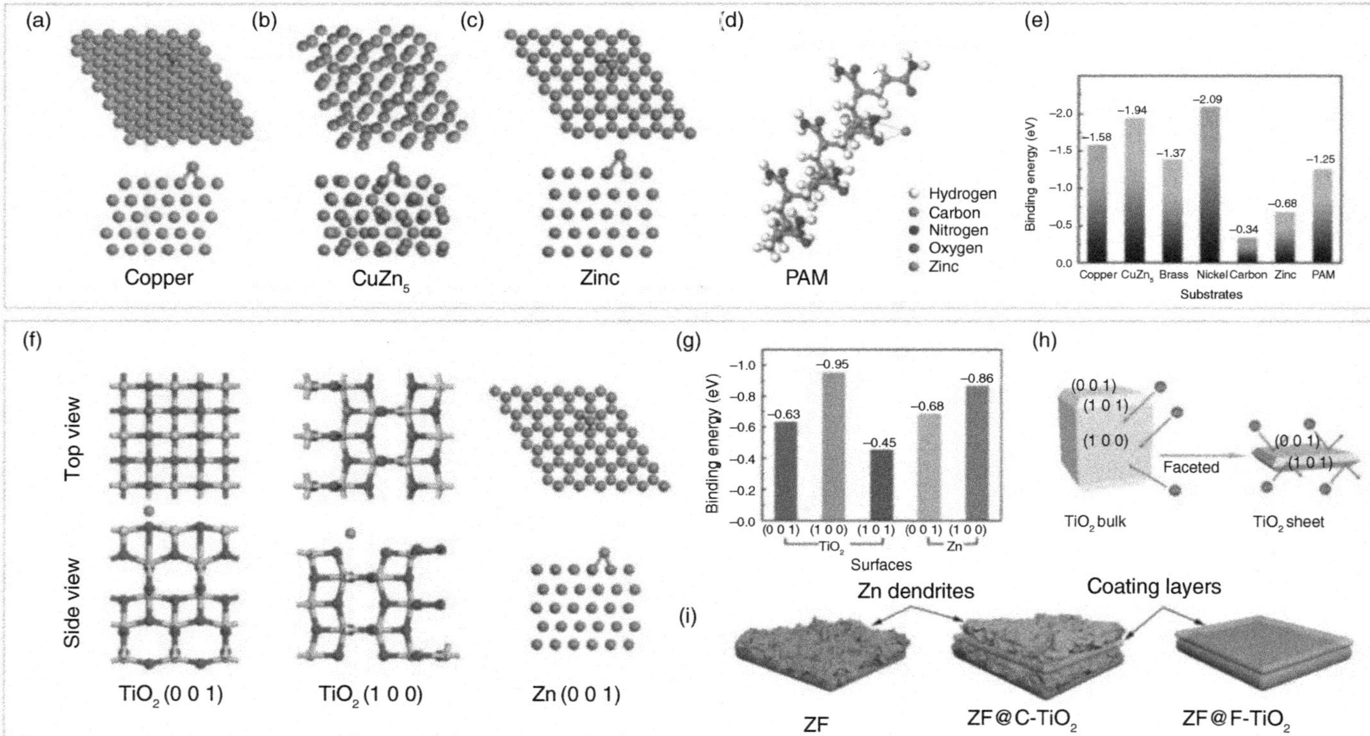

Figure 8.19 Calculation models of interactions of Zn with (a) Cu, (b) $CuZn_5$, (c) Zn, (d) PAM, and (e) corresponding adsorption energies. (f) Calculation models of Zn adsorbed on the TiO_2 and Zn metal surfaces with different facets and (g) corresponding adsorption energies. (h) Schematic illustration of the interactions between Zn atom and different TiO_2 surfaces. (i) Schematic illustration of the Zn plating process with the different surface exposure. Source: (a–e) Zhang et al. [83]/Reproduced with permission of John Wiley & Sons. (f–i) Zhang et al. [84]/Springer Nature / Licensed under CC BY 4.0.

of Zn atom on TiO_2 (100) plane was −0.95 eV, higher than that of Zn atom on TiO_2^- (100) plane (−0.68 eV and −0.86 eV), indicating that Zn was more inclined to be deposited on TiO_2^- (100) plane than on Zn surface (Figure 8.19b). This priority caused the Zn dendrites to grow on the layer, rendering the protective layer ineffective. In contrast, the (001) and (101) planes of TiO_2 exhibited relatively low Zn affinity (−0.63 and −0.45 eV) because of the exposure of low-coordination Ti to these sites and the stronger repulsion of the Zn atoms. From the above analysis, it can be seen that surface orientation plays a key role in zincophilicity, and a suitable protective layer can be obtained by controlling specific surface exposure.

8.3.2.2 Calculations for Structural Evolution With Zn^{2+} Insertion/Extraction

In Section 8.2, in situ techniques were used to study the structural evolution of active materials during charging/discharging. In this section, the study of predicting tectonic evolution with theoretical calculations will be further discussed in detail. Chen et al. deduced the structural evolution of C4Q before and after absorption of Zn^{2+} through DFT calculation [85]. Firstly, the active sites (redox centers) of C4Q electrode were deduced by electrostatic potential (ESP) method to determine the exact location of Zn^{2+} adsorption (Figure 8.20). In organic chemistry, ESP is used to predict electrophilic or nucleophilic reactions; the positive charge regions of ESP tend to be nucleophilic, while the negative charge regions of ESP tend to be electrophilic. Since the discharge process involves the absorption of Zn^{2+}, the more negative the potential current is, the more attractive the discharge reaction is. The ESP results displayed that the regions near the carbonyl group had more negative ESP values and were considered highly active sites, which agreed with the experimental results. For C4Q structure, ESP values of upper and lower carbonyl groups were lower than those of carbonyl groups on both sides, which was more conducive to Zn^{2+} absorption. The extremely short distance between Zn and O indicated a strong interaction between Zn^{2+} and the carbonyl groups at the top and bottom. Notably, the structure and volume of C4Q did not change much after the third Zn^{2+} absorption, which was attributed to the low polarization of the aqueous Zn-C4Q batteries.

In the experiment, only carbonyl group could be identified as the active central site. At the same time, the specific absorption position and structural evolution of Zn^{2+} after absorption were not clear but could be clarified by theoretical calculation. The above analysis indicated that theoretical calculation can not only verify the experimental results but also further understand the mechanism, which is of great significance for the development of AZIBs.

8.3.2.3 Calculations of Zn^{2+} Diffusion Kinetics

The quantitative characterization of Zn^{2+} diffusion kinetics is described in Section 8.1.2. The Zn^{2+} diffusion kinetics will be further analyzed by theoretical calculation in this section. The effect of charge density distribution on Zn^{2+} diffusion kinetics has been investigated by many researchers. For example, Guo et al. established the interface models of Zn and ZnS (002) planes for DFT calculations to study the effect of the ZnS-protective layer on Zn^{2+} [86]. The DFT results displayed that the bonds formed between S atoms (from ZnS) and Zn atoms (from Zn metal)

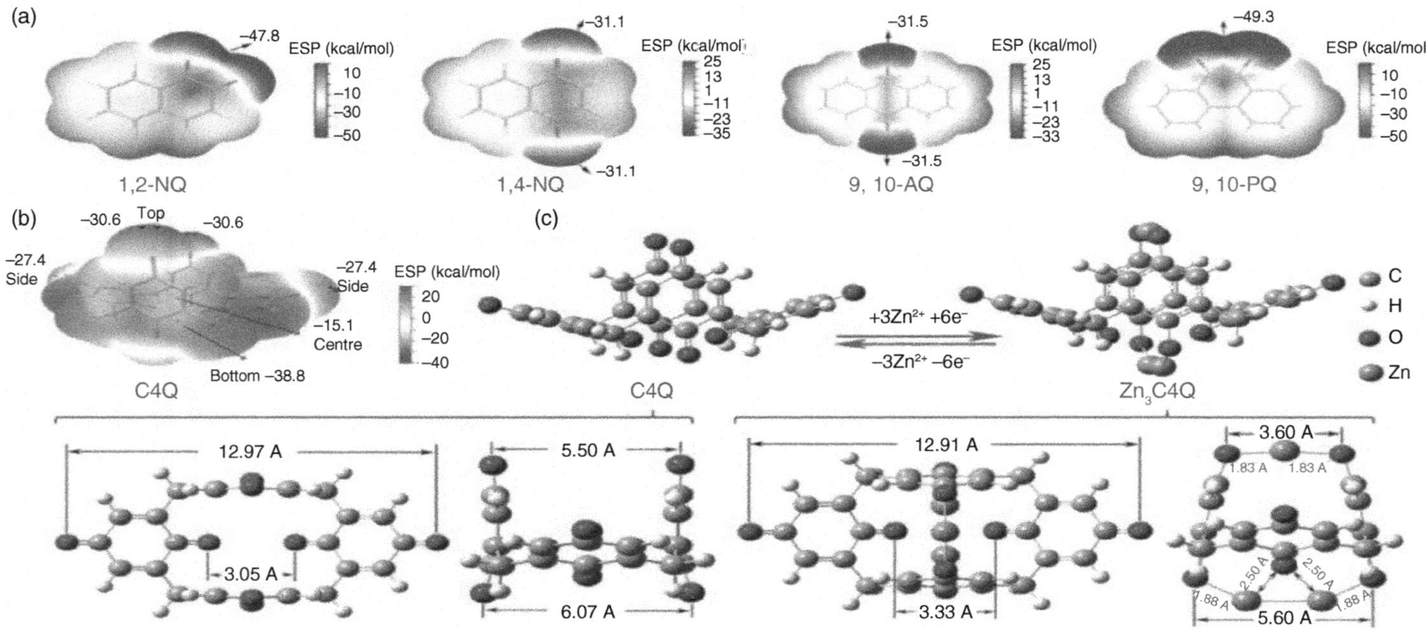

Figure 8.20 ESP-mapped molecular van der Waals surface of (a) 1,2-NQ, 1,4-NQ, 9,10-AQ, 9,10-PQ, and (b) C4Q, respectively. (c) Optimized configurations of C4Q before and after Zn^{2+} uptake. Bottom: Corresponding configurations at different viewpoints. Source: Adapted from Bhalla et al. [85].

changed the charge distribution at the interface, resulting in an imbalance charge distribution at the interface (Figure 8.21a–c). The imbalanced charge distribution not only promoted the diffusion of Zn^{2+} at the ZnS@Zn interface, but also enhanced the adhesion of the ZnS-protective layer to Zn metal.

Similarly, Zhou et al. also used charge density distribution calculations to explore the factors affecting the solvation structure in order to improve the Zn^{2+} diffusion kinetics [87]. The geometrically optimized oxygen element (from ZnO-3D) can generate an additional charge concentration on the surface (Figure 8.21d). In addition, Zn^{2+} insertion into Zn@ZnO-3D possessed a low energy barrier, which can be verified by the reduction of Gibbs free energy. Thus, this results in static conditions preferentially attracting Zn^{2+} over solvation or hydration conditions, reducing the energy consumption of the tight solvation sheath of free water near the Zn metal and thus effectively preventing H_2 precipitation.

Li et al. presented a water–ethylene glycol (EG)–$ZnSO_4$ mixed electrolyte for low-temperature Zn batteries. They systematically studied the effect of the solvation structure of the modified electrolyte on the diffusion kinetics of Zn^{2+} [88]. The simulated electrolyte structure showed that the solvation cluster of Zn^{2+} was synergistic with SO_4^{2-}, H_2O, and EG in the mixed electrolyte (Figure 8.21e–j). With the increase in EG/water ratio, the coordination number of SO_4^{2-} near Zn^{2+} increases slightly. In the first hydration layer, the coordination number of H_2O gradually decreased. At the same time, that of EG gradually increased, indicating that the introduction of EG destroyed the solvation of Zn^{2+} with H_2O to a large extent.

Meanwhile, the binding energies of Zn^{2+} to SO_4^{2-}, H_2O, and EG were ordered as $Zn^{2+}–H_2O < Zn^{2+}–EG < Zn^{2+}–SO_4^{2-}$, which indicated that Zn^{2+} was more likely to coordinate with EG than with H_2O, leading to rapid exchange of EG around Zn^{2+} and thus rapid diffusion of Zn^{2+} in the mixed electrolyte. In addition, the ESP of Zn^{2+}-$5H_2O$ solvation state was significantly reduced after introducing EG into the electrolyte. Therefore, the electrostatic repulsion can be reduced, which is conducive to the rapid migration of Zn^{2+}.

A fast ionic conductor, $NaTi_2(PO_4)_3$ (NTP), was designed by Zhu et al. as a protective layer for stabilizing Zn anode and systematically evaluated Zn^{2+} diffusion kinetics by calculating the Zn^{2+} migration energy barrier [89]. As shown in Figure 8.21k–m, the band gaps of NTP, TiP_2O_7 (TPO), and $Zn_3(PO_4)_2$ (ZPO) were between 0.5 and 4.5 eV, indicating the three phosphates with low conductivity. Compared with TPO and ZPO with higher Zn^{2+} migration barriers (0.86 and 1.30 eV, respectively), the calculated Zn^{2+} migration barrier was as low as 0.52 eV in NTP, indicating that the Zn^{2+} diffusion kinetics were the fastest in NTP. The lower the Zn^{2+} migration barrier, the faster the Zn^{2+} migration rate and charge rate, which is conducive to improving Zn-leaching/plating kinetics.

8.4 Conclusion

Characterization techniques are important tools for in-depth understanding of battery materials, which play a crucial role in understanding their physical and

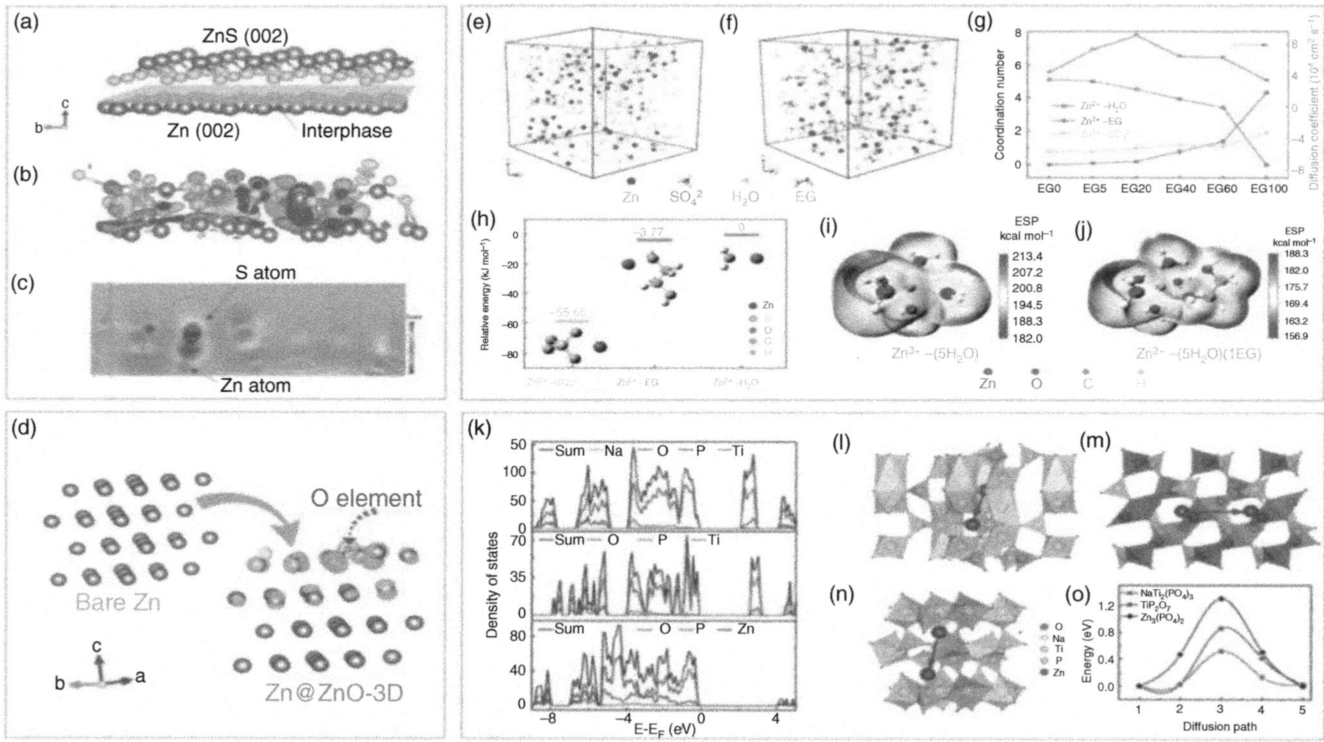

Figure 8.21 (a) Interfacial model of Zn and ZnS (002) surfaces of the ZnS@Zn anode. (b) Electron density difference map at the interface of ZnS@Zn, and (c) its slice to demonstrate the unbalanced charge distribution. (d) Differential charge density distribution of the Zn@ZnO-3D. Snapshots of (e) EG0 and (f) EG40 during molecular dynamics simulations. (g) Coordination numbers of SO_4^{2-}, H_2O, and EG around Zn^{2+} in the series of electrolytes with different EG proportions. (h) The binding energies for Zn^{2+} with SO_4^{2-}, H_2O, and EG. ESP maps of (i) the Zn^{2+}–H_2O system and (j) added-EG system. (k) The density of states of NTP, TPO, and ZPO. Schematic illustrations of Zn^{2+} migration pathways for (l) NTP, (m) TPO, and (n) ZPO. (o) Migration energy barriers of Zn^{2+} in NTP, TPO, and ZPO, respectively. Source: (a–c) Hao et al. [86]/John Wiley & Sons. (d) [87]; Copyright 2020, Royal Society of Chemistry. (e–j) Reproduced with permission Chang et al. [88]; Copyright 2020, Royal Society of Chemistry. (k–o) Liu et al. [89]/Reproduced with permission of John Wiley & Sons.

chemical properties. In this chapter, the main characterization techniques for investigating AZIBs are reviewed, and the recently advanced process in both ex situ and in situ techniques are highlighted. Advanced scattering techniques, such as XRD, have proven to be powerful methods for providing valuable insights into the structural evolution and phase transitions of electrode materials during electrochemical processes. Meanwhile, various spectral techniques, such as XAS, FTIR, and Raman, can also provide valuable information about surface and electronic structures, which can provide valuable information about the reaction of AZIBs. In addition, microscopic techniques, such as SEM, TEM, STEM, and AFM, can provide visualized information about morphological changes and phase transformations during the charge/discharge processes. More importantly, the combination of EDS, STEM-EELS, and other microscopic and spectral techniques makes it possible to study not only the morphology but also the structure, elemental composition, and coordination chemistry of electrode materials from multiple angles. The latest progress of these in situ techniques and operating conditions is also reviewed. In the near future, these in situ techniques and performance tests will be more widely used and further developed in AZIB research community, and information about them needs to be constantly updated. In addition, the researchers developed a range of analytical techniques for simulation and theoretical calculations. This chapter summarizes the research progress of AZIB analysis techniques in recent years.

References

1 Armand, M. and Tarascon, J.M. (2008). *Nature* 451: 652–657.
2 Tarascon, J.M. and Armand, M. (2001). *Nature* 414: 359–367.
3 Dunn, B.S., Kamath, H., and Tarascon, J.M. (2011). *Science* 334: 928–935.
4 Wang, L., Song, J., Qiao, R. et al. (2015). *Journal of the American Chemical Society* 137: 2548–2554.
5 Choi, J.W. and Aurbach, D. (2016). *Nature Reviews Materials* 1: 16013.
6 Li, W., Dahn, J.R., and Wainwright, D.S. (1994). *Science* 264: 1115–1118.
7 Naveed, A., Yang, H., Shao, Y. et al. (2019). *Advanced Materials* 31: 1900668.
8 Yang, H., Zhou, W., Chen, D. et al. (2022). *Energy & Environmental Science* 15: 1106–1118.
9 Li, C., Sun, Z., Yang, T. et al. (2020). *Advanced Materials* 32: 2003425.
10 Shi, X., Xu, G., Liang, S. et al. (2020). *Energy & Environmental Science* 13: 503–510.
11 Rubel, O., Tran, T.N.T., Gourley, S. et al. (2022). *The Journal of Physical Chemistry C* 126: 10957–10967.
12 Wang, J., Zhang, B., Cai, Z. et al. (2022). *Science Bulletin* 67: 716–724.
13 Zhang, N., Cheng, F., Liu, J. et al. (2017). *Nature Communications* 8: 405.
14 Hao, J., Li, X., Zhang, S. et al. (2020). *Advanced Functional Materials* 30: 2001263.
15 Li, L., Jia, T., Pei, X. et al. (2022). *Ionics* 28: 283–293.

16 Soundharrajan, V., Sambandam, B., Kim, S. et al. (2020). *Energy Storage Materials* 28: 407–417.
17 Meng, H., Ran, Q., Dai, T.-Y. et al. (2022). *Nano-Micro Letters* 14: 128.
18 Findlay, S.D., Shibata, N., Ikuhara, Y. et al. (2017). *Microscopy Today* 25: 36–41.
19 Shadike, Z., Zhao, E., Zhou, Y.-N. et al. (2018). *Advanced Energy Materials* 8: 1702588.
20 Xu, W., Sun, C., Zhao, K. et al. (2019). *Energy Storage Materials* 16: 527–534.
21 Zhu, S., Dai, Y., Li, J. et al. (2022). *Science Bulletin* 67: 1882–1889.
22 Danis, L., Gateman, S.M., Kuss, C. et al. (2017). *ChemElectroChem* 4: 6.
23 Liu, X., Wang, D., and Wan, L. (2015). *Science Bulletin* 60: 839–849.
24 Liu, Y., Zhi, J., Sedighi, M. et al. (2020). *Advanced Energy Materials* 10: 2002578.
25 Zhang, Y., Zhu, M., Wang, G. et al. (2021). *Small Methods* 5: 2100650.
26 Li, B., Yang, K., Ma, J. et al. (2022). *Angewandte Chemie International Edition* 1: e202212587.
27 Su, Y., Wang, X., Zhou, S. et al. (2022). *Energy Storage Materials* 53: 909–916.
28 Zhao, Z., Wang, R., Peng, C. et al. (2021). *Nature Communications* 12: 6606.
29 Canette, A. and Briandet, R. (2014). *Encyclopedia of Food Microbiology (Second Edition)* (ed. C.A. Batt and M.L. Tortorello), 676–683. Oxford: Academic Press.
30 Qiu, H., Du, X., Zhao, J. et al. (2019). *Nature Communications* 10: 5374.
31 Kabzinski, J., Budde-Meiwes, H., Rahe, C., and Sauer, D.U. (2016). *Journal of The Electrochemical Society* 163: A995.
32 Qin, R., Wang, Y., Zhang, M. et al. (2021). *Nano Energy* 80: 105478.
33 Yang, J.-L., Li, J., Zhao, J.-W. et al. (2022). *Advanced Materials* 34: 2202382.
34 Zheng, J., Huang, Z., Zeng, Y. et al. (2022). *Nano Letters* 22: 1017–1023.
35 Wang, R., Wu, Q., Wu, M. et al. (2022). *Nano Research* 15: 7227–7233.
36 Yan, H., Li, S., Nan, Y. et al. (2021). *Advanced Energy Materials* 11: 2100186.
37 Yang, Z., Zhang, Q., Li, W. et al. (2023). *Angewandte Chemie International Edition* 62: e202215306.
38 Wapner, K., Stratmann, M., and Grundmeier, G. (2006). *Electrochimica Acta* 51: 3303–3315.
39 Wu, B., Guo, B., Chen, Y. et al. (2023). *Energy Storage Materials* 54: 75–84.
40 Zhao, R., Yang, Y., Liu, G. et al. (2021). *Advanced Functional Materials* 31: 2001867.
41 Zhang, C., Holoubek, J., Wu, X. et al. (2018). *Chemical Communications* 54: 14097–14099.
42 Yang, H., Chang, Z., Qiao, Y. et al. (2020). *Angewandte Chemie International Edition* 59: 9377–9381.
43 Zhang, Q., Ma, Y., Lu, Y. et al. (2020). *Nature Communications* 11: 4463.
44 Chen, Z., Mo, F., Wang, T. et al. (2021). *Energy & Environmental Science* 14: 2441–2450.
45 Bie, Z., Yang, Q., Cai, X. et al. (2022). *Advanced Energy Materials* 12: 2202683.
46 Jha, P., Pandey, O., and Singh, K. (2015). *Journal of Molecular Structure* 1083.
47 Wu, Z., Li, M., Tian, Y. et al. (2022). *Nano–Micro Letters* 14: 110.
48 Sun, P., Ma, L., Zhou, W. et al. (2021). *Angewandte Chemie International Edition* 60: 18247–18255.

49 Hao, J., Yuan, L., Ye, C. et al. (2021). *Angewandte Chemie International Edition* 60: 7366–7375.

50 Petibois, C., Rigalleau, V., Melin, A.-M. et al. (1999). *Clinical Chemistry* 45: 1530–1535.

51 Wan, F., Zhang, L., Dai, X. et al. (2018). *Nature Communications* 9: 1656.

52 Wang, L., Fan, G., Liu, J. et al. (2021). *Chinese Chemical Letters* 32: 1095–1100.

53 Jin, Y., Han, K.S., Shao, Y. et al. (2020). *Advanced Functional Materials* 30: 2003932.

54 Lv, Y., Zhao, M., Du, Y. et al. (2022). *Energy & Environmental Science* 15: 4748–4760.

55 Zhou, A., Zhang, J., Chen, M. et al. (2022). *Advanced Materials* 34: 2270328.

56 Tang, C., Singh, G., Housel, L.M. et al. (2021). *ECS Meeting Abstracts* 2090: MA2021–MA2001.

57 Zhang, N., Cheng, F., Liu, Y. et al. (2016). *Journal of the American Chemical Society* 138: 12894–12901.

58 Kumar, S., Verma, V., Chua, R. et al. (2020). *Batteries & Supercaps* 3: 619–630.

59 Li, M., Sciacca, R., Maisuradze, M. et al. (2021). *Electrochimica Acta* 400: 139414.

60 Rudowicz, C., Stefaniuk, I., Dziembaj, R. et al. (2007). *Research on Chemical Intermediates* 33: 853–862.

61 Povia, M., Sottmann, J., Portale, G. et al. (2018). *The Journal of Physical Chemistry C* 122: 5917–5923.

62 Ai, W., Du, Z., Fan, Z. et al. (2014). *Carbon* 76: 148–154.

63 Wang, J. and Sun, X. (2012). *Energy & Environmental Science* 5: 5163–5185.

64 Fukushima, T., Matsuda, Y., Hashimoto, H., and Arakawa, R. (2002). *Journal of Power Sources* 110: 34–37.

65 Guo, Z., Fan, L., Zhao, C. et al. (2022). *Advanced Materials* 34: 2105133.

66 Pan, R., Zheng, A., He, B. et al. (2022). *Nanoscale Horizons* 7: 1501–1512.

67 Zheng, J., Zhao, Q., Tang, T. et al. (2019). *Science* 366: 645–648.

68 Seo, D.M., Afroz, T., Allen, J.L. et al. (2014). *The Journal of Physical Chemistry C* 118: 25884–25889.

69 Liu, T., Hong, J., Wang, J. et al. (2022). *Energy Storage Materials* 45: 1074–1083.

70 Jing, P., Wei, W., Luo, W. et al. (2020). *Inorganic Chemistry Communications* 117: 107953.

71 Yang, B., Quan, H., Gao, J. et al. (2021). *Journal of Alloys and Compounds* 889: 161680.

72 Li, X., Feng, J., Wen, N. et al. (2021). *Solid State Ionics* 364: 115632.

73 Wu, T.-H., Lin, Y.-Q., Althouse, Z.D., and Liu, N. (2021). *ACS Applied Energy Materials* 4: 12267–12274.

74 Zhang, Z., Said, S., Smith, K. et al. (2021). *Journal of Materials Chemistry A* 9: 15355–15362.

75 Wang, S., Wang, Z., Yin, Y. et al. (2021). *Energy & Environmental Science* 14: 4077–4084.

76 Zhao, K., Fan, G., Liu, J. et al. (2022). *Journal of the American Chemical Society* 144: 11129–11137.

77 Yang, Z., Lv, C., Li, W. et al. (2022). *Small* 18: 2104148.

78 Yang, Z., Zhang, Q., Xie, C. et al. (2022). *Energy Storage Materials* 47: 319–326.

79 Zeng, Y., Zhang, X., Qin, R. et al. (2019). *Advanced Materials* 31: 1903675.

80 Yang, Q., Guo, Y., Yan, B. et al. (2020). *Advanced Materials* 32: 2001755.

81 Shen, C., Li, X., Li, N. et al. (2018). *ACS Applied Materials & Interfaces* 10: 25446–25453.

82 Yin, Y., Wang, S., Zhang, Q. et al. (2020). *Advanced Materials* 32: 1906803.

83 Zhang, Q., Luan, J., Fu, L. et al. (2019). *Angewandte Chemie International Edition* 58: 15841–15847.

84 Zhang, Q., Luan, J., Huang, X. et al. (2020). *Nature Communications* 11: 3961.

85 Bhalla, S., Melnekoff, D.T., Aleman, A. et al. *Science Advances* 7: eabg9551.

86 Hao, J., Li, B., Li, X. et al. (2020). *Advanced Materials* 32: 2003021.

87 Xie, X., Liang, S., Gao, J. et al. (2020). *Energy & Environmental Science* 13: 503–510.

88 Chang, N., Li, T., Li, R. et al. (2020). *Energy & Environmental Science* 13: 3527–3535.

89 Liu, M., Cai, J., Ao, H. et al. (2020). *Advanced Functional Materials* 30: 2004885.

9

Conclusion and Future Perspectives

Aqueous zinc-ion batteries (AZIBs) hold great promise for practical applications in large-scale energy storage due to their cost-effectiveness, environmental friendliness, intrinsic safety, and relatively high gravimetric energy and power densities. In this work, we adopt a holistic perspective to systematically review and discuss in depth the basic mechanisms, development, and promising strategies of AZIBs over the past ten years. However, a major obstacle to the commercialization of rechargeable AZIBs is their low energy density, caused by their low areal capacity. Typically, the utilization of zinc anodes is limited based on uncontrollable zinc dendrite growth, hydrogen evolution, and side reactions. Cathode materials suffer from problems such as dissolution, irreversible by-product formation, and surface passivation, which pose threats to battery stability and reversibility, especially at low current densities and thicker cathodes. Furthermore, most of the current research on AZIBs is still based on small systems (coin cells) at the laboratory stage, which cannot fully reflect the issues they encounter in practical applications. Therefore, it is necessary to scale up the evaluation system of AZIBs to realize integrated devices with kWh or Ah levels (e.g. pouch cells). Currently, no single strategy is perfect for each component of AZIBs, and a strategy can optimize several key properties but may adversely affect others. To meet the application requirements of specific scenarios, components and strategies with different characteristics need to be reasonably combined. Strategies focused on high energy densities should avoid using methodologies that might worsen zinc-ion transport. The construction of advanced AZIBs for large-scale energy storage requires joint efforts from the anode, cathode, electrolyte, separator, and cell configuration.

Uniform zinc deposition/stripping and suppression of parasitic reactions are the keys to solving severe zinc anode issues and improving anode reversibility. In-situ and ex-situ formed artificial interfacial coatings have been shown to induce uniform electric and ionic fields and suppress zinc dendrites. Among them, ex situ interfacial layers may cause battery failure during cycling due to their thick and fragile nature. The in-situ artificial solid interfacial layers with the self-healing ability are encouraged to be further studied. Note that strategies involving artificial solid interfacial layers should also mitigate their negative impact on device power density and cost. The structural design of the current collector is also a feasible strategy, and its future research direction should focus on increasing the areal loading of

Aqueous Zinc Ion Batteries: Fundamentals, Materials, and Design, First Edition.
Haiyan Wang, Qi Zhang, Yixin Li, and Yougen Tang.

zinc, thereby increasing the areal capacity of the cells. The low areal capacity (less than 0.5 mAh cm^{-2}) and utilization of zinc anodes are currently one of the main factors hindering the practical application of zinc metal anodes. As the theoretical areal capacity of the commonly used zinc foil with a thickness of 100 mm is about 58.5 mAh cm^{-2}, the utilization of zinc is less than 1%. The utilization of zinc anode is closely correlated to the capacity fading and energy density of practical batteries. Furthermore, optimizing the capacity ratio (N/P) between the negative and positive electrodes should be taken into consideration. For instance, the specific energy density at an N/P ratio of one (ideal condition) is almost four times that of an N/P ratio of 10, with the same positive electrode. However, the current lab-grade mass loadings of these cathode materials are usually below 2 mg cm^{-2}, resulting in an areal capacity of less than 1 mAh cm^{-2}, which is far less than the practical value. Therefore, the fundamental understanding of zinc dendrite growth, parasitic reactions, and electrode–electrolyte interface mechanisms is still insufficient, and more in-depth investigations are urgently needed. Meanwhile, the reasonable utilization of zinc anode and N/P ratio is necessary to accelerate the practical application of zinc metal anode.

The main obstacle hindering the further development of AZIBs is the lack of ideal cathode materials to meet all the requirements, including large capacity, high voltage, fast Zn-ion-insertion/extraction kinetics, good structural stability, and compatibility with electrochemical stable windows. Current strategies can be divided into two types: (i) reducing intermolecular interactions between zinc ions and electrode materials, such as pre-intercalation of foreign ions and heteroatom doping; (ii) modulating the morphology of materials, including surface coating, pore size engineering, size regulation, and defect engineering. Pre-intercalation of foreign ions proved to be an effective strategy for enhancing structural integrity and contributing to fast zinc-ion deintercalation kinetics. However, synergistic effects between different atoms and substrates need to be further investigated. Taking manganese-based cathodes as an example, the zinc-ion storage mechanism is still controversial. Different mechanisms have been reported so far, involving intercalation/extraction of zinc ions, co-intercalation/extraction of hydrogen and zinc ions, chemical transition reactions, the combination of intercalation and transition, and dissolution/deposition mechanisms. A thorough understanding of the reaction mechanism by combining advanced in-situ characterization techniques would facilitate future exploration of novel cathode materials and modification of known cathode materials. The future focus of novel cathode materials should be on high voltage, high capacity, and cathode-free batteries to promote the development of high energy density AZIBs.

The capacity fading of AZIBs is mostly caused by the consumption of electrolytes rather than the short circuit induced by zinc dendrites. The water-in-salt and molecular crowding electrolytes can effectively change the zinc-ion solvent sheath structure and suppress the water activity, thereby broadening the battery-operating voltage and improving reversibility. However, the decomposition of active materials and salt precipitation are still unavoidable in these electrolytes due to water evaporation. In addition to reducing the production cost, the high-concentration salt strategy is also required to improve the long-term cycling stability of AZIBs. Quasi-solid gel electrolytes are safer than liquid electrolytes because they can suppress zinc dendrites and corrosion reactions and improve the reversibility of zinc anodes. In future studies, in-depth research should be devoted to interfacial regulation and electrode/electrolyte compatibility, except for enhancing the mechanical strength and ionic conductivity of the gel electrolyte. Overall, for the future development of both aqueous and solid-state electrolytes for AZIBs, special attention should be paid to the fundamental understanding of the electrolyte degradation mechanism as well as the interactions between the electrolyte and other battery components.

The separator is a crucial component in rechargeable AZIBs. The main role of the separator is to impede the connection between the positive and negative electrodes to avoid short circuits and act as an electrolyte reservoir for ion transport. The ideal separator should possess abundant electrolyte uptake for low internal resistance, extremely thin thickness with strong mechanical strength, and electrochemically stable structure. However, the most widely used glass fiber separators cannot meet the aforementioned performance requirements at the same time. The exploration of new high-performance separators and the modification of conventional separators are urgent. Novel functional separators should be stable, safe, and sustainable. Cellulose is one of the most abundant, thermally and chemically stable, and sustainable materials. These key qualities make cellulose an attractive candidate for sustainable membranes. More work is required for future research and development of cellulose-related and composite materials as sustainable separators. For conventional separators, modification strategies in state-of-the-art lithium-ion batteries can be used as references, such as plasma treatment, UV light irradiation, electron beam irradiation, and polymer coating. Furthermore, along with the development of novel separators and an in-depth understanding of the critical role of separators in battery systems, it is expected that functional separators will not only advance existing battery systems but also inspire the development of new battery technologies.

Currently, the main research on integrated devices for AZIBs is conducted in coin cells, in which the size and areal capacity of the cells are limited (0.5–1.7 cm^{-2} and 0.1–1 mAh cm^{-2}). Upon developing the potential battery prototypes on a laboratory scale, it is equally important to establish medium and large battery modules for industrial production and commercial applications. It is unclear whether high-performance AZIBs suitable for small coin cells can also work well in large cells, especially when the electrode area and thickness are increased. To meet future energy market requirements, different commercially available cell configurations, such as pouch and prismatic cells, or novel cell devices, are supposed to be assembled to evaluate the cycling stability, reversible capacity, and Coulombic efficiency of modular cells. Therefore, researchers on AZIBs are encouraged to focus on solving scientific issues and emphasizing applied engineering issues by designing practical batteries and converting laboratory-scale batteries into commercially viable technologies for ensuring a sustainable future of AZIBs' energy storage.

Index

Aqueous Zinc Ion Batteries: Fundamentals, Materials, and Design, First Edition.
Haiyan Wang, Qi Zhang, Yixin Li, and Yougen Tang.